CELL METABOLISM – CELL HOMEOSTASIS AND STRESS RESPONSE

Edited by **Paula Bubulya**

Cell Metabolism – Cell Homeostasis and Stress Response
Edited by Paula Bubulya

Published by InTech
Janeza Trdine 9, 51000 Rijeka, Croatia

Publishing Process Manager Igor Babic

Technical Editor Teodora Smiljanic

Cover Designer InTech Design Team

Additional hard copies can be obtained from orders@intechopen.com

Cell Metabolism – Cell Homeostasis and Stress Response, Edited by Paula Bubulya
p. cm.
ISBN 978-953-307-978-3

For EU product safety concerns:
IN TECH d.o.o., Prolaz Marije Krucifikse Kozulić 3, 51000 Rijeka, Croatia,
info@intechopen.com or visit our website at intechopen.com.

Meet the editor

Paula A. Bubulya, PhD, has been an Associate Professor at the Department of Biological Sciences, Wright State University, Dayton, Ohio, USA, since 2005. She received her PhD in Biological Sciences from the University of Toledo in Toledo, Ohio in 1998, for her dissertation on cadherin-catenin interactions in adherens junctions. Dr Bubulya was a postdoctoral fellow studying cell biology, nuclear organization and microscopy at Cold Spring Harbor Laboratory from 1999 till 2005, and currently teaches both undergraduate and graduate courses in Cell Biology at Wright State University. In line with her passion for involving students in laboratory research, Dr Bubulya's research team at Wright State University is comprised primarily of undergraduate, MS and PhD students. Members of the lab study the connection between gene expression and nuclear organization in mammalian cells, and gain expertise in cellular imaging and in situ approaches to examine gene expression. The long-term goal of Dr Bubulya's research program is to determine how nuclear domains called nuclear speckles organize into discrete, yet dynamic nuclear domains, how key components of nuclear speckles help to coordinate pre-mRNA transcription and processing, and how the function of speckle components impacts mitosis. Dr Bubulya's research is funded by the NIH (NIGMS) and she is a member of the American Society for Cell Biology.

Contents

Preface

All organisms possess molecular mechanisms for responding to environmental stress induced under conditions such as limited nutrient availability, chemical exposure or phototoxicity. All organisms, from single cells to complex multicellular societies, activate biochemical pathways in response to specific stress cues in order to promote their survival. Ongoing research aims to understand both similarities and differences in these pathways among a wide variety of organisms. As an example, molecular pathways in lower eukaryotic organisms such as yeast can be placed in the context of broader stress response networks and serve as working models for understanding human cellular pathways. Likewise, comparisons can be made among different cell types within an organism, or between different organisms to understand common molecular responses to a given stressor.

The chapters in this book address stress response in yeast, plants and humans. The topics discussed include acetic acid-induced stress response in yeast, damaging effects of reactive oxygen species in chloroplasts, stress-induced gene expression in plants, plant defenses that detoxify and preserve integrity of plant tissues, interaction of metabolons with cytoskeleton to enhance cell survival during stress, toxic exposure that promotes cancer onset, irreversible damage of tumor cells, the use of Gene Ontology annotations to integrate human signaling pathways, and a wide array of approaches for imaging cellular metabolism.

Paula A. Bubulya
Wright State University, Dayton, Ohio
USA

1

Oligoglucan Elicitor Effects During Plant Oxidative Stress

Abel Ceron-Garcia[2], Irasema Vargas-Arispuro[1], Emmanuel Aispuro-Hernandez[1] and Miguel Angel Martinez-Tellez[1]
[1]*Centro de Investigación en Alimentación y Desarrollo, Hermosillo, Sonora*
[2]*Centro de Investigación y Asistencia en Tecnología y Diseño del Estado de Jalisco, Parque de Investigación e Innovación Tecnológica (PIIT), Apodaca, Nuevo León*
México

1. Introduction

Molecular oxygen is essential for the existence of life of aerobic organisms including plants. However, Reactive Oxygen Species (ROS), which include the superoxide anion ($O_2{\bullet^-}$), hydroxyl radical ($\bullet OH$), perhydroxyl radical ($\bullet O_2H$) and hydrogen peroxide (H_2O_2), are generated in all aerobic cells as byproducts of normal metabolic processes. In general, under various conditions of environmental stress, plant cells show an increase in ROS levels leading to oxidative stress. Indeed, oxidative stress is a major cause of cell damage in plants exposed to environmental stress. Plants under the effect of biotic (senescence, pathogen attack) and/or abiotic factors (heat, chilling, drought, salinity, chemical compounds, mechanical damage) may increase ROS levels, and their accumulation produce a disruption of the redox homeostasis.

Plants employ an efficient ROS scavenging system based on enzymatic (superoxide dismutase, SOD; catalase, CAT; ascorbate peroxidase, APX) and non-enzymatic antioxidants (carotenoids, tocopherols, glutathione, phenolic compounds) to counteract ROS adverse effects against important macromolecules like lipids, proteins and nucleic acids, which are necessary for cell structure and function. However, the catalytic activity of these antioxidant systems could be negatively affected by several stress conditions due to abiotic and biotic factors; a very common situation for plants in fields or commercial stocks. The efforts of farm growers to bring up healthy crops and sufficient yields could be reinforced with the scientific experience and development of novel techniques focused in plant physiology and crop protection by means of the elicitation of plant defense responses against any kind of stress.

Multiple biological responses in plants including controlled ROS overproduction during phytopathogen attack, changes in ionic fluxes across lipid membranes, phosphorylation of proteins, transcription factors activation and up-/down-regulation of defense related genes have been demonstrated when using oligogalacturonides and some oligoglucans derivatives from plants and fungi cell wall. The study of these elicitors is essential for designing strategies to reduce negative effects of oxidative stress in plants. Therefore, the objective of this chapter was to review the oxidative stress generated in plants and its relationship with the elicitation of defense responses carried out by oligosaccharides, and particularly, by oligoglucans.

2. Oxidative stress and reactive oxygen species

Oxidative stress is defined as the rapid production of $O_2\bullet^-$ and / or H_2O_2 in response to various external stimuli (Wojtaszek, 1997) therefore their disturbance between production and elimination of the host cell. The decrease in catalytic activity of the plant antioxidant system is also a reason for oxidative stress to appear (Shigeoka et al., 2002). The balance of the antioxidant system may be disturbed by a large number of abiotic stresses such as bright light, drought, low and high temperatures and mechanical damage (Tsugane et al., 1999). The presence of heavy metals in the field, like pollution by lead (Pb) induces oxidative stress that damages cells and their components such as chloroplasts, in addition to altering the concentration of different metabolites including soluble proteins, proline, ascorbate and glutathione, and antioxidant enzymes (Reddy et al., 2005). On the other hand, processes related to the deterioration of fruits and vegetables, either by attack of pathogens, senescence or changes in the storage temperature are factors that increase ROS levels, leading to further economic losses (Reilly et al., 2004).

In plants, ROS are byproducts of diverse metabolic pathways localized in different cell compartments (chloroplasts, mitochondria and peroxisomes, mainly). Under physiological conditions, ROS are eliminated or detoxified by different components of enzymatic or non-enzymatic antioxidant defense system (Alscher et al., 2002). However, when plants are under the effect of single or multiple biotic and/or abiotic factors, the catalytic action of various antioxidants is negatively affected, allowing ROS accumulation that turns oxidative stress into an irreversible disorder (Qadir et al., 2004).

A common feature among different types of ROS is their ability to cause oxidative damage to proteins, lipids and DNA. However, depending on its intracellular concentration, ROS can also function as signaling molecules involved in the regulation and defense responses to pathogens, but mainly at very low concentrations (Apel & Hirt, 2004). It is proposed that ROS affect stress responses in two different ways. ROS act on a variety of biological molecules, causing irreversible damage leading to tissue necrosis and in extreme cases, death (Girotti, 2001). On the other hand, ROS affect the expression of several genes and signal transduction pathways related to plant defense (Apel & Hirt, 2004).

3. Antioxidant system in plants

The chloroplast is the cellular compartment associated with photosynthetic electron transport system and is a generous provider of oxygen, which is a rich source of ROS (Asada, 1999). In a second place, peroxisomes (glyoxisomes) and mitochondria are another ROS generating places inside the cell. A large number of enzymatic and non-enzymatic antioxidants have evolved to detoxify ROS and/or prevent the formation of highly reactive and damaging radicals such as hydroxyl radical ($\bullet OH$). Non-enzymatic antioxidants include ascorbate, glutathione (GSH), tocopherol, flavonoids, alkaloids, carotenoids and phenolic compounds. There are three key enzymatic antioxidants for detoxification of ROS in chloroplasts, superoxide dismutase (EC 1.15.1.1, SOD), ascorbate peroxidase (EC 1.11.1.11, APX) and catalase (EC 1.11.1.6, CAT). SOD catalyzes the dismutation of two molecules of $O_2\bullet^-$ in O_2 and H_2O_2. On the other hand, using ascorbate as electron donor, the enzyme APX reduces H_2O_2 to H_2O. The formation of hydroxyl radicals by $O_2\bullet^-$ and H_2O_2 can be controlled by the combination of dismutation reactions carried out by enzymes SOD, APX and CAT (Tang et al., 2006) (Figure 1).

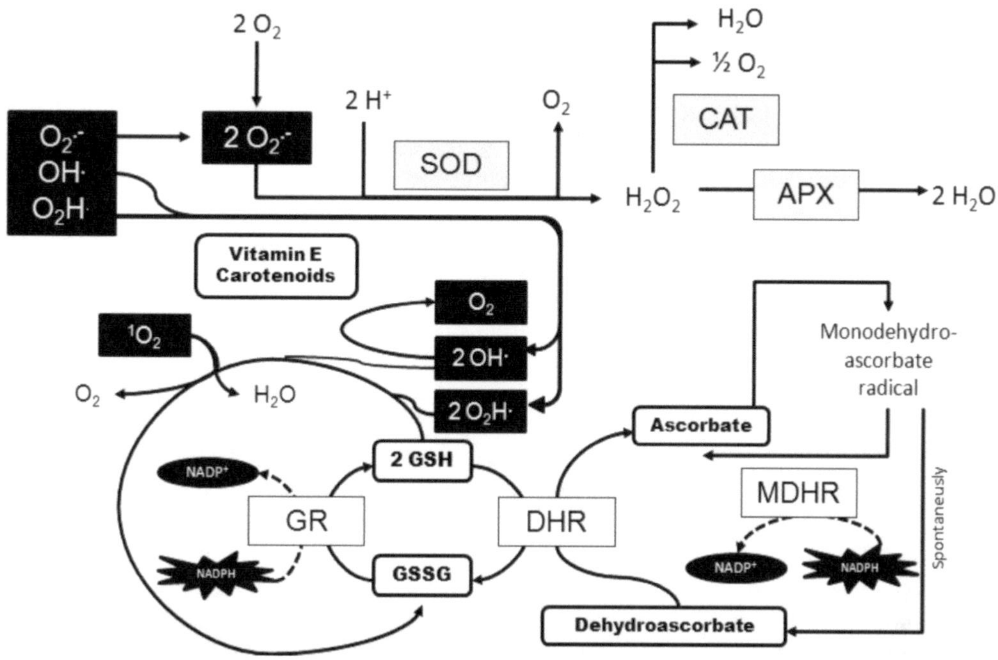

Fig. 1. Enzymatic and non-enzymatic antioxidant system in plants. Superoxide dismutase (SOD), catalase (CAT) and ascorbate peroxidase (APX) are the proteins responsible for eliminating ROS. While the elimination of ROS by non-enzymatic processes is carried out by vitamin E, carotenoids, ascorbate, oxidized glutathione (GSH) and reduced (GSSG). Enzymes that promote the elimination of ROS via the ascorbate-glutathione cycle are monodehydroascorbate reductase (MDHR), dehydroascorbate reductase (DHR) and glutathione reductase (GR) (Modified from Halliwell, 2006).

Superoxide Dismutase is a major ROS scavenging enzyme found in aerobic organisms. In plants, three types of SOD were distinguished on the basis of its active site cofactor: manganese SOD (MnSOD), copper / zinc SOD (Cu / ZnSOD) and iron SOD (FeSOD) (Reilly et al., 2004). CAT is a tetramer containing 4 heme groups, located mainly in peroxisomes (Apel & Hirt, 2004) and eliminates H_2O_2. It is proposed that CAT plays a role in mediating signal transduction where H_2O_2 acts as second messenger, possibly via a mechanism related to salicylic acid (Leon et al., 1995). On the other hand, APX enzyme has been found in higher plants, algae and some cyanobacteria, but not in animals. It is necessary for plants to have high levels of ascorbate to maintain functionally viable the endogenous antioxidant action of this enzyme (Shigeoka et al., 2002). APX activity in plants has increased in response to various stress conditions such as drought, ozone, chemicals, salinity, heat, infection (López et al., 1996; Mittler & Zilinskas, 1994). The sequencing of *Arabidopsis thaliana* genome has revealed the presence of 9 genes of APX (The Arabidopsis Genome Initiative, 2000). This fact shows how relevant the antioxidant enzymes-coding genes are in plants, as well as their down or up-regulated expression during stress conditions.

Different APX isoenzymes have been identified in plant cells: cytosolic (Ishikawa et al., 1995), peroxisomal (Ishikawa et al., 1998), two chloroplasmatic APX (in the stroma and thylakoid) (Ishikawa et al., 1996) and mitochondrial (De Leonardis et al., 2000). Each one, with a specific role as antioxidant enzyme, being activated or inhibited in response to different cellular signals as a consequence of biotic or abiotic stresses. The cytosolic APX isoenzyme has been considered one of the most important enzymes in defense against H_2O_2. Because of its cellular localization is the first to receive the signals produced during stress, acting very quickly to prevent severe damage to the cell and/or whole tissue. It has been reported the characterization of cDNAs encoding for cytosolic APX from various plants such as pea (Mittler & Zilinskas, 1992), Arabidopsis (Jespersen et al., 1997), rice (Morita et al., 1999), spinach (Webb & Allen, 1995) , tobacco (Orvar & Ellis, 1995) and potato (Kawakami et al., 2002; Park et al., 2004). However, the information about the genomic organization of the cytosolic APX is scarce, since there is only complete information of APX genes for tomato (Gadea et al., 1999) and pea (Mittler & Zilinskas, 1992).

4. Defense responses in plants during oxidative stress

During oxidative metabolic processes, ROS are generated at controllable levels and they play a key role in facilitating the defense of plants. This can be summarized in the following points: (1) strengthening the cell wall by structural carbohydrate modifications in linkages, (2) the induction of defense-related genes encoding protein-related proteins like glucanase, chitinase or protein inhibitors, and (3) causing cell death in a particular region of the plant (Reilly et al., 2004). During the defense response against pathogens, ROS are produced by the plant cell by increasing the activities of NADPH oxidase enzymes bound to plasma membranes, peroxidase attached to the cell wall and amino oxidase in the apoplast (Hammond-Kosack & Jones, 2000). The strengthening of the cell wall plays an important role in defense mechanisms against penetration by fungal pathogens (Bolwell et al., 2001). During defense responses by the attack of pathogens, plants produce higher levels of ROS while decreasing the detoxifying capacity, then the accumulation of ROS and activation of programmed cell death (PCD) happens. The suppression of ROS removal mechanisms is crucial for the establishment of the PCD. The production of ROS in the apoplast alone without the detoxification of ROS does not result in the induction of PCD (Delledonne et al., 2001).

Reactive Oxygen Species are among the major signaling molecules in the cell. These molecules are small and can diffuse a short distance, and there are several mechanisms for its production, many of which are fast and controllable. H_2O_2 generation occurs locally and systemically in response to mechanical damage or wounding (Orozco-Cardenas & Ryan, 1999). Other research shows that H_2O_2 acts as a second messenger mediating the systemic expression of several defense-related genes in tomato plants (Orozco-Cardenas et al., 2001).

5. Biological active elicitors

An elicitor can be defined as a molecule which, when introduced in low concentrations in a biological system, initiates or promotes the synthesis of biologically active metabolites (Radman et al., 2003). The type and structure of elicitors varies greatly, so there is no universal elicitor (Radman et al., 2003). Various elicitors have been purified: oligosaccharides, proteins, glycoproteins and lipophilic compounds (Coté & Hahn, 1994). The oligosaccharides are the most studied elicitors today. There are four types of oligosaccharides: oligoglucans, oligochitin, oligochitosan (predominantly from fungal source) and oligogalacturonides from plants (Coté & Hahn, 1994) (Figure 2). In the same way that the fungal and plant oligosaccharides have been studied, the oligosaccharides obtained from algae and animals have presented a great potential as signaling molecules (Delattre et al., 2005).

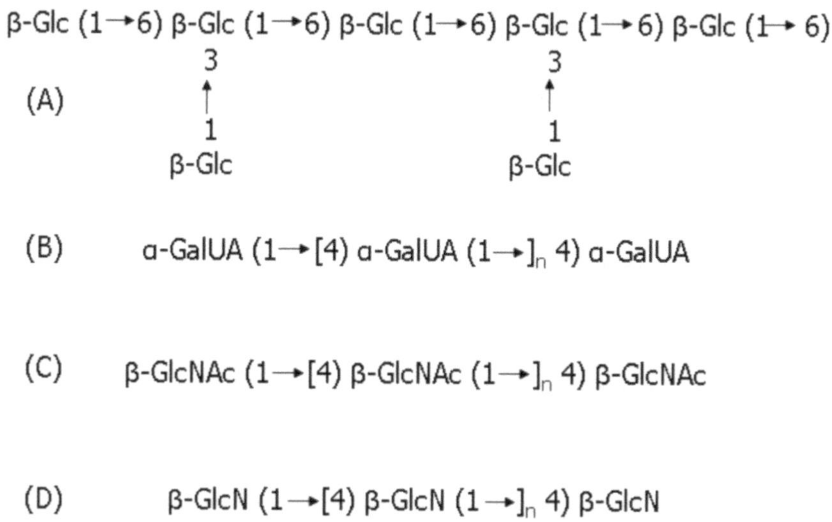

Fig. 2. Major oligosaccharides recognized by plants: (A) oligoglucans, (B) oligogalacturonide, (C) chitin-oligomer (D) chitosan-oligomer. Glc, glucose; GalUA, galacturonic acid; GlcNAc, N-acetyl glucosamine; GlcN, N-glucosamine.

5.1 Biochemical responses elicited by oligosaccharides

Of the major biochemical responses (Radman et al., 2003) that occur when a plant or cell culture is confronted with an elicitor are:

- Elicitor recognizing by plasma membrane receptor
- Changes in the flow of ions across the membrane
- Rapid changes in protein phosphorylation patterns
- Activation of NADPH oxidase enzyme complex responsible for ROS production and cytosolic acidification
- Reorganization of the cytoskeleton
- Accumulation of defense-related proteins
- Cell death at the site of infection (hypersensitive response)
- Structural changes in the cell wall (lignification, callose deposition)
- Transcriptional activation of defense related genes
- Synthesis of jasmonic acid and salicylic acid as second messengers
- Systemic acquired resistance

5.2 Oligoglucans

In the search for active oligosaccharides, at first it was considered the fungi kingdom, and specially biotrophic or necrotrophic fungi such as pests, because they cause important damage in plants, fruits and vegetables. But these organisms are the cue to reinforce the defense mechanisms of plants. When the plant-pathogen interaction occurs, several signaling receptor are activated by fungi or plant cell wall fragments, and then a biological response could be the main factor determining the survival or decline of plants. Many fungal pathogens have β-glucans as major components of their cell walls, which are recognized by different plant species (Yoshikawa et al., 1993). The Albersheim working group, at the middle of 70's, was the first to extract glucans elicitors of phytoalexins (a natural antimicrobial compound) in soybean from the mycelial walls of *Phytophthora megasperma* by heat treatment. These fungal wall structures were analyzed by Sharp et al., (1984) detailing the primary structure of an active glucan from *Phytophthora megasperma* f. sp. *glycinea* (Pmg) obtained by partial acid hydrolysis, finding that the hepta-β-glucoside elicitor was the active subunit.

Partial characterization of the fraction with elicitor activity from Pmg walls showed β-glucans with terminal residues 1-3 (42%), 1-6 (2%) and 1-3, 1-6 (27 %) glycosidic bonds (Sharp et al., 1984; Waldmüller et al., 1992). They observed that the obtention method of the cell wall fragments influenced the type of links present in the fungal elicitor. If the elicitor is released naturally or by heat treatment, then elicitors differ greatly from those glucans obtained by partial acid hydrolysis. While naturally released glucans have β-(1-3, 1-6) ramifications, β-(1-6) links are in greater proportion when glucans are released from acid hydrolysis (Waldmüller et al., 1992).

5.3 Oligoglucan receptors in plants

The recognition of elicitors by plants could be possible if the oligoglucan-receptor interaction occurs (Yoshikawa et al., 1993). In plants, receptors of fungal elicitors are found on the cell surface, while bacterial receptors are found within the cell (Ebel & Scheel, 1997). Other binding sites for oligosaccharides, glycopeptides, peptides and proteins are located on the cell surface and in the membranes (Cosio et al., 1990). Hence, many defense responses could be activated against pathogens, if the correct single or complex mixtures of elicitors are applied in healthy or unhealthy plants.

Binding proteins have been reported in soybean membranes for the hepta-β-glucosides (1-3, 1-6) and their branching fractions (Cosio et al., 1992). Other binding sites for yeast glycopeptides have been reported in tomato cells (Basse et al., 1993), for chitin-oligosaccharides these binding proteins have been found in tomato, rice (Baureithel et al., 1994) and parsley cells (Nürnberger et al., 1994). On the other hand, induction of phytoalexins by fungal β-glucans showed good correlation with the presence or absence of high affinity binding sites in several Fabaceae family plants (Cosio et al., 1996). A key method for assessing the presence of receptors on the membranes is through homogeneous ligand binding assays in isolated membranes (Yoshikawa et al., 1993). The radiolabeled ligand competition experiments using non-derivatized hepta-β-glucan as a competitive agent showed the existence of specific binding in at least four (alfalfa, bean, lupin and pea) of six species of Fabaceae family plants analyzed (Cosio et al., 1996).

The active oligoglucans can be isolated from the cell wall of algae and phytopathogenic fungi (Shinya et al., 2006). The oligoglucan laminarin is a β-(1-3)-glucan branching β-(1-6) glucose, which significantly stimulates defense responses in various crops including tobacco. The best known fungal elicitor is the heptaglucan (penta-β-(1-6) glucose with two branches β-(1-3) glucose) that was isolated from the cell walls of *Phytophthora megasperma*. This oligoglucan elicits defense responses in soybean cell cultures but not in cell cultures of tobacco or rice (Cheong & Hahn, 1991; Klarzinsky et al., 2000, Yamaguchi et al., 2000). A branched oligoglucan isolated from *Pyricularia oryzae* induces phytoalexins in rice but not in soybean (Yamaguchi et al., 2000). Linear oligoglucans were active in tobacco (Klarzinsky et al., 2000), but not in rice (Yamaguchi et al., 2000) or soybean plants (Cheong & Hahn, 1991). Another oligoglucans obtained from the cell walls of *Colletotrichum lindemuthianum* produce oxidative damage, common plant response to the invasion of pathogens, has been extensively studied in cell cultures of *Phaseolus vulgaris* (Sudha & Ravishankar, 2002). This clearly explains the great diversity of oligoglucans and the various biological effects that can be generated in the plant or crop to be evaluated. Clearly these facts show that the successful recognition for this kind of elicitor depends on specific plant receptors among plant species, even within families.

5.4 Oligoglucans action mechanism in plants

At the present time, only few reports about the action mechanisms of oligoglucans have been described. These reports focused in the final steps of the defense response, mainly during fungal attack, while other abiotic factors such as stress by uncontrollable temperatures (heat or cooling) have been less addressed. In order to address these issues, Doke *et al.*, (1996) proposed a mechanism of oxidative damage in plant cells in response to elicitors derived from fungal cell wall. The invasive fungal elicitor molecule (oligoglucan or, if the elicitation is mediated by pectic oligogalacturonic from plants) is recognized by the plasma membrane receptor (peripherial or transmembrane proteins), this recognition stimulates Ca^{2+} influx through Ca^{2+} channels. The increase in free Ca^{2+} in the cell acts as a second messenger, together with the activation of calmodulin (CaM) to activate protein kinases and protein factors by phosphorylation. Then the activated NADPH oxidase provides electrons through the oxidation of NADPH, and the electron transport system reduces O_2 molecules generating the radical $O_2^{\bullet -}$ (Figure 3).

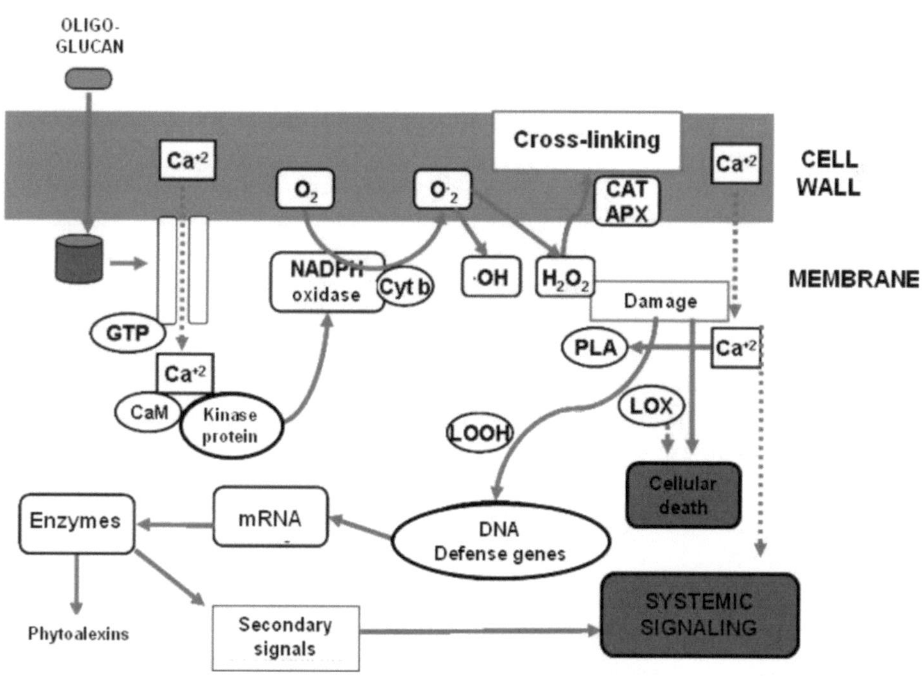

Fig. 3. Oligoglucans action mechanism in plants (modified Doke *et al.*, 1996).

6. Fungal glucans and their relationship with the enzymatic antioxidant system in cold stressed plants

Every day, the non-desirable climate change effects are present in our agriculture and the worldwide food production suffers the adverse consequences. Therefore, crop yields fell around fifty percent for several crops (Wahid et al., 2007). Several environmentally agencies report increments or reductions in temperature along the year. It is crucial to find an environmental friendly solution to challenge against low crop yields.

Under thermal stress (heat or chilling temperatures), important metabolic and physiologic plant processes are interrupted. As a consequence, protein aggregation and denaturalization in chloroplasts and mitochondria, destruction of membrane lipids, production of toxic compounds and the ROS overproduction (Howarth, 2005) are the most common responses of plant cells. Those are some reasons of the destructive effects of this kind of abiotic stress.

There are several pre- and postharvest treatments to deal with thermal stress like genetic modifications, thermal conditioning treatments of seeds and fruits or triggering early defense systems in plants by exogenous elicitation (Falcón-Rodríguez et al., 2009; Islas-Osuna et al., 2010). Our work team, evaluated the triggering of some important antioxidant enzymes in squash (*Cucurbita pepo* L.) seedlings at low temperature by the spraying of a novel mixture of fungal glucans isolated from *Trichoderma harzianum* by chemical and/or enzymatic fungal cell wall hydrolysis (Cerón-García et al., 2011). Two of the most active antioxidant enzymes, catalase and ascorbate peroxidase, were triggered by the exogenous elicitation with fungal oligoglucans in cold-stressed squash seedlings. Both antioxidant enzymes are the main active H_2O_2 detoxificant elements in the plant cell. Antioxidant enzymatic system in plants became unstable under thermal stresses, mainly by the inhibition of the catalytic activities during extreme temperatures. However, the elicitation with fungal glucans restored the deficiency of the antioxidant enzymatic system.

7. Conclusion

Biotic and abiotic factors may have a negative effect on plants, favoring the accumulation of ROS to generate further oxidative stress. Multiple biochemical responses are clearly generated by the use of oligoglucans as elicitors of defense responses against oxidative stress. The recognition of elicitors may vary depending on their characteristics, on the plant species or even for a particularly tissue, where specific receptors enables the generation of secondary signals that promote the most active plant defense against various biotic and/or abiotic factors by strengthening the antioxidant system, the accumulation of antimicrobial compounds such as phytoalexins and the activation of plant defense-related genes. Since there is little research on plant-oligoglucan interactions, so many questions remain unanswered.

8. Acknowledgment

Abel Ceron-García thanks the fellowship from Consejo Nacional de Ciencia y Tecnología (CONACyT). The authors would like to thank Olivia Briceño-Torres, Francisco Soto-Cordova and Socorro Vallejo-Cohen for technical assistance. We also thank Emmanuel Aispuro-Hernández for critical reading of the manuscript.

9. References

Alscher, R.G.; Erturk, N. & Heath, L.S. (2002). Role of superoxide dismutases (SODs) in controlling oxidative stress in plants. *Journal of Experimental Botany*, Vol.53, No. 372. pp. 1331-1341. http://jxb.oxfordjournals.org/cgi/content/abstract/53/372/1331.

Apel, K. & Hirt, H. (2004). Reactive oxygen species: Metabolism, oxidative stress, and signal transduction. *Annual Review in Plant Biology*, Vol.55, pp. 373-399. ISBN/ISSN 1543-5008.

Asada, K. (1999). The water–water cycle in chloroplasts: scavenging of active oxygen and dissipation of excess photons. *Annual Review in Plant Physiology and Plant Molecular Biology*, Vol.50, pp. 601-639. DOI: 10.1146/annurev.arplant.50.1.601.

Basse, C.W.; Fath, A. & Boller, T. (1993). High affinity binding of a glycopeptide elicitor to tomato cells and microsomal membranes and displacement by specific glycan suppressors. *The Journal of Biological Chemistry*, Vol.268, pp.14724-14731. ISSN 0021-9258.

Baureithel, K.; Félix, G. & Boller, T. (1994). Specific high affinity binding of chitin fragments to tomato cells and membranes. *The Journal of Biological Chemistry*, Vol.269, pp. 17931-17938. ISSN 0021-9258.

Bolwell, G.P.; Page, A.; Pislewska, M. & Wojtaszek, P. (2001). Pathogenic infection and the oxidative defences in plant apoplast. *Protoplasma*, Vol.217. pp. 20-32. ISBN/ISSN 0033-183X.

Cerón-García, A.; Gonzalez-Aguilar, G.A.; Vargas-Arispuro, I.; Islas-Osuna, M.A. & Martinez-Tellez, M.A. (2011). Oligoglucans as Elicitors of an Enzymatic Antioxidant System in Zucchini Squash (*Cucurbita pepo* L.) Seedlings at Low Temperature. *American Journal of Agricultural and Biological Sciences*, Vol.6, No. 1. pp. 52-61. ISSN 1557-4989.

Cheong, J.J. & Hahn, M.G. (1991). A specific, high affinity binding site for the hepta-β-glucoside elicitor exists in soybean membranes. *The Plant Cell*, Vol.3, pp. 137-147. ISSN 1040-4651.

Cosio, E.G.; Feger, M.; Miller, C.J.; Antelo, L. & Ebel, J. (1996). High-affinity binding of fungal β-glucan elicitors to cell membranes of species of the plant family Fabaceae. *Planta*, Vol.200, pp. 92-99. DOI: 10.1007/BF00196654.

Cosio, E.G.; Frey, T. & Ebel, J. (1992). Identification of a high-affinity binding protein for a hepta-β-glucoside phytoalexin elicitor in soybean. *European Journal of Biochemistry*, Vol.204, pp. 1115-1123. DOI: 10.1111/j.1432-1033.1992.tb16736.x.

Cosio, E.G.; Frey, T.; Verduyn, R.; Van Boom, J. & Ebel, J. (1990). High-affinity binding of a synthetic heptaglucoside and fungal glucan phytoalexin elicitors to soybean membranes. *FEBS Letters*, Vol.271, pp. 223-226. DOI: 10.1016/0014-5793(90)80411-B.

Coté, F. & Hahn, M.G. (1994). Oligosaccharins: Structure and signal transduction. *Plant Molecular Biology*, Vol.26, pp. 1379-1411. DOI: 10.1007/BF00016481.

De Leonardis, S.; Dipierro, N. & Dipierro, S. (2000). Purification and characterization of an ascorbate peroxidase from potato tuber mitochondria. *Plant Physiology and Biochemistry*, Vol.38, pp. 773-779. DOI: 10.1016/S0981-9428(00)01188-8.

Delattre, C.; Michaud, P.; Lion, J. & Courtois, J. (2005). Production of glucuronan oligosaccharides using a new glucuronan lyase activity from a *Trichoderma* sp. strain. *Journal of Biotechnology*, Vol.118, pp. 448-457. ISBN/ISSN 0168-1656.

Delledonne, M.; Marocco, A. & Lamb, C. (2001). Signal interactions between NO and reactive oxygen intermediates in the plant hypersensitive disease resistance response. *Proceedings of the National Academy of Sciences of the United States of America,* Vol.98, pp. 13454-13459. DOI: 10.1073/pnas.231178298.

Doke, N.; Miura, Y.; Sanchez, L.M.; Park, H.J.; Noritake, T.; Yoshioka, H. & Kawakita, K. (1996). The oxidative burst protects plants against pathogen attack: mechanism and role as an emergency signal for plant bio-defense - a review. *Gene,* Vol.179, pp. 45-51. ISBN/ISSN 0378-1119.

Ebel, J. & Scheel, D. (1997). Signals in host-parasite interactions. In: *The Mycota V Part A. Plant Relationships.* G C Carroll, T Tudzynski (Eds). pp. 85-105. Springer-Verlag. Berlin. Heidelberg.

Falcón-Rodríguez, A.B.; Cabrera, J.C.; Ortega, E. & Martinez-Tellez, M.A. (2009). Concentration and physicochemical properties of chitosan derivatives determine the induction of defense responses in roots and leaves of tobacco (*Nicotiana tabacum*) plants. *American Journal of Agricultural and Biological Sciences,* Vol.4, pp. 192-200. ISSN 1557-4989.

Gadea, J.; Conejero, V. & Vera, P. (1999), Developmental regulation of a cytosolic ascorbate peroxidase gene from tomato plants. *Molecular Genomics and Genetics,* Vol.262, pp. 212-219. DOI: 10.1007/s004380051077.

Girotti, A.W. (2001). Photosensitized oxidation of membrane lipids: reaction pathways, cytotoxic effects and cytoprotective mechanisms. *Journal of Photochemistry & Photobiology,* Vol.63, pp. 103-113. DOI: 10.1016/S1011-1344(01)00207-X.

Halliwell, B. (2006). Reactive species and antioxidants. Redox biology in fundamental theme of aerobic life. *Plant Physiology,* Vol.141, pp. 312-322. www.plantphysiol.org/cgi/doi/10.1104/pp.106.077073.

Hammond-Kosack, K. & Jones, J.D.G. (2000). Responses to plant pathogens. In: *Biochemistry and Molecular Biology of Plants.* B.B. Buchanan, W. Gruissem, R.L. Jones (Eds). pp. 1102-1156. American Society of Plant Physiologist. ISBN 0-943088-37-2. Rockville, MD.

Howarth, C.J. (2005). Genetic Improvements of Tolerance to High Temperature, In: Abiotic *stresses: Plant resistance through breeding and molecular approaches,* Ashraf, M. & Harris, P.J.C. pp. 725. Howarth Press Inc., ISBN: 1-56022-965-9. New York, USA.

Ishikawa, T., Sakai, K, Takeda, T. & Shigeoka, S. (1995). Cloning and expression of cDNA encoding a new type of ascorbate peroxidase from spinach. *FEBS Letters,* Vol.367, pp. 28-32. DOI: 10.1016/0014-5793(95)00539-L.

Ishikawa, T.; Sakai, K.; Yoshimura, K.; Takeda, T. & Shigeoka, S. (1996). cDNAs encoding spinach stromal and thylakoid-bound ascorbate peroxidase, differing in the presence or absence of their 3'-coding regions. *FEBS Letters,* Vol.384, pp. 289-293. DOI: 10.1016/0014-5793(96)00332-8

Ishikawa, T.; Yoshimura, K.; Sakai, K.; Tamoi, M.; Takeda, T. & Shigeoka, S. (1998). Molecular characterization and physiological role of a glyoxisome-bound ascorbate peroxidase from spinach. *Plant Cell Physiology,* Vol. 30, pp. 23-34. ISSN 0032-0781.

Islas-Osuna, M.A., N.A. Stephens-Camacho, C.A. Contreras-Vergara, M. Rivera Dominguez, E. Sanchez Sanchez, M.A. Villegas-Ochoa and G.A. Gonzalez Aguilar, 2010. Novel postharvest treatment reduces ascorbic acid losses in mango (*Mangifera indica* L.) Var. Kent. Am. J. Agric. Biol. Sci., 5: 342-349. ISSN: 15574989.

Jespersen, H.; Kjaersgard, I.; Ostergaard, L. & Welinder, K. (1997). From sequence analysis of three novel ascorbate peroxidases from *Arabidopsis thaliana* to structure, function and evolution of seven types of ascorbate peroxidase. *Biochemical Journal,* Vol.326, pp. 305-310. PMCID: PMC1218670.

Kawakami, S.; Matsumoto, Y.; Matsunaga, A.; Mayama, S. & Mizuno, M. (2002). Molecular cloning of ascorbate peroxidase in potato tubers and its response during storage at low temperature. *Plant Science,* Vol.163, pp. 829-836.

Klarzinsky, O.; Plesse, B.; Joubert, J.M.; Yvin, J.C.; Kopp, M.; Kloareg, B. & Fritig, B. (2000). Linear β-1,3 glucans are elicitors of defense responses in tobacco. *Plant Physiology,* Vol.124, pp. 1027-1037.

Leon, J.; Lawton, M. & Raskin, I. (1995). Hydrogen peroxide stimulates salicylic acid biosynthesis in tobacco. *Plant Physiology,* Vol.108, pp. 1673-1678.

López, F.; Vansuyt, G.; Case-Delbart, F. & Fourcroy, P. (1996). Ascorbate peroxidase activity, not the mRNA level, is enhanced in salt stressed *Raphanus sativas* plants. *Physiological Plantarum,* Vol.97, pp. 13-20.

Mittler, R. & Zilinskas, B.A. (1992). Molecular cloning and characterization of a gene encoding pea cytosolic ascorbate peroxidase. *The Journal of Biological Chemistry,* Vol.267, pp. 21802-21807. ISSN 0021-9258.

Mittler, R. & Zilinskas, B.A. (1994). Regulation of pea cytosolic ascorbate peroxidase and other antioxidant enzymes during the progression of drought stress and following recovery from drought. *The Plant Journa,l* Vol.5, pp. 397-405. DOI: 10.1111/j.1365-313X.1994.0

Morita, S.; Kaminaka, H.; Masumura, T. & Tanaka, K. (1999). Induction of rice cytosolic ascorbate peroxidase mRNA by oxidative stress; the involvement of hydrogen peroxide in oxidative signal. *Plant and Cell Physiology,* Vol.1999, No.40, pp. 417-422. ISSN 0032-0781.

Nürnberger, T.; Nennstiel, D.; Jabs, T.; Sacks, W.R.; Hahlbrock, K. & Scheel, D. (1994). High affinity binding of a fungal oligopeptide elicitor to parsley plasma membranes triggers multiple defense responses. *Cell,* Vol.78, pp. 449-460. DOI: 10.1016/0092-8674(94)90423-5.

Orozco-Cardenas, M.L.; Narvaez-Vasquez, J. & Ryan, C.A. (2001). Hydrogen peroxide acts as a second messenger for the induction of defense genes in tomato plants in response to wounding, systemin, and methyl jasmonate. *The Plant Cell,* Vol.13, pp. 179-191. DOI: 10.1105/tpc.13.1.179

Orozco-Cardenas, M.L. & Ryan, C.A. (1999). Hydrogen peroxide is generated systemically in plant leaves by wounding and systemin via the octadecanoid pathway. *Proceedings of the National Academy of Sciences of the United States of America,* Vol.96, pp. 6553-6557. DOI: 10.1073/pnas.96.11.6553

Orvar, B. & Ellis, B. (1995). Isolation of a cDNA encoding cytosolic ascorbate peroxidase in Tobacco. *Plant Physiology,* Vol.108, pp. 839-840. PMCID: PMC157414.

Park, S.Y.; Ryu, S.H.; Jang, I.C.; Kwon, S.Y.; Kim, J.G. & Kwak, S.S. (2004). Molecular cloning of a cytosolic ascorbate peroxidase cDNA from cell cultures of sweetpotato and its expression in response to stress. *Molecular Genetics and Genomics,* Vol.271, No. 3. pp. 339-346. DOI 10.1007/s00438-004-0986-8

Qadir, S.; Qureshi, M.I.; Javed, S. & Abdin, M.Z. (2004). Genotypic variation in phytoremediation potential of *Brassica juncea* cultivars exposed to Cd-stress. *Plant Science,* Vol.167, pp. 1171-1181. DOI: 10.1016/j.plantsci.2004.06.018

Radman, R.; Saez, T.; Bucke, C. & Keshavarz, T. (2003). Elicitacion of plant and microbial cell systems. *Biotechnology Applied Biochemistry,* Vol.37, pp. 91-102. ISBN/ISSN 0885-4513.

Reddy, A.M.; Kumar, S.G.; Jyothsnakumari, G.; Thimmanaik, S. & Sudhakar, C. (2005). Lead induced changes in antioxidant metabolism of horsegram (*Macrotyloma uniflorum* (Lam.) Verdc.) and bangalgram (*Cicer arietinum* L.). *Chemosphere,* Vol.60, pp. 97-104.

Reilly, K.; Gomez-Vasquez, R.; Buschmann, H. & Beeching, J.R. (2004). Oxidative stress responses during cassava post-harvest physiological deterioration. *Plant Molecular Biology,* Vol.56, pp. 625-641.

Sharp, J.K.; Valent, B. & Albersheim, P. (1984). Purification and partial characterization of a β-Glucan fragment that elicits phytoalexin accumulation in soybean. *The Journal of Biological Chemistry,* Vol.259, No. 18. pp. 11312-11320.

Shigeoka, S.; Ishikawa, T.; Tamoi, M.; Miyagawa, Y.; Takeda, T. & Yoshimura, K. (2002). Regulation and function of ascorbate peroxidase isoenzymes. *Journal of Experimental Botany,* Vol.53, No. 372. pp. 1305-1319. http://jxb.oxfordjournals.org/cgi/content/abstract/53/372/1305

Shinya, T.; Ménard, R.; Kozone, I.; Matsuoka, H.; Shibuya, N.; Kauffmann, S.; Matsuoka, K. & Saito, M. (2006). Novel β-1,3-, 1,6-oligoglucan elicitor from *Alternaria alternata* 102 for defense responses in tobacco. *FEBS Journal,* Vol.273, No. 11. pp. 2421-2431. ISBN/ISSN 1742-4658.

Sudha, G. & Ravishankar, G.A. (2002). Involvement and interaction of various signaling compounds on the plant metabolic events during defense response, resistance to stress factors, formation of secondary metabolites and their molecular aspects. *Plant Cell, Tissue and Organ Culture,* Vol.71, pp. 181-212.

Tang, L.; Kwon, S.Y.; Kim, S.H.; Kim, J.S.; Choi, J.S.; Cho, K.Y.; Sung, C.K.; Kwak, S.S. & Lee, H.S. (2006). Enhanced tolerance of transgenic potato plants expressing both superoxide dismutase and ascorbate peroxidase in chloroplasts against oxidative stress and high temperature. *Plant Cell Report,* Vol.25, No. 12. pp. 1380-1386. DOI 10.1007/s00299-006-0199-1.

The Arabidopsis Genome Initiative. (2000). Analysis of the genome sequences of the flowering plant *Arabidopsis thaliana*. *Nature,* Vol.408, pp. 796-815.

Tsugane, K.; Kobayashi, K.; Niwa, Y.; Ohba, Y.; Wada, K. & Kobayashi, H. (1999). A recessive Arabidopsis mutant that grows enhanced active oxygen detoxification. *Plant Cell,* Vol.11, pp. 1195-206. PMC: 144266.

Wahid, A.; Gelani, S.; Ashraf, M. & Foolad, M.R. (2007). Heat tolerance in plants: An overview. *Environmental & Experimental Botany,* Vol.61, pp. 199-223. ISBN/ISSN 0098-8472.

Waldmüller, T.; Cosio, E.G.; Grisebach, H. & Ebel, J. (1992). Release of highly elicitor-active glucans by germinating zoospores of *Phytophthora megasperma glycinea*. *Planta,* Vol.188, pp. 498-505. DOI: 10.1007/BF00197041.

Webb, R. & Allen, R. (1995). Isolation and characterization of a cDNA for spinach cytosolic ascorbate peroxidase. *Plant Physiology,* Vol.108, pp. 1325. PMC: 157502.

Wojtaszek, P. (1997). Oxidative burst: an early plant response to pathogen infection. *Biochemical Journal,* Vol.322, pp. 681–692. PMC 1218243.

Yamaguchi, T.; Yamada, A.; Hong, N.; Ogawa, T.; Ishii, T. & Shibuya, N. (2000). Differences in the recognition of glucan elicitor signals between rice and soybean: beta-glucan fragments from the rice blast disease fungus *Pyricularia oryzae* that elicit phytoalexin biosynthesis in suspension-cultured rice cells. *The Plant Cell,* Vol.12, No. 5. pp. 817-826. http://www.plantcell.org/cgi/content/abstract/12/5/817.

Yoshikawa, M.; Yamaoka, N. & Takeuchi, Y. (1993). Elicitors: Their significance and primary modes of action in the induction of plant defense reactions. *Plant Cell Physiology,* Vol.34, No. 8. pp. 1163-1173. ISSN 0032-0781.

2

Regulation of Gene Expression in Response to Abiotic Stress in Plants

Bruna Carmo Rehem[1], Fabiana Zanelato Bertolde[1] and Alex-Alan Furtado de Almeida[2]
[1]*Instituto Federal de Educação, Ciência e Tecnologia da Bahia (IFBA)*
[2]*Universidade Estadual de Santa Cruz*
Brazil

1. Introduction

The multiple adverse conditions but not necessarily lethal, that occur sporadically as either permanently in a location that plants grow are known as "stress." Stress is usually defined as an external factor that carries a disadvantageous influence on the plant, limiting their development and their chances of survival. The concept of stress is intimately related to stress tolerance, which is the plant's ability to confront an unfavorable environment. Stress is, in most definitions, considered as a significant deviation from the optimal conditions for life, and induces to changes and responses in all functional levels of the organism, which are reversible in principle, but may become permanent.

The dynamics of stress include loss of stability, a destructive component, as well as the promotion of resistance and recovery. According to the dynamic concept of stress, the organism under stress through a series of characteristic phases. *Alarm phase:* the start of the disturbance, which is followed by loss of stability of structures and functions that maintain the vital activities. A very rapid intensification of the stressor results in an acute collapse of cellular integrity, before defensive measures become effective. The alarm phase begins with a stress reaction in which the catabolism predominates over anabolism. If the intensity of the stressor does not change the restitution in the form of repair processes such as protein synthesis or synthesis of protective substances, will be quickly initiated. This situation leads to a *resistance phase,* in which, under continuous stress, the resistance increases (hardening). Due to the improved stability, normalization occurs even under continuous stress (adaptation). The resistance may remain high for some time after the disturbance occurred. If the state of stress is too lengthy or if the intensity of the stress factor increases, a state of exhaustion can occur at the *final stage,* leaving the plant susceptible to infections that occur as a consequence of reduced host defenses and leading to premature collapse or still a chronic damage may occur, leading to plant death. However, if the action of the stressor is only temporary, functional status is restored to its original level. If necessary, any injury caused can be repaired during the restitution (Larcher, 1995).

The characteristics of the state of stress are manifestations nonspecific, which represent firstly an expression of the severity of a disturbance. A process can be considered nonspecific if it can not be characterized as a pattern, whatever the nature of the stressor.

Examples of non-specific indications of the state of stress are: increased respiration, inhibition of photosynthesis, reduction in dry matter production, growth disorders, low fertility, premature senescence, leaf chlorosis, anatomical alterations and decreased intracellular energy availability or increased energy consumption due to repair synthesis. The cell responses to stress include changes in cell cycle and division, changes in the system of vacuolization, and changes in cell wall architecture. All this contributes to accentuate tolerance of cells to stress. Biochemically, plants alter metabolism in several manners, to accommodate environmental stress (Hirt & Shinozaki, 2004).

Currently, all plant life is being threatened by rapid environmental changes. The gases associates to global warming as CO_2 and methane have a enormous impact on global environmental conditions, resulting in extreme changes in temperatures and weather patterns in many regions of the world (Hirt & Shinozaki, 2004). In contrast to animals, plants are sessile organisms and can not escape from environmental changes. The greenhouse effect also affects the ozone layer causing the levels of ultraviolet (UV) are much larger to reach the ground (Hirt & Shinozaki, 2004). Besides resulting in an increase in the registers of the occurrence of diseases in humans such as skin cancer. The greenhouse effect also affects the ozone layer causing the levels of ultraviolet (UV) are much larger to reach the ground. Another concern is the intense use of chemical fertilizers and artificial irrigation in agriculture. In many areas of the world, these practices have increased soil salinity. Under these conditions, resistance to abiotic stress corresponds to a more required to be found in several plant species (Hirt & Shinozaki, 2004). In short, the factors discussed above, together with the increasing use of agricultural land cultivated is one of the biggest challenges for the future humanity with regard to agriculture and conservation of genetic diversity in plant species.

2. Water stress

Water has a key role in all physiological processes of plants, comprising between 80 and 95% of the biomass of herbaceous plants. If water becomes insufficient to meet the needs of a particular plant, this will present a water deficit. The water deficit or drought is not caused only by lack of water but also the environment in low temperature or salinity. These different tensions negatively affect plant productivity (Hirt & Shinozaki, 2004).

Plants developed different mechanisms to adapt their growth in conditions where water is limited. These adjustments depend on the severity and duration of drought, as well as the development phase and morphology and anatomy of plants. The cellular response includes the action of solute transporters such as aquaporin, activators of transcription, some enzymes, reactive oxygen species and protective proteins. Two main strategies can be taken to defend the damage caused by dehydration: synthesis of molecules of protection to prevent damage and a repair mechanism based on rehydration in order to neutralize the damage. In the classic signaling pathways, environmental stimuli are captured by receptor molecules (Hirt & Shinozaki, 2004).

The main response that distinguishes tolerant plants of sensitive plants to drought stress is the marked intracellular accumulation of osmotically active solutes in tolerant plants. This mechanism, known as osmotic adjustment, is the ability of many species adjusts their cells by decreasing the osmotic potential and water potential in response to drought or salinity without a decrease in cell turgor.

In plants, dehydration activates a protective response to prevent or repair cell damage. The plant hormone, abscisic acid (ABA) has a central role in this process. The ABA is considered a "stress hormone" because plants respond to environmental challenges such as water and salt stress with changes in the availability of ABA, as well as being an endogenous signal required for adequate development. Dehydration in plants leads to increased levels of ABA, which in turn induces the expression of several genes involved in defense against the effects of water deficit. High levels of ABA cause complete closure of stomata and alteration of gene expression. Stomatal closure reduces water loss through transpiration (Hirt & Shinozaki, 2004). The ABA signaling is composed of multiple cellular events, including the regulation of turgor and differential gene expression.

Plants have developed several mechanisms to adapt their growth to the availability of water. The movement of water molecules is determined by water potential gradient across the plasma membrane, which in turn is influenced by the concentration of solute molecules inside and outside the plant cell. Fluctuations in water availability and flows of transmembrane extracellular solute disrupt cellular structures, altering the composition of the cytoplasm and modulate cell function (Hirt & Shinozaki, 2004).

One effect of the signal transduction cascade of dehydration is the activation of transcription factors, which each activates a set of target genes, including those necessary for the synthesis of protective molecules. Transcription factors that are activated by dehydration are differentially expressed in tissues. Dehydration causes high level of expression of many genes, among which the most prominent are the so called late embryogenesis abundant genes (LEA) (Hirt & Shinozaki, 2004). The last step in the signaling cascade in response to dehydration is the activation of genes responsible for synthesis of compounds that serve to protect cellular structures against the deleterious effects of dehydration. Plants that are able to survive in drought conditions have taken a variety of different strategies. There are three important mechanisms to allow the plants to resist dehydration: the accumulation of solutes, elimination of reactive oxygen species and synthesis of proteins with protective functions (Hirt & Shinozaki, 2004).

In many species, dehydration leads to the accumulation of a variety of compatible solutes. Compatible solutes are soluble molecules of low molecular weight that are not toxic and do not interfere with cellular metabolism. The chemical nature of solutes differ among plant species. They include betaines, including glycine betaine, amino acids (especially proline) and sugars such as mannitol, sorbitol, sucrose or trehalose. These compounds help to maintain turgor during dehydration, increasing the number of particles in solution. Furthermore, can modulate membrane fluidity and protein by keeping it hydrated, allowing the stabilization of its structure (Hoekstra et al., 2001).

One consequence of dehydration is an increase in the concentration of reactive oxygen intermediates (ROI) (Mittler, 2002). ROI cause irreversible damage to membranes, proteins, DNA and RNA. However, a low concentration of ROI is vital to the plant cells, they are essential components in defense signaling to stress. When the ROI concentration increases because of dehydration, prevention of damage to competitors is essential for survival. The accumulation of ROI is largely controlled by intrinsic antioxidant systems that include the enzymatic action of superoxide dismutase, peroxidases and catalases.

The analysis of differential gene expression and analysis of global patterns of gene expression using macro and microarray approaches have identified a broad spectrum of transcripts whose expression is modified in response to dehydration (Fowler & Thomashow, 2002; Kreps et al., 2002; Seki et al., 2002). These studies have provided a fairly comprehensive overview of the types of transcripts modulated by dehydration plant. They showed that at least hundreds of genes are affected by dehydration.

3. Oxidative stress

For plants, as for all aerobic organisms, oxygen is required for normal growth and development, but continuous exposure to oxygen can result in cellular damage and ultimately death. This is because molecular oxygen is continually reduced within cells by various forms of reactive oxygen species (ROS), especially the free radical superoxide anion (O_2^-) and hydrogen peroxide (H_2O_2), which react with many cellular components resulting in acute or chronic damage resulting in cell death (Scandalios, 2002). Oxidative stress results from disequilibrium in the generation and removal of ROS within cells. In plant cells, ROS are generated in large quantities by both constitutive and inducible pathways, but in normal situations, the cellular redox balance is maintained through the action of a great variety of antioxidant mechanisms that evolved to remove ROS.

The calcium ions may also be related to oxidative stress and the antioxidant system in plants. Oxidative stress, many enzymes are involved in the mechanisms of protecting the protoplasm and cell integrity. This defense includes antioxidant enzymes able to remove or neutralize free radicals and intermediate compounds that enable their production. Among these enzymes, highlight the peroxidase (POX), catalase (CAT) and superoxide dismutase (SOD). The mechanisms of elimination of reactive oxygen involving SOD, whose synthesis is induced probably by increased production of O_2^-. In the process, SOD converts O_2^- to hydrogen peroxide (H_2O_2) and then peroxidase and catalase removes hydrogen peroxide formed. Hydrogen peroxide and superoxide radicals can exert deleterious effects in cells, acting on lipid peroxidation of membranes, as well as damaging their DNA.

Several environmental stresses and endogenous stimuli can disrupt the redox balance by increasing ROS production or reduced antioxidant activity, with continued oxidative stress. In response toto increased ROS is induced the expression of genes encoding antioxidant proteins and the genes that encode proteins involved in a variety of cellular processes of rescue. ROS are produced during photosynthesis and respiration, as a byproduct of metabolism, or by specific enzymes. Cells are equipped with a variety of effective antioxidant mechanisms to eliminate ROS. Transcriptome analyses indicate that the expression of many genes is regulated by ROS. These antioxidants include genes that encode the rescue of cell defense proteins and signaling proteins. ROS can lead to programmed cell death, stomatal closure, and gravitropism (Hirt & Shinozaki, 2004).

Oxygen is normally reduced by four electrons to produce water, a reaction catalyzed by cytochrome oxidase complex and the electron transport chain of mitochondria. It is relatively unstable and can be converted back to molecular form of oxygen or H_2O_2, either spontaneously or through a reaction catalyzed by the enzyme superoxide dismutase (SOD). H_2O_2 in particular, acts as a signaling molecule with regulated synthesis, specific effects and presents a series of removal mechanisms (Hirt & Shinozaki, 2004).

The evolution of photosynthesis and aerobic metabolism led to the development of processes of generation of ROS in chloroplasts, mitochondria and peroxisomes. It seems likely that the antioxidant mechanisms have evolved to combat the negative effects of these ROS (Scandalios, 2002). As environmental pressures increase the generation of ROS, would have been the evolutionary pressure for selection of ROS signaling mechanisms inducing genes encoding antioxidant proteins and cellular defense. This role of "defense" of ROS and these proteins may be one reason that leads to induction of cellular defense, where many genes show a common response to various environmental stresses and oxidative stress, allowing for acclimatization and tolerance (Bowler & Fluhr, 2000). Functions of protection against ROS may also have been responsible for the evolution of enzymes such as NADPH oxidase, where the reaction seems to be the key ROS generation, in which the enzyme activity can be regulated by environmental stresses. Thus, abiotic stress not only increases the generation of ROS through non-specific mechanisms, but also trigger the signaling of defense mechanisms that start with the induction of ROS production, continue with the induction of defense responses and end with removal of ROS to restore the redox status and cell survival.

Oxidative stress causes the intracellular environment becomes more electropositive, which may induce a change in the redox environment and thus interfere with signaling pathways. ROS are generated both electron transport and enzymatic sources. The generation of ROS occurs through the process of electron transport in chloroplasts and mitochondria. H_2O_2 is generated by various enzymatic reactions, from specific enzymes such as NADPH oxidase (Hirt & Shinozaki, 2004). Plant cells are rich in antioxidants, where the activity and location of these can affect the concentration of H_2O_2.

Gene expression in response to oxidative stress may be coordinated through the interaction of transcription factors (TF) with cis-elements common to the entity regulatory regions of these genes. Certainly, the increase of ROS in cellular compartments such as mitochondria or chloroplasts results in new profiles of transcription (Hirt & Shinozaki, 2004).

4. Flooding stress

The temporary or continuous flooding of the soil resulting from high rainfall, intensive practice of large-scale irrigation farms or soils with inadequate drainage (Kozlowski, 1997). In normal drainage, the soil contains air-filled pores that contain content similar to oxygen from the atmosphere (20%) (Pezeshki, 1994). Excess water replaces the air in these pores, extremely restricting the flow of oxygen in the soil, creating a condition of hypoxia (low O_2 availability) or, in more severe cases, anoxia (lack of O_2) (Peng et al., 2005). The gas diffusion becomes extremely slow in soils saturated with water, about 10,000 times slower than in air (Armstrong, 1979).

Under natural conditions, the flooding changes numerous physical and chemical properties of soil through processes of biological reduction, resulting from depletion of available oxygen (O_2), increasing the availability of P, Mn and Fe, and decreased availability of Zn and Cu and the formation of hydrogen sulfide and organic acids (Camargo et al., 1999). This soil is also characterized by accumulating a larger amount of CO_2 (Jackson, 2004) and stimulate organic matter decomposition (Kozlowski, 1997; Pezeshki, 2001; Probert & Keating, 2000). The phytotoxic compounds that accumulate in flooded soils can be produced both by plant roots (ethanol and acetaldehyde) and the metabolism of anaerobic microorganisms (methane, ethane, unsaturated acids, aldehydes, ketones), and ethylene may be produced by by plants and microorganisms (Kozlowski, 1997).

Flooding can devastate vegetation species poorly adapted to this kind of stress (Jackson, 2004). The stress tolerance of hypoxia or anoxia can vary in hours, days or weeks depending on the species, the organs directly affected the stage of development and external conditions (Vartapetian & Jackson, 1997). The duration and severity of flooding may be influenced not only by the rate of influx of water, but also by the rate of water flow around the root zone and the absorption capacity of soil water (Jackson, 2004).

One major effect of flooding is the privation of O_2 in the root zone, attributed to the slow gas diffusion in soil saturated with water and O_2 consumption by microorganisms (Folzer et al., 2005). Higher plants are aerobic and O_2 supplies depend on the environment to support respiration and several other reactions of oxidation and oxygenation of vital (Vartapetian & Jackson, 1997). The O_2 as participate of aerobic respiration final electron acceptor in oxidative phosphorylation, generation of ATP and regeneration of NAD^+, and in several crucial biosynthetic pathways as the synthesis of chlorophyll, fatty acids and sterols (Dennis et al., 2000). Under hypoxia, glycolysis and fermentation can exceed the aerobic metabolism and become the only way to produce energy (Sousa & Sodek, 2002). The main products of fermentation in plant tissues are lactate, ethanol and alanine derived from pyruvate, the end product of glycolysis (Drew, 1997). Evidence suggests that cytosolic acidosis causes lactic fermentation and therefore the maintenance of cytosolic pH is important for the survival of plants under waterlogged conditions (Dennis et al., 2000).

The initial responses of many plants to flooding correspond to wilt and to stomatal closure, starting from one or two days of exposure of roots to this stress, accompanied by decreases in photosynthetic rate (Chen et al., 2002). These changes promote a decline in growth of stems and roots and can damage the roots and death of many species of plants (Kozlowski, 1997). The saturation of the soil causes significant decreases in total plant biomass and biomass allocation to roots and changes in biomass allocation pattern in many woody species and herbaceous (Pezeshki, 2001; Rubio et al., 1995).

The plant damage induced by hypoxia, have been attributed to physiological dysfunction, which include the change in the relationship between carbohydrates, minerals, water and hormones, as well as the reduction or alteration of several metabolic pathways (Kennedy et al., 1992). Initially, the plant under stress by hypoxia shows decreases in the rate of CO_2 uptake by leaves (Kozlowski, 1997). Some authors suggest that stomatal closure may be associated with a decrease in hydraulic conductivity of roots (Davies & Flore, 1986; Kozlowski, 1997), as well as the transmission of hormonal signals from roots to shoots. Among the hormones involved in signal transmission are the ABA and cytokinin (Else et al., 1996). During stages of prolonged flooding, the progressive decrease in photosynthetic rate is attributed to changes in enzyme carboxyl groups and loss of chlorophyll (Drew, 1997; Kozlowski, 1997). The decrease in activity of ribulose -1,5-bisphosphate carboxylase-oxygenase (RUBISCO), the enzyme responsible for assimilation of CO_2 in the biochemical phase of photosynthesis, contributing to losses in photosynthetic capacity (Pezeshki, 1994).

It is known that water temperature, light, O_2 and nutrient availability are the main abiotic factors that control the growth of woody plants (Kozlowski, 1997). The excess or shortage of any of these factors, such as water, causes significant deviations in the optimum conditions for growth for a given species, generating a stress condition. Such a condition, depending on the level of specialization of the organism, the amplitude and duration of stress, may be reversible or become permanent (Lichtenthaler, 1996). Excess water in the root zone of terrestrial plants may be injurious or even lethal because it blocks the transfer of O_2 and other gases between soil and atmosphere (Drew, 1997). The responses of plants to flooding vary according to several factors, among which may include the species, genotype and age of the plant, the properties of water and duration of flooding (Kozlowski, 1997).

Plant growth and primary productivity of ecosystems are ultimately dependent on photosynthesis (Pereira et al., 2001). Either environmental stressor that, somehow, can interfere with the photosynthetic rate, will affect the net gain of dry matter and, therefore, growth (Pereira, 1995). Growth and development of most vascular plant species are restricted by flooding, particularly when they are completely submerged and may result in death (Jackson & Colmer, 2005). Generally, waterlogged soils affect the growth of aerial part of many woody species, suppressing the formation of new leaves, retarding the expansion of leaves and internodes that formed before flooding and reduce growth of stem diameter of species not tolerant to flooding, causing senescence and premature leaf abscission (Kozlowski, 1997). The plant response to flooding during the growing phase, including injury, inhibiting seed germination, of vegetative and reproductive growth and changes in plant anatomy (Kozlowski, 1997).

Morphological changes, such as the formation of hypertrophied lenticels, aerenchyma and adventitious roots, observed during O_2 deficiency in the soil are key to increasing the availability of O_2 in the tissues of plants. The lenticels participate in the uptake and diffusion of O_2 to the root system and the liberation of potentially toxic volatile products such as ethanol, acetaldehyde and ethylene (Medri, 1998). The best strategy of flooding tolerance is the supply of internal aeration, increased with the formation of hypertrophied lenticels, which are the main points of entry of O_2 in the plants, associated with the appearance of intercellular air spaces (White & Ganf, 2002). According Topa & McLeod (1986), the increase of these air spaces allows for an efficient entry of O_2 causing the lenticels assume the role of gas exchange in hypoxic conditions. The formation of hypertrophic lenticels occurs in submerged portions of stems and roots of various woody angiosperms and gymnosperms (Kozlowski, 1997), and involves both increased activity felogênica as the elongation of cortical cells (Klok et al., 2002). In addition, participating in the uptake and diffusion of O_2 to the root system and release potentially toxic volatile products such as ethanol, acetaldehyde and ethylene (Medri, 1998).

It can be observed the formation of aerenchyma in stems and roots of aquatic species tolerant to flooding, which usually occurs by cell separation during development (esquizogeny) or by lysis of cortical cells and cell death (lysogeny) (He et al., 1994; Drew, 1997). The point of view adaptive aerenchyma provides a low resistance to diffusion of air inside the submerged tissue, promoting survival of plants to flooding (Drew, 1997). The formation of aerenchyma, recorded during the stress by O_2 deficiency in soil is associated with the accumulation of ethylene. Roots under flooded soil containing high concentrations of ethylene, compared with roots in normal, and its precursor (acid-1-aminocyclopropane-1-carboxylic acid - ACC), and high activity of ACC synthase and ACC oxidase (He et al., 1994; He et al., 1996). Ethylene induces the activity of different enzymes such as cellulases, and hydrolases xyloglucanases, enzymes, cell wall loosening related to the formation of aerenchyma (He et al., 1994; Drew, 1997).

In the submerged portion of the plant there is a dead roots and the production of adventitious roots on the root system portions of the original stem. These roots induced by flooding, are usually thickened and exhibit more intercellular spaces of the roots growing in well drained soils (Hook et al., 1971). The induction of adventitious roots has been reported in a wide variety of plant species tolerant and non-tolerant, but usually occurs in species tolerant to flooding (Kozlowski & Pallardy, 1997). According Kozlowski (1997), the adventitious roots are produced from the original roots and submerged portions of stems. According to the author in terms of flooding the induction of adventitious root formation can be reported in both angiosperms and gymnosperms in tolerant and non tolerant to this type of stress (Kozlowski, 1997). For Chen et al. (2002), the adventitious roots are important in plants with high root hypoxia, since they are responsible for obtaining O_2 needed for their development. The increased number of adventitious roots can be accompanied by an increment of damage and death of the original roots (Chen et al., 2002). Among the root adaptations to flooding can also cite the development of aerenchyma, induced by increasing endogenous levels of ethylene (Mckersie, 2001). This tissue serves as an air transport system in aquatic plants and can develop into plants that grow in hydromorphic soils. The intercellular spaces are developed primarily by disintegration of cells due to an increase in cellulase activity, or by increasing the intercellular spaces when there is lack of O_2 and hence increase in ethylene production (Fahn, 1982). In some plants, such stress induces abnormal formation of the wood and increase the proportion of parenchymatous tissues of xylem and phloem (Kozlowski, 1997).

Flooding can also cause a decline in the growth of petioles and leaf stomatal conductance (Domingo et al., 2002). Moreover, the saturation of the soil (i) interfere in the allocation of photoassimilates in woody and herbaceous plants, root can decrease metabolism and oxygen demand (Chen et al., 2002); (ii) inhibits the initiation of flower buds and the increase in fruit species not tolerant to flooding; (iii) induces abscission of flowers and fruits; (iv) reduces the quality of fruits due to the reduction of size, changing its appearance and interfering in its chemical composition (Kozlowski, 1997).

Other responses of plants to flooding include: (i) decreased permeability of the root and the absorption of water and mineral nutrients, the death and suppression of root metabolism; (ii) the epinasty, leaf chlorosis and necrosis, and (iii) decrease in fruit production. On the other hand, several morpho-physiological responses are driven by differential expression of a large number of genes induced by conditions of hypoxia or anoxia (Vartapetian & Jackson, 1997; Kozlowski, 1997; Holmberg & Büllow, 1998; Vantoai et al., 1994; Klok et al., 2002).

The decreased availability of O_2 also affects different processes of plant genetics (Blom & Voesenek, 1996; Kozlowski, 2002; Drew, 1997). Saab & Sachs (1995) observed in maize under conditions of flooding, the 1005 induction of the gene that encodes a homologue of the xyloglucan endotransglicosilase (x and t), an enzyme potentially involved in cell wall loosening (Peschke & Sachs, 1994). This gene, which is among the first to be induced by flooding, does not encode enzymes of glucose metabolism. Believed to be associated with the onset of the structural changes induced by flooding (Saab & Sachs 1995), because the substrate of XET and t are the xyloglucans that are part of the cell wall structure. Saab &

Sachs (1996) found that gene induction in regions of primary root and mesocotyl exhibiting signs of development of aerenchyma in flooded soil, suggesting an association between this enzyme and the structural changes induced by this type of stress.

The plants survive under conditions of anoxia, replacing the aerobic metabolism by anaerobic (Drew, 1997). However, before the metabolic adjustments are made, the stress can be perceived and signaled to induce the appropriate cellular and molecular responses (Dat et al., 2004). Studies on plants under conditions of O_2 deficiency in the soil have focused the role of calcium (Ca^{2+}) as a marker (Subbaiah et al. 1998; Subbaiah et al., 2000). Molecular processes are modulated by Ca^{2+} via high-affinity proteins such as calmodulin (CaM) and its isoforms in plants (Zielinski, 1998). The complex active Ca^{2+}-CaM can regulate the activity of many target molecules associated with plant responses to stress (Snedden & Fromm, 2001). Many CaM have been identified in herbaceous plants, but for woody plants have been little information. Folzer et al. (2005) were the first to identify a family of CaM in woody plant, *Quercus petraco* Liebl., demonstrating that the isolated isoforms exhibit organ-specific distribution and differential expression in the plant when the plant is subjected to flooding of the soil, suggesting that each isoform plays a role and specific activation or modulation of target enzymes directly associated with the stress response (Subbaiah et al., 2000).

Changes in gene expression in plants, induced by lack of O_2 in the soil, levels occur in transcriptional, translational and post-translational (Sachs et al., 1980). These changes result in immediate suppression of protein synthesis pre-existing, with simultaneous selective synthesis of transition of polypeptides (TPs, 33 kDa) and, after prolonged exposure to this condition of the plant, selective synthesis of anaerobic proteins (ANPs) (Sachs et al., 1980). The ANPs have been extensively studied, among which the enzymes are involved in sucrose consumption, glycolysis and fermentation in ethanol, lactate and alanine (Drew, 1997; Vartapetian & Jackson, 1997; Dennis et al., 2000). On the other hand, little is known about the genes or TPs rapid induction (1-2 h of anoxia). According to Dennis et al. (2000), the first stage (0-4 h) of the responses of plants to O_2 deficiency was the rapid induction or activation of markers of transduction components that promote the second stage (4-24 h) metabolic and structural adaptations, including the induction of genes required to maintain a continuous production of energy. In the third stage (24-48 h), critical to the survival of O_2 tension, there is the formation of aerenchyma in the roots; genes activated by the stage 1 and or 2 and accumulation of the hormone ethylene (Drew, 1997). Thus the TPs are associated with the initial signal, which would allow the plant to survive anoxia.

Flooding decreases the absorption of N, P and K and, in some species, this type of stress alters the partition of carbohydrates for the production of xylem cells and the cell wall thickening (Kozlowski, 1997). The decrease of macronutrients and decrease of nutrients found in the leaves of plants not tolerant to flooding can be attributed to mortality of roots, decrease of mycorrhizae, root metabolism, transpiration and water conductivity (Domingo et al., 2002). The mechanisms presented by plants tolerant to water stress over which survive periods of flooding are complex (Pezeshki, 2001).

The physical properties of water affect the leaf gas exchange when soils are submerged in water (Vartapetian & Jackson). The primary adaptation of plants to flooding of the substrate is the ability to absorb air O_2 into tissues, increasing its concentration in these tissues and favor the formation of hypertrophic lenticels, aerenchyma and adventitious roots

(Kozlowski, 1997). The transport of O_2 is necessary for the maintenance of aerobic respiration mainly in roots that are under hypoxia or anoxia (Pezeshki, 2001).

Under stress conditions, if the carbon assimilation is dependent on stomatal closure, the internal concentration of CO_2 in the leaf (Ci) may be low, resulting in limitations of photosynthetic activity (Ashraf, 2003). In a relatively long period of time, the non-stomatal limitations of photosynthesis are strongly associated with changes in the Calvin cycle enzymes and degradation of photosynthetic pigments, which, in turn, are directly related to the decrease in efficiency carboxylic acid and the quantum yield apparent photosynthesis (α) of plants under flooding (Pezeshki, 1994). In addition, the flooding stress promotes the reduction of plant transpiration rates, resulting from changes in stomatal conductance, because in this situation, the route may be of little use apoplastic hydraulic resistance and hence increased (Steudle & Peterson, 1998; Jones, 1998).

5. Salt stress

Soil salinity is a major abiotic stress that adversely affects crop productivity. The saline soil is characterized by toxic levels of chlorides and sulphates of sodium. The problem of soil salinity is enhanced by irrigation, improper drainage, sea water in coastal areas, and the accumulation of salt in arid and semi-arid. Considerable efforts have been invested to unravel the mechanisms of salt tolerance in plants. The success of breeding programs aimed at improving crop productivity is limited by lack of understanding of the molecular basis of salt tolerance (Hirt & Shinozaki, 2004). Excessive soil salinity could slow the differentiation of primary xylem and secondary xylem development. The salt stress can also bring down the rate of multiplication and cell elongation, limiting the length of the leaf and, consequently, plant growth.

Sodium (Na) is an essential micronutrient for some plants, but most crops are sensitive to this element. The salinity is detrimental to plant growth because it causes nutritional restrictions, reducing the absorption of phosphorus, potassium nitrate and calcium, increases the cytotoxicity of ions and results in an osmotic stress. Under salinity, excess Na + ions, these interfere with the function of protein hydration. The ionic toxicity, osmotic stress and nutritional problems under conditions of salinity can lead to tolerance mechanisms to salt, which can be grouped into: (i) cellular homeostasis (including ion homeostasis and osmotic adjustment), (ii) control of damage caused by stress (repair and detoxification), (iii) metabolic imbalances, which result in oxidative stress (Zhu, 2001b) and (iv) growth regulation. The cellular ionic homeostasis under salinity is achieved through the following strategies: (i) exclusion of Na^+ cell plasma membrane, (ii) use of Na^+ for osmotic adjustment to the partitioning of Na^+ in the vacuole. Thus, regulation of ion transport systems is essential for plant tolerance to salt. The cytoplasmic ion homeostasis, by excluding Na + from the excess cytoplasm may require the plant to synthesize compatible osmolytes to reduce the osmotic potential, which is necessary for the absorption of water under saline stress (Hirt & Shinozaki, 2004).

Excess Na^+ and Cl^- can lead to conformational changes in the structure of proteins and/or changes in the electrical potential of the plasma membrane, while the osmotic stress leads to loss of turgor and cell volume changes. Thus, the excess of ions (Na^+ and Cl^-) and osmotic stress induce changes in turgor can act as inputs to salt stress signaling. The loss of turgor caused by osmotic stress leads to the synthesis and accumulation of ABA, which in turn regulates part of the cellular response to osmotic stress under salinity. ABA regulates the water balance of the cell by regulating stomatal and genes involved in biosynthesis of osmolytes as conferring tolerance to dehydration through LEA genes (Zhu, 2002).

Under high Na^+ concentration, it can enter cells through non-specific ion channels, causing membrane depolarization. A change in the polarization of the membrane can also indicate salt stress, known to activate Ca^{2+} channels (Sanders et al., 1999). The loss of turgor leads to a change in cell volume and retraction of the plasma membrane of the cell wall. Sodium is essential for a few C4 plants that import of pyruvate by mesophyll chloroplasts of the cotransporter Na^+/pyruvate.

Cytosolic Ca^{2+} oscillations occur within 50-10 seconds of salt stress. The three major families of proteins sensitive to signs of Ca^{2+} in plants are (Harmon et al. 2000): (i) calmodulin (CAM), which have no enzymatic activity, but they do signal transduction for protein interaction CAM, (ii) calcium dependent protein kinases (CDPKs), and (iii) SOS3 and SOS3 proteins with affinity for calcium (SCaBPs) (Guo et al 2001). Genes involved in the biosynthesis of osmoprotetores are regulated under drought and salt stress (Zhu 2002). The cytosolic signals induced by stress caused by excess calcium are perceived by the calcium-sensing protein, SOS3. The SOS3 activates SOS2, a protein kinase to be/thr. Salinity induces the biosynthesis and accumulation of the stress hormone in plants, the ABA (Jia et al 2002) and also induces accumulation of ROS (Hernandez et al., 2001). The ROS-mediated signaling under salt stress, occurs through the mitogen-activated protein kinases (MAPKs) (Pei et al. 2000).

The salt stress induces a decrease in the ratio of K^+/Na^+, and thus an enemy of cellular biochemical processes. In addition, the K^+ provides osmotic potential required for water uptake by plant cells. Thus, the K^+ uptake is crucial for maintaining cell turgor and the biochemical processes under salinity. In plants, Na^+ competes with K^+ uptake under salinity conditions (Hirt & Shinozaki, 2004).

6. Cold stress

One of the most severe environmental challenges for plants is the low temperature. The photoinhibition induced by low temperature process is completely reversible and may represent a strategy for protecting the damage caused by light energy absorbed and not used. When the photoinhibition is reversible, it can be considered as a protective and regulatory process. The photoinhibition is more pronounced at low temperatures.

The initial signaling events in response to cold hardening include changes in the membrane and cytoskeleton of cells, which is followed by increased levels of extra-and intracellular Ca^{2+}- and activation of protein kinase cascades, leading to the activation of transcription factors.

Different plant species vary greatly as to the ability to tolerate cold stress. The low temperature not only affects plant growth and distribution, but also causes serious damage to many plants. Tropical species sensitive to this kind of stress can be damaged even at temperatures significantly higher than the freezing temperature of the tissues. Injuries are caused by a deficiency of metabolic processes, changes in membrane properties, changes in protein structure and interactions of macromolecules as well as inhibition of enzymatic reactions. Tolerant plants are able to survive freezing at temperatures slightly below zero, but are severely damaged after ice formation in tissues. On the other hand, frost-tolerant plants are able to survive to varying levels of low temperatures; the actual degree of cold tolerance depends on species, developmental stage and duration of stress.

The exposure of plants to temperatures below freezing results in formation of extracellular ice, water loss and cellular dehydration. Therefore, the cold tolerance is strongly correlated with tolerance to dehydration (eg, caused by drought or high salinity). The dehydration freezing-induced can cause various disturbances in membrane structure. Although the cellular dehydration induced by freezing is the central cause of the damage caused by such stress, additional factors also contribute to damage to the plants. The growth of ice crystals can cause mechanical damage to cells and tissues and low temperatures can cause dehydration, protein denaturation and disruption of macromolecular complexes. Common responses in various types of stress are the production of reactive oxygen species (ROS), which can cause damage to different macromolecules in cells. Low temperatures can cause excessive production of ROS and therefore tolerance to cold also correlates with effective systems for elimination of ROS in response to oxidative stress (Hirt & Shinozaki, 2004).

Temperate plants respond to low temperature, turning a cold acclimation that confers tolerance to freezing temperatures. This acclimation process is accompanied by changes in expression of several genes in response to stress as well as in controlling the synthesis of proteins and metabolites that protect cellular structures from the adverse effects of freezing and cold-induced cellular dehydration. Changes in gene expression in response to cold are controlled by a set of transcription factors responsive to stimuli of low temperature (Hirt & Shinozaki, 2004). Changes in temperature can induce a signal transduction cascade by activating the expression of target genes in response to cold. In plants, CBF1 is a transcriptional regulator for cold-responsive genes. One of the major classes of plasma membrane receptors consists of the receptor protein kinases. The plasma membrane fluidity is directly affected by changes in temperature and therefore may be involved in detecting low temperature (Hirt & Shinozaki, 2004).

In winter the acclimation process in woody plants usually occurs in two steps. Initially, the reduction of the photoperiod occur for a critical value there is a pause in growth, development of dormancy and leads to a moderate increase in freezing tolerance. The second phase of acclimation is triggered by a subsequent exposure to low temperatures (Hirt & Shinozaki, 2004).

The extreme cold leads to an acclimation response that requires the synergistic action of two factors. The perception of the photoperiod, presumably involves the phytochrome A (Phya) (Olsen et al., 1999), which is the critical component for acclimatization (Li et al., 2003). In many annual and herbaceous plants in winter, the temperature is the only one capable of unleashing the full acclimatization, regardless of photoperiod. However, recent studies have indicated that phytochrome-mediated processes may also have an important role in the process of acclimation in herbaceous plants (Kim et al., 2002). Furthermore, photosynthesis is controlled by acclimation, because this process requires energy supplied by photosynthesis (Wanner & Junttila, 1999). The close association with other stress freezing results in water deficit, such as drought or high salinity. There is a temporary increase in the level of ABA during cold acclimation. This hormone in the acclimation process, by applying in plants leads to an increased tolerance to freezing.

Cold adaptation is a polygenic characteristic controlled by several genes. The control of expression of these genes leads to a series of physiological, cellular and molecular changes, including changes in membrane lipid composition, accumulation of compatible solutes, changes in levels of hormones and antioxidants, and synthesis of new proteins (Xin & Browse, 2000). Genes expressed in response to cold are divided into two distinct main categories: (i) genes that encode enzymes or structural components of cells and believed to be involved in the direct protection of cells against damage caused by freezing (Thomashow, 1999) and (ii) genes that encode transcription factors and other regulatory proteins, it is believed that regulate the responses to low temperatures, or transcriptionally postranscriptionally (Viswanathan & Zhu, 2002). The cold response genes appear to exhibit a complex temporal pattern of expression controls involving both transcriptional and post-transcriptional (Hughes & Dunn, 1996). The cold response genes appear to exhibit a complex temporal pattern of expression controls involving both transcriptional and post-transcriptional (Zhu, 2001a). The genes that encode transcription factors, such as CBF, DREB, ABF, AREB, are involved in gene regulation in response to cold. The activity of these TFs is modulated by cold temperatures. Temperature variations can be recognized in any part of cell, but the cellular components that are most directly affected by changes in temperature are the membranes and proteins.

Calcium is also acting as a second messenger in signal transduction in response to cold. A transient increase in levels of cytosolic Ca^{2+} has been shown in response to cold. The levels of $Ca2^{+-}$ Cytosolic change in response to a variety of different stimuli than the cold, such as light, growth regulators, wind and touch (Gilroy & Trewavas, 1994).

7. Heat stress

Response to heat stress occurs in organisms as diverse as bacteria, fungi, plants and animals, and is characterized by the sudden increase in body temperature and the synthesis of a set of proteins called heat shock proteins (HSPs). HSPs include several families of evolutionarily conserved proteins such as hsp100, hsp90, hsp70, HSP60 and small HSPs (sHSPs). A common response to heat stress is the thermotolerance. How thermotolerant cells express high levels of HSPs, these proteins have been associated with the development of thermotolerance. A common feature of the response to heat stress is that an initial exposure to mild heat stress provides resistance against a subsequent lethal dose of the usual heat stress. This phenomenon is known as acquired thermotolerance (Hirt & Shinozaki, 2004).

High level of temperature leads to extensive denaturation and aggregation of cellular proteins, which, if not controlled, can lead to cell death. Through its activity HSPs help cells deal with damage induced by heat. During stress, the function of these proteins is to prevent aggregation and promote proper renaturation of denatured proteins, they also play important roles in normal conditions. The main role of HSPs under normal conditions is to assist in synthesis, transport and proper folding of target proteins (Hirt & Shinozaki, 2004).

In nature, temperature changes may occur more quickly than other stress-causing factors. Plants due to their inability to translocation are subject to large fluctuations in both diurnal and seasonal temperature, and must therefore adapt to heat stress quickly and efficiently. The response to heat stress is characterized by inhibition of transcription and translation, increased expression of HSPs and induction of thermotolerance. If stress is severe, the signaling pathways leading to cell death by apoptosis are also activated. As molecular chaperones, HSPs provide protection for cells against the harmful effects of heat stress and improve survival. The higher expression of HSPs is regulated by transcription factors heat shock (HSFS). Although knowledge about the expression and function of HSPs has already been acquired, understanding the regulation of these mechanisms is still limited (Hirt & Shinozaki, 2004).

One striking feature of plants is that they contain a highly complex multigene family that encodes HSFS and HSPs. The HSF gene in plants is composed of a conserved DNA molecule, an oligomerization domain and an activation domain. The HSFS of higher eukaryotes are converted from a monomer of a trimeric form in response to stress. The trimeric active form binds to DNA and activates transcription factors. Although all HSPs function as molecular chaperones, each family of HSPs has a unique mechanism of action. The relative importance in stress tolerance of the different families of HSPs varies from one organism to another (Hirt & Shinozaki, 2004).

The hsp100 family of proteins present in prokaryotes and eukaryotes, with sizes ranging from 75 to 100 kDa. Hsp100 proteins are divided into two main classes: 1 class represented by proteins that contain two ATP binding sites, and the 2 class that contains proteins with only one ATP binding site (Miernyk, 1999). An important feature of the hsp100 protein is its ability to promote dissociation of protein aggregates in a manner dependent on ATP, as opposed primarily to prevent the unfolding and aggregation of proteins, as is attributed to other chaperones (Parsell et al., 1994). Hsp100 proteins have been identified in several plant species, and its expression analysis revealed that both are induced by stress (Agarwal et al., 2001).

The molecular chaperone Hsp90 is essential for eukaryotic cells, with key functions in signal transduction, cell cycle control, degradation and transport of proteins. Hsp90 genes are found in the inner compartments of plastids, mitochondria and endoplasmic reticulum (ER). The occurrence of multiple proteins hsp90 in the cytoplasm and other family members in different subcellular compartments suggests a number of specific functions for these proteins. While hsp90 is an abundant protein in normal conditions, its highest expression is observed in response to high temperatures, suggesting a protective role for them in conditions of heat stress (Krishna & Gloor, 2001).

Hsp70 family members are found in the cytosol of eukaryotes, and within the mitochondria, ER and plastids of eukaryotic cells (Lin et al., 2001). In higher eukaryotes, including plants, some hsp70 is constitutively expressed (Hsc70), while others are induced by stress (Hartl & Hayer-Hartl, 2002). The structure of hsp70 is composed of an NH_2-ATPase domain of approximately 45 kDa domain and a COOH-terminal peptide bond of approximately 25 kDa. In higher eukaryotes, the functions of hsp70 co-occur and post-translationally. HSP70 family members can be expressed in response to stress from heat or cold, on maturity and seed germination.

SHSPs are a group of proteins ranging in size from 15-42 kDa that can be synthesized in prokaryotic and eukaryotic cells in response to thermal stress. The patterns of expression and function of sHSPs suggest that its synthesis is correlated with thermotolerance. The functions of sHSPs extend beyond those associated with protection against heat stress; they are also synthesized under normal conditions in specific stages of development such as germination, embryogenesis, pollen development and maturation of fruits (Sun et al., 2002).

The chaperonins comprise a diverse family of molecular chaperones that are present in the cytoplasm, in plastids and mitochondria of eukaryotes. They occur in two distinct subgroups, type I and type II (Hirt & Shinozaki, 2004). The chaperonins type I are located in the chloroplast and mitochondria and are expressed as chaperonins 60 (Cpn60). The plastidic Cpn60 is composed of two types of subunit, α and β.

During the prolonged thermal stress in plants, sHSPs aggregates produce orderly cytoplasmic complex of 40 mm in diameter. The heat stress granules (HSGS) are unique and are synthesized in all plant species. The HSGS comprise mainly sHSPs cytosolic of classes I and II. The HSGS represent the local storage and protection for cleaning of mRNPs, which are released after removal of stress (Nover et al., 1989). During heat stress in the long term, the sHSP protein complexes are stored temporarily in HSGS (Löw et al., 2000).

High temperature exposure is often accompanied by high light exposure, it is important also to consider how high intensity lighting leads to plant stress. The strong radiation introduces a number of the leaf photochemical energy greater than the ability to use this energy in photosynthesis, overloading the photosynthetic processes and, ultimately, resulting not only in a quantum low utilization, but also on a low efficiency assimilatory (photoinhibition). High light intensity can, under aerobic conditions, catalyze the generation of oxygen species on highly damaging to cellular integrity and functionality. The photosynthetic capacity is impaired, with effects on plant growth and productivity.

The term photoinhibition is often used generically by reference to events of photoprotection and photodamage. Photoinhibitory process often comes down to photodamage, a drop in quantum efficiency caused by damage to the photosynthetic apparatus. Recovery from photoinhibition requires the presence of low-intensity light. In general, photoinhibition causes a decrease in the efficiency of photosynthetic O_2 release and induces changes in photochemical reactions associated with chlorophyll *a*. Plants sensitive to photoinhibition, biochemical and physiological changes occur that result in loss of membrane integrity (the lamellae of thylakoids) chloroplastidics, including its lipid and protein domains. The photoinhibition is a process dependent on temperature and light, resulting in the decreased efficiency of utilization of energy of photons captured in photochemical reactions in photosystems I and II.

When subjected to high light conditions, many higher plants, especially those of tropical origin, suffer from inhibition of photosynthesis due to the inability of tissues chlorophyll dissipate excess radiant energy, resulting in photoinhibition. Photoinhibitory process may be reversible or not. Only characterizes the occurrence of photoinhibition occurs when a prolonged downturn in the quantum efficiency of PSII. The reaction center of PSII is the primary target of photoinhibitory process, where certain subunits of proteins are quickly broken. The photosynthetic damage can be repaired by its own mechanisms, intrinsic to each species according to their degree of sensitivity to cooling temperatures. The resilience of the damage photoinhibitory, even for those species less sensitive, is greater the lower the extent of damage.

Two types of photoinhibition are identified: (a) dynamic photoinhibition, which occurs under moderate excess of light, is caused by deviation of absorbed light energy in the direction of heat dissipation, which leads to a decrease in quantum efficiency, but this decrease is temporary quantum efficiency and can return to its initial value higher; (b) chronic photoinhibition, which results from exposure to high levels of excess light, which damage the photosynthetic system and decrease the quantum efficiency and maximum photosynthetic rate, as opposed to dynamic photoinhibition, such effects are relatively long duration, persisting for weeks or months.

Extremely high temperatures can occur in soils exposed to the sun without vegetal cover, or even rock formations. Species closely related to the same genus may differ in relation to this ability, and even different organs and tissues of the same individual have different capacities of resistance to high irradiance. Typical differences in resistance, related to the conditions of the distribution and the geographical origin of the species are developed in the course of evolution.

The successful establishment of plants under conditions of high irradiance stress depends, in fact, the ability of different species to capture and use in an efficient, light. The less efficient for the capture of radiation by the pigment complex, the plant is more resistant compared to the strong radiation. As a first measure of protection, the inputs of radiation is deviated directly from photosystem via fluorescence and, especially, in the form of heat. Another protective mechanism by which energy can be diverted in a cyclic process is the metabolism of glycolate, or plants MAC, reassimilation internal CO_2, when the stomata are permanently closed due to drought.

Three types of resistance to high irradiance can be highlighted: (i) sensitive species: species which include all species that suffer injuries at 30-40 ° C or a maximum of 45 ° C (algae, lichens and most terrestrial species); (ii) relatively resistant eukaryotic organisms: dry sunny spots are usually able to acquire a rustification in relation to high irradiance, including plants that can survive in temperatures of 50-60 ° C for at least 30 minuntes; however, the range between 60-70 °C is the typical temperature limit for the survival of highly differentiated cells and organisms; (iii) prokaryotic organisms tolerant to high irradiance, some thermophilic prokaryotic organisms (cyanobacteria, bacteria and archaea hyperthermophiles) can withstand extremely high temperatures. All these organisms have cell membranes, proteins and nucleic acids highly resistant.

An extremely high temperature can destroy the photosynthetic pigments and thylakoid structures (photodestruction). The photodestruction is responsible for the decline in photosynthetic capacity of leaves senescence. The effect of heat depends on its duration; it follows the rule of the dose, which indicates that little heat for a long period causes so much injury as an intense heat for a short period. Photosynthesis and respiration are inhibited at high temperatures, but the photosynthetic rates fall before respiratory rates. Photosynthesis at elevated temperatures can not replace the carbon used as a substrate for respiration. As a result, the carbohydrate reserves decrease and lose the fruit sugars. Heat stress can alter the rate of metabolic reactions that consume or produce protons, it can affect the activity of proton pumping ATPases. This would cause an acidification of the cytosol, which would cause additional metabolic disturbances during stress.

8. Stress by heavy metals

Heavy metals are defined as metals with a density greater than 5 g cm^{-3}. However, only a limited number of these elements is soluble under physiological conditions and therefore may become available to living cells. Among them, the factors used for the metabolism of plants as micronutrients or trace elements (Fe, Mo, Mn, Zn, Ni, Cu, V, Co, W, Cr) and become toxic when in excess, as well as other biological functions not known and high toxicity, as As, Hg, Ag, Sb, Cd, Pb and U (Hirt & Shinozaki, 2004).

Regulatory limits for heavy metals in the environment are defined by national legislation. The concentrations of heavy metals in soils are regional differences and could exceed regulatory limits in 10 to 50 times (Haag-Kerwer et al., 1999). The soil covering the rocks of ore naturally contains heavy metals in amounts that are toxic to most species of plants. In such places, specialized plant communities called "chemotypes" evolved offering opportunities to investigate traces of resistance to heavy metals. There is a growing concern about the increased release of heavy metals in the environment. The sources of heavy metals include traffic, garbage and sewage sludge. Emissions of dust, aerosols, ashes and metal processing industries lead the spread of heavy metals in rural areas. In agricultural soils, pollution by heavy metals is a growing problem because of soil contamination with municipal sewage sludge and intensive use of phosphate fertilizers containing cadmium as a contaminant.

The long-term biological and retention of heavy metals in soil favor its accumulation in the food chain with potentially negative effects on human health. The bioavailability of heavy metals in plants is specific and depends on the demand for specific metals as micronutrients and plant capacity to actively regulate the mobilization of metals by exuding organic acids or protons in the rhizosphere. In addition, soil properties influence the chemical mobility of metals, thus regulating its release to the soil solution. The ability of plants to extract metals from the soil, the internal allocation of the metal in the plant cell and detoxification mechanisms are areas of research that has attracted increasing attention (Hirt & Shinozaki, 2004).

The absorption of heavy metals in plant cells is modulated by biotrophic interactions and the inherent characteristics of plants, such as its ability to retain heavy metals in the roots, for example, by binding to cell wall components. A common response to exposure to heavy metals is a significant reduction in plant growth (Sanita di Toppi & Gabrielli, 1999). Normal growth is the result of cell division, elongation and differentiation also including programmed cell death in certain tissues such as xylem. The excess of heavy metals affects root functions on various levels and causes the accumulation of abscisic acid (ABA). In the whole plant, roots are the main site access for heavy metals. In general, a large fraction of cadmium (Cd) or copper (Cu) is retained by the roots and only a relatively small amount (about 10%) are transported to the shoot (Liao et al., 2000). The cytokinins act as antagonists to Cd, indicating that the internal hormonal status can critically affect plant tolerance to heavy metals.

Cu is a micronutrient essential for the catalytic activity of many enzymes. His capture and transport are regulated and mediated by specific transporters and escorts. The Cu serves as an intermediary for the flag receiving the hormone ethylene. Excess Cu is detected by binding to transcription factors, thus activating an arsenal of defense against abiotic stresses, including increased expression of metallothioneins, phytochelatins and antioxidants which help to remove the Cu "free" and to restore homeostasis ionic and cellular redox. These free metals are potentially dangerous, and therefore its uptake and cellular concentration must be regulated. The Cu has a high affinity for peptide, carboxylic and phenolic groups. Therefore, Cu is usually present in living cells. It is believed that the superoxide dismutase (SOD), which contain Cu / Zn, Fe or Mn in its reaction center, play a dual role in preventing metal toxicity on the one hand, they clean the O_2^- radical, thus maintaining lower concentrations of reactive oxygen species and on the other, they seem to be involved in preventing the accumulation of free metal (Hirt & Shinozaki, 2004).

Differently from Cu, no specific uptake system for Cd is known. Once accumulated in the body, trigger discharges of Cd in the activation of signaling pathways as well as a sequence of biochemical reactions and morpho-physiological changes that can cause programmed cell death in various tissues and organs (Souza et al., 2010). According to these authors, cell death induced by Cd can present apoptotic features such as chromatin condensation, nuclear DNA cleavage oligonucleossomos and formation of apoptotic bodies. Cd enters cells through transporters with broad specificity for metals, and probably also through calcium channels. It is toxic due to its high reactivity with sulfur and compounds derived from this type of stress causes depletion of antioxidant systems and stimulates the production of H_2O_2 by enzymes. The CD has more affinity with thiol groups of other metal micronutrients (Schützendübel & Polle, 2002). This feature is probably also the main basis for its toxicity. Cadmium inhibits HS-structural, redox regulated enzymes in living organisms (Hall, 2002). CDs can also be linked to other functional groups containing nitrogen or oxygen.

Exposure to Cd leads to oxidative damage as lipid peroxidation and protein carbonylation (Romero-Pueras et al., 2002). The excess of Cd in plant tissue can stimulate the formation of free radicals and reactive oxygen species (ROS), causing severe oxidative stress (Souza et al., 2010), in addition, the Cd disturbs the cellular redox balance. One of the most important answers to the CD, and also for other heavy metals, is an initial transient depletion of GSH, which is probably due to an increased demand for this precursor for the synthesis of PC (Schützendübel et al., 2001). Cd suppresses cell expansion. In shoots, Cd inhibits the proton pump responsible for maintenance of turgor (Aidid & Okamoto, 1992). This is likely to occur in other parts of the plant as well. Moreover, the roots exposed to Cd increased the production of ethylene, a hormone that inhibits cell expansion. The Cd also leads to significant accumulation of H_2O_2 which causes hardening of the cell wall (Ros Barcelo, 1997). Therefore, inhibition of root growth is probably a pleiotropic effect caused by direct inhibition of enzymes important for Cd interference in cell signaling.

Cu and Cd activate the formation of phytochelatins (PCs) and metallothioneins (MT), both compounds with functions sequestration of heavy metals (Cobbett & Goldsbrough, 2002). MTs are a family of small proteins. The promoter regions of MT transport elements MRE (metal responsive elements: GCGCGCA), leading to the accumulation of MT because of exposure to heavy metals (Cobbett & Goldsbrough, 2002). PCs are produced enzymatically from the tri-peptide precursor GSH (glycine cysteinyl γ-glutamyl). In response to heavy metals γECS (synthase γ-glutamyl transferase cisteinyl) and is transcriptionally activated (Noctor et al., 1998) and is transcriptionally activated (Lee & Korban, 2002) leading to accumulation of PC (Rauser, 1999). MTs may contribute to control the concentration of metals "free" and reactive oxygen species can activate the defenses.

Cd induces the biosynthesis of ABA and ethylene in the roots (Chen, et al., 2001). These are signs that lead to stress responses in the shoot. Ethylene inhibits the growth of cells and plays a role in cell signaling. In situations of stress by excess Cd, the absorption of water in the roots is disturbed, the hydraulic conductivity decreases and thus the water supply to the shoot decreases (Marchiol et al., 1996). The transport of Cd to the shoot is driven by perspiration and can be reduced through the application of ABA (Salt et al., 1995). The excess of Cu leads to a decline in the efficiency of water use and an accumulation of proline (Vinit-Dunant et al., 2002). The biosynthesis of proline was also found in plants stressed with Cd (Talanova et al., 2000).

9. Mechanical Stress

Plants respond to various mechanical disturbances, such as weight changes, snow, ice, wind and rain. Thygmotropism directional growth is determined by a stimulus and describes phenomena such as the ability of roots to grow around objects in the soil (Porter et al., 2009). The change in physiology, morphology and composition of plants resulting from prolonged stimulation by mechanical disturbances is referred to as thygmomorphogenesis. The most common morphological changes associated with thigmomorphogenesis is the development of compact plants decreased due to stretching, bending and friction caused by the shoot in the wind, animals or other plant parts. However, it is possible to observe significant differences in responses thigmomorphogenetics between and within species, with an indication of genetic diversity in the regulation of this phenomenon (Porter et al., 2009).

These thigmomorphogenetics changes may lead to increases in the width of the plants and decreased sensitivity to these various stresses. The most pronounced effects found in conditions of high rates of turbulent wind or water flow (Onguso et al., 2006). Plants react to mechanical stress according to species, cultivar, habit and growth. The morphological responses to mechanical stress are characterized by: decreasing the ratio between growth and stem diameter growth and decrease or increase in stem diameter. The responses also include increased production of xylem and flexed at the point decreases leaf area.

10. Conclusion

The evolution of knowledge about the mechanisms of stress tolerance in various species, through the study of functional genomes and proteomes, has provided valuable information for the development of genotypes that can tolerate periods of stress without productivity is substantially impaired. The dissemination of new molecular techniques such as DNA microarrays, which allow simultaneous analysis of thousands of genes and implicate metabolic pathways activated or deactivated under specific conditions, will promote the viewing of hundreds of interactions that occur in the context transcriptional and proteomic analysis in response to stress events. Confirmation of such changes by real-time PCR then allows the precise quantification of mRNA levels of genes of interest under different conditions. These techniques allow the design strategies aimed at increasing tolerance to environmental stress conditions. These strategies have been through traditional breeding methods, facilitated by the use of molecular markers linked to individual genes or linked to quantitative trait locus (QTL) of importance, or through the use of genetic engineering. Genes identified by tolerance mechanisms have shown potential to be used in studies of plant transformation.

11. References

Agarwal, M.; Katiyar-Agarwal, S.; Sahi, C., Gallie; D.R. & Grover, A. (2001). *Arabidopsis thaliana* Hsp100 proteins: kith and kin. *Cell Stress & Chaperones*, Vol. 6, pp. 219-224, ISSN 1355-8145.

Aidid, S.B. & Okamoto, H. (1992). Effects of lead, cadmium and zinc on the electric membrane potential at the xylem /symplast interface and cell elongation of Impatiens balsamina. *Environmental Experimental Botany*, Vol. 32, pp. 439-448, ISSN 0098-8472.

Armstrong, W. (1979). Aeration in higher plants. *Advances in Botanical Research*, Vol. 7, pp. 225–232.

Ashraf, M. (2003). Relationships between leaf gas exchange characteristics and growth of differently adapted populations of Blue panicgrass (*Panicum antidotale* Retz) under salinity or waterlogging. *Plant Science*, Vol. 165, pp. 69-75, ISSN 0168-9452.

Blom, C.W.P.M. & Voeseneck, L.A.C.L. (1996). Flooding: the survival strategies of plants. *Tree*, Vol.11, No.7, pp. 290-295, ISSN 0931-1890.

Bowler, C. & Fluhr, R. (2000). The role of calcium and activated oxygens as signals for controlling cross-tolerance. *Trends Plant Science*, Vol. 5, pp. 241-245, ISSN 1360-1385.

Camargo, F.A.O.; Santos, G.A. & Zonta, E. (1999). Alterações eletroquímicas em solos inundados. *Ciência Rural*, Vol.29, No.1, pp.171-180, ISSN 0103-8478.

Chen, C.T.; Chen, L.M.; Lin, C.C. & Kao, C.H. (2001). Regulation of proline accumulation in detached rice leaves exposed to excess copper. *Plant Science*, Vol. 160, pp.283-290, ISSN 0168-9452.

Chen, H.; Qualls, R.G. & Miller, G.C. (2002). Adaptive responses of *Lepidium latifolium* to soil flooding: biomass allocation, adventitious rooting, aerenchyma formation and ethylene production. *Environmental Experimental Botany*, Vol. 48, pp. 119-128, ISSN 0098-8472.

Cobbett, C. & Goldsbrough, P.B. (2002). Phytochelatins and metallothioneins:roles in Heavy Metal Detoxification and Homeostasis. *Annual Review Plant Biology*, Vol.53, pp.159–182, ISSN 1543-5008.

Dat, J.F.; Capelli, N.; Folzer, H.; Bourgeade, P. & Badot, P.M. (2004). Sensing and ignalling during plant flooding. *Plant Physiology and Biochemistry*, Vol. 42, pp. 273–282, ISSN 0981-9428.

Davies, F.S. & Flore, J.A. (1986). Short-term flooding effects on gas exchange and quantum yield of rabbiteye blueberry (*Vaccinium ashei* Reade). *Plant Physiology*, Vol. 81, pp. 289-292, ISSN 0032-0889.

Dennis, E.S.; Dolferus, R.; Ellis, M.; Rahman, M.; Wu, Y.; Hoeren, F.U.; Grover, A.; Ismond, K.P.; Good, A.G. & Peacock, W.J. (2000). Molecular strategies for improving waterlogging tolerance in plants. *Journal of Experimental Botany*, Vol. 51, pp. 89-97, ISSN 0022-0957.

Domingo, R.; Pérez - Pastor, A. & Ruiz - Sánches, M.C. (2002). Physiological responses of apricot plants grafted on two different rootstocks to flooding conditions. *Journal of Plant Physiology*, v. 159, p. 725-732, ISSN 0176-1617.

Drew, M.C. (1997). Oxygen deficiency and root metabolism: injury and acclimation under hypoxia and anoxia. *Annual Review of Plant Physiology and Plant Molecular Biology*, Vol. 48, pp. 223-250, ISSN 1040-2519.

Else, M.A.; Tiekstra, A.E.; Croker, S.J.; Davies, W.J. & Jackson, M.B. (1996). Stomatal closure in flooded tomato plants involves abscisic acid and a chemically unidentified anti-transpirant in xylem sap. *Plant Physiology*, Vol. 112, pp. 239-247, ISSN 0032-0889.

Fahn, A. (1982). *Plant Anatomy* (3ª ed.), Pergamon Press. 544p.

Folzer, H.; Capelli, N.; Dat, J. & Badot, P.-M. (2005). Molecular cloning and characterization of calmodulin genes in young oak seedlings (*Quercus petraea* L.) during early flooding stress. *Biochimica et Biophysica Acta,* Vol. 1727, pp. 213- 219, ISSN 0167-4781.

Fowler, S. & Thomashow, M.F. (2002). Arabidopsis transcriptome profiling indicates that multiple regulatory pathways are activated during cold acclimation in addition to the CBF cold response pathway. *Plant Cell,* Vol. 14, pp.1675-1690, ISSN 1040-4651.

Gilroy, S. & Trewavas, A. (1994) A decade of plant signals. *BioEssays,* Vol. 16, pp.677-682, ISSN 0265-9247.

Guo, Y.; Halfter, U.; Ishitani, M. & Zhu, J-K. (2001). Molecular characterization of functional domains in the protein kinase SOS2 that is required for plant salt tolerance. *Plant Cell,* Vol.13, pp.1383-1400, ISSN 1040-4651.

Haag-Kerwer, A.; Schäfer, H.J.; Heiss, S.; Walter, C. & Rausch, T. (1999). Cadmium exposure in *Brassica juncea* causes a decline in transpiration rate and leaf expansion without effect on photosynthesis. *Journal of Experimental Botany,* Vol.50, pp.1827-1835, ISSN 0022-0957.

Hall, J.L. (2002). Cellular mechanisms for heavy metal detoxification and tolerance. *Journal of Experimental Botany,* Vol.366, pp.1-11, ISSN 0022-0957.

Harmon, A.C.; Gribskov, M. & Harper, J.F. (2000). CDPKs—a kinase for every Ca^{2+} signal? *Trends Plant Science,* Vol.5, pp. 154–159, ISSN 1360-1385.

Hartl, F.U. & Hayer-Hartl, M. (2002). Molecular chaperones in the cytosol: from nascent chain to folded protein. *Science,* Vol.295, pp.1852-1858, ISSN 0036-8075.

He, C.J.; Drew, M.C. & Morgan, P.W. (1994). Induction of enzime associated with lysigenous aerenchyma formation in roots of *Zea mays* during hypoxia or nitrogen-starvation. *Plant Physiology,* Vol. 105, pp. 861-865, ISSN 0032-0889.

He, C.J.; Finlayson, S.A.; Drew, M.C.; Jordan, W.R. & Morgan, P.W. (1996). Ethylene biosynthesis during aerenchyma formation in roots of *Zea mays* subjected to mechanical impedance and hypoxia. *Plant Physiology,* Vol. 112, pp. 1679-1685, ISSN 0032-0889.

Hernandez, J.A.; Ferrer, M.A.; Jiménez, A.; Barceló, A.R. & Sevilla, F. (2001). Antioxidant systems and $O_2{\cdot}^-/H_2O_2$ production in the apoplast of pea leaves. Its relation with salt-induced necrotic lesions in minor veins. *Plant Physiology,* Vol.127, pp.817 - 831, ISSN 0032-0889.

Hirt, H. & Shinozaki, K. (2004). *Plant responses to abiotic stress.* Berlin Heidelberg: Springer.

Hoekstra, F.A.; Golovina, E.A. & Buitink, J. (2001). Mechanisms of plant desiccation tolerance. *Trends Plant Science,* Vol. 6, pp. 431-438, ISSN 1360-1385.

Holmberg,N. & Büllow, L. (1998). Improving stress tolerance in plants by gene transfer. *Trends in plant science* (reviews), Vol.3, No. 2, pp. 61-66, ISSN 1366-2570.

Hook, D.D.; Brown, C.L. & Kormanik, P.P. (1971). Inductive flood tolerance in swamp tupelo (*Nyssa sylvatica* var. biflora (Walt.) Sarg.). *Journal of Experimental Botany,* Vol. 22, p. 78-89, ISSN 0022-0957.

Hughes, M.A. & Dunn, M.A. (1996). The molecular biology of plant acclimation to low temperature. *Journal of Experimental Botany,* Vol.47, pp.291-305, ISSN 0022-0957.

Jackson, M.B. (2004). The impact of flooding stress on plants and crops. 2005, Available from: <http://www.plantstress.com/Articles/waterlogging_i/waterlog_i.htm>.

Jackson, M.B. & Colmer, T.D. (2005). Response and adaptation by plants to flooding stress. *Annals of Botany,* Vol. 96, pp. 501-505, ISSN 0305-7364.

Jia, W.; Wang, Y.; Zhang, S. & Zhang, J. (2002). Salt-stress-induced ABA accumulation is more sensitively triggered in roots than in shoots. *Journal of Experimental Botany,* Vol.53, pp.2201-2206, ISSN 0022-0957.

Jones, H.G. (1998). Stomatal control of photosynthesis and transpiration. *Journal of Experimental Botany,* Vol. 49, pp. 387-398, ISSN 0022-0957.

Kennedy, R.A.; Rumpho, M.E. & Fox, T.C. (1992). Anaerobic metabolism in plants. *Plant Physiology,* Vol. 100, pp. 1-6, ISSN 0032-0889.

Kim, H-J.; Kim, Y-K.; Park, J-Y & Kim, J. (2002). Light signaling mediated by phytochrome plays an important role in cold-induced gene expression through the Crepeat/dehydration responsive element (C/DRE) in *Arabidopsis thaliana. Plant Journal,* Vol. 29, pp.693-704, ISSN 0960-7412.

Klok, J.E.; Wilson, I.W.; Chapman, C.S.; Ewing, M.R.; Somerville, C.S.; Peacock, W.J.; Dolferus, R. & Dennis, E.S. (2002). Expression profile analysis of low-oxygen response in *Arabidopsis* root cultures. *Plant Cell,* Vol. 14, pp. 2481–2494, ISSN 1040-4651.

Kozlowski, T.T. (1997). Responses of woody plants to flooding and salinity. *Tree Physiology,* Mon. 1, pp. 1–29, 1997.

Kozlowski, T.T. (2002). Acclimation and Adaptive Responses of Woody Plants to Environmental Stresses. *The Botanical Review,* Vol. 68, No. 2, pp. 270-334, ISSN 0006-8101.

Kozlowski, T.T. & Pallardy, S.G. (1997). *Growth control in woody plants.* San Diego: Academic Press.

Kreps, J.A.; Wu, Y.; Chang, H-S.; Zhu, T.; Wang, X. & Harper, J.F. (2002). Transcriptome changes for Arabidopsis in response to salt, osmotic, and cold stress. *Plant Physiology,* Vol. 130, pp.2129-2141, ISSN 0032-0889.

Krishna, P. & Gloor, G. (2001). The Hsp90 family of proteins in *Arabidopsis thaliana. Cell Stress & Chaperones,* Vol. 6, pp.238-246, ISSN 1355-8145.

Larcher, W. (1995). *Physiological plant ecology.* Berlin: Springer-Verlag, 506p.

Lee, S. & Korban, S.S. (2002) The trancriptional regulation of Arabidopsis thaliana (L.) Heynh. *Planta,* Vol.215, pp.689-693, ISSN 0032-0935.

Li, C.; Viherä-Aarnio, A.; Puhakainen, T.; Junttila, O.; Heino, P. & Palva, E.T. (2003). Ecotype dependent control of growth, dormancy and freezing tolerance under seasonal changes in *Betula pendula* Roth. *Trees,* Vol.17, pp.127-132, ISSN 0931-1890.

Liao, M.T.; Hedley, M.J.; Woolley, D.J.; Brooks, R.R. & Nichols, M.A. (2000). Copper uptake and translocation in chicory (*Cichorium intybus* L. cv. Grasslands Puna) and tomato (*Lycopersicon esculentum* Mill. cv. Rondy) plants grown in NFT system. I. Copper uptake and distribution in plants. *Plant and Soil,* Vol.221, pp.135-142, ISSN 0032-079X.

Lichtenthaler, H.K. (1996). Vegetation stress: an introduction to stress concepts in plants. *Journal of Plant Physiology,* Vol.148, pp.4-14, ISSN 0176-1617.

Lin, B-L.; Wang, J-S.; Liu, H-C.; Chen, R-W.; Meyer, Y.; Barakat, A. & Delseny, M. (2001). Genomic analysis of the Hsp70 superfamily in *Arabidopsis thaliana*. *Cell Stress & Chaperones*, Vol.6, pp.201-208, ISSN 1355-8145.

Löw, D.; Brändle, K.; Nover, L. & Forreiter, C. (2000). Cytosolic heat-stress proteins Hsp17.7 class I and Hsp17.3 class II of tomato act as molecular chaperones in vivo. *Planta*, Vol. 211, pp.575-582, ISSN 0032-0935.

Marchiol, L.; Leita, L.; Martin, M.; Peresotti, A. & Zerbi, G. (1996). Physiological responses of two soybean cultivars to cadmium. *Journal of Environmental Quality* Vol. 25, pp.562-566, ISSN 0047-2425.

Mckersie, B.D. (2001). *Plant environment interactions and stress physiology*. Module 2. p. 83-460.

Medri, M.E. (1998). Aspectos morfo-anatômicos e fisiológicos de *Peltophorum dubium* (Spr.) Taub. submetida ao alagamento e à aplicação de etrel. Revista Brasileira de Botânica, Vol. 21, pp. 261-267, ISSN 0100-8404.

Miernyk, J.A. (1999). Protein folding in the plant cell. *Plant Physiology*, Vol.121, pp.695-703, ISSN 0032-0889.

Mittler, R. (2002). Oxidative stress, antioxidants and stress tolerance. Trends Plant Science, Vol.7, pp.405-410, ISSN 1360-1385.

Noctor, G.; Arisi, A.C.M.; Jouanin, L.; Kunert, K.J. & Rennenberg, H. (1998). Glutathione: biosynthesis, metabolism and relationship to stress tolerance explored in transformed plants. *Journal of Experimental Botany*, Vol.49, pp.623-647, ISSN 0022-09.

Nover, L.; Scharf, K-D. & Neumann, D. (1989). Cytoplasmic heat shock granules are formed from precursor particles and are associated with a specific set of mRNAs. *Molecular and Cellular Biology*, Vol.9, pp.1298-1308, ISSN 0270-7306.

Olsen, J.E.; Junttila, O.; Nilsen, J.; Eriksson, M.E.; Martinussen, I.; Olsson, O.; Sandberg, G. & Moriz, T. (1997). Ectopic expression of oat phytochrome A in hybrid aspen changes critical daylength for growth and prevents cold acclimatization. *Plant Journal*, Vol.12, pp.1339-1350, ISSN 0960-7412.

Onguso, J.M.; Mizutani, F. & Hossain, A.B.M.S. (2006). The effect of trunk electric vibration on the growth, yield and fruit quality of peach trees (*Prunus persica* [L.] Batsch). *Scientia Horticulturae*, Vol.108, pp. 359–363, ISSN 0304-4238.

Parsell, D.A.; Kowal, A.S.; Singer, M.A. & Lindquist, S. (1994). Protein disaggregation mediated by heat stress protein 104. *Nature*, Vol.372, pp.475-478, ISSN 0028-0836.

Pei, Z.M.; Murata, Y.; Benning, G.; Thomine, S.; Klusener, B.; Allen, G.J.; Grill, E. & Schroeder, J.I. (2000). Calcium channels activated by hydrogen peroxide mediate abscisic acid signaling in guard cells. *Nature*, Vol.406, pp.731-734, ISSN 0028-0836.

Peng, H.; Lin, T.; Wang, N. & Shih, M. (2005). Differential expression of genes encoding 1-aminocyclopropane-1-carboxylate synthase in *Arabidopsis* during hypoxia. *Plant Molecular Biology*, Vol. 58, pp. 15-25, ISSN 0167-4412.

Pereira, J.S. (1995). Gas exchange and growth. In: *Ecophysiology of Photosynthesis*, Shulze, E.D., Caldwell, M.M. (eds.), pp. 147-181, Berlin: Springer-Verlag.

Pereira, J.N.; Ribeiro, J.F. & Fonseca, C.E.L. (2001). Crescimento inicial de *Piptadenia gonoacantha* (Legiminosae, Mimosoideae) sob inundação em diferentes níveis de luminosidade. *Revista Brasileira de Botânica*, Vol. 24, No. 4, pp. 561-566, ISSN 0100-8404.

Peschke, V.M. & Sachs, M.M. (1994). Characterization and expression of anaerobically induced maize transcripts. *Plant Physiology*, Vol. 104, pp. 387-394, ISSN 0032-0889.

Pezeshki, S.R. (1994). Responses of baldcypress (*Taxodium distichum*) seedlings to hypoxia: leaf protein content, ribulose-1,5-bisphosphate carboxilase/oxigenase activity and photosynthesis. *Photosynthetica*, Vol. 30, pp. 59-68, ISSN 0300-3604.

Pezeshki, S.R. (2001). Wetland plant responses to soil flooding. *Environmental Experimental Botany*, Vol. 46, pp. 299-312, ISSN 0098-8472.

Porter, B.W.; Zhu, Y.J.; Webb, D.T. & Christopher, D.A. (2009). Novel thigmomorphogenetic responses in Carica papaya: touch decreases anthocyanin levels and stimulates petiole cork outgrowths. *Annals of Botany*, Vol.103, pp. 847–858, ISSN 0305-7364.

Probert, M.E. & Keating, B.A. (2000). What soil constraints should be included in crop and forest models? *Agriculture, Ecosystems & Environment*, Vol. 82, pp. 273–281, ISSN 0167-8809.

Rauser, W. (1999). Structure and function of metal chelators produced by plants. *Cell Biochem Biophys* Vol.31, pp.19-48.

Romero-Pueras, M.C.; Palma, J.M.; Gomez, L.A.; del Rio, L.A. & Sandalio, L.M. (2002). Cadmium causes oxidative modification of proteins in plants. *Plant Cell Environmental*, Vol.25, pp.677-686.

Ros Barcelo, A. (1997). Lignification in plant cell walls. *International Review Cytology*, Vol.176, pp.87-132.

Rubio, G.; Casasola, G. & Lavado, R.S. (1995). Adaptation and biomass production of two grasses in response to waterlogging and soil nutrient enrichment. *Oecologia*, Vol. 102, pp. 102–105, ISSN 0029-8549.

Saab, I.N. & Sachs, M.M. (1995). Complete cDNA and genomic sequence encoding a flooding-responsive gene from maize (*Zea mays* L.) homologous to xyloglucan endotransglycosylase. *Plant Physiology*, Vol. 108, pp. 439-440, ISSN 0032-0889.

Sachs, M.M.; Freeling, M. & Okimoto, R. (1980). The anaerobic proteins of maize. *Cell*, Vol. 20, pp. 761-767, ISSN 0092-8674.

Salt, D.E.; Prince, R.C.; Pickering, I.J. & Raskin, I. (1995). Mechanism of cadmium mobility and accumulation in Indian Mustard. *Plant Physiology*, Vol.109, pp.1472-1433, ISSN 0032-0889.

Sanders, D.; Brownlee, C. & Harper, J.F. (1999). Communicating with calcium. *Plant Cell*, Vol.11, pp.691-706, ISSN 1040-4651.

Sanita di Toppi, L. & Gabrielli, R. (1999). Response to cadmium in higher plants. *Environmental Experimental Botany*, Vol.41, pp.105-130, ISSN 0098-8472.

Scandalios, J.G. (2002). The rise of ROS. *Trends in Biochemical Sciences*, Vol.27, pp.483-486, 0968-0004.

Schützendübel, A. & Polle, A. (2002). Plant responses to abiotic stresses:heavy metal-induced oxidative stress and protection by mycorrhization. *Journal of Experimental Botany*, Vol.53, pp.1351-1365, ISSN 0022-0957.

Schützendübel, A.; Schwanz, P.; Teichmann, T.; Gross, K.; Langenfeld-Heyser, R.; Godbold, D. & Polle, A. (2001). Cadmium-induced changes in antioxidative systems, H_2O_2 content and differentiation in pine (*Pinus sylvestris*) roots. *Plant Physiology*, Vol.127, pp.887-898, ISSN 0032-0889.

Seki, M.; Narusaka, M.; Ishida, J.; Nanjo, T.; Fujita, M.; Oono, Y.; Kamiya, A.; Nakajima, M.; Enju, A.; Sakurai, T.; Satou, M.; Akiyama, K.; Taji, T.; Yamaguchi-Shinozaki, K.; Carninci, P.; Kawai, J.; Hayashizaki, Y. & Shinozaki, K. (2002). Monitoring the expression profiles of 7000 Arabidopsis genes under drought, cold and high-salinity stresses using a fulllength cDNA microarray. *Plant Journal*, Vol.31, pp.279-292, ISSN 0960-7412.

Snedden, W.A. & Fromm, H. (2001). Calmodulin as a versatile calcium signal transducer in plants. *New Phytologist*, Vol.151, pp. 35–66, ISSN 0028-646X.

Sousa, C. A. F. & Sodek, L. (2002). The metabolic response of plants to oxygen deficiency. *Brazilian Journal of Plant Physiology*, Vol. 14, No. 2, pp. 83-94, ISSN 1677-0420.

Souza, V. L.; Almeida, A-A. F.; Lima, S. G. C.; Cascardo, J. C. M.; Silva, D. C.; Mangabeira, P. A. O. & Gomes, F. P. (2011). Morphophysiological responses and programmed cell death induced by cadmium in *Genipa americana* L. (Rubiaceae). *Biometals*, Vol. 24, pp. 59–71, ISSN 0966-0844.

Steudle, E. & Peterson, C.A. (1998). How does water get through roots? *Journal of Experimental Botany*, Vol. 49, No. 322, pp. 775-788, ISSN 0022-0957.

Subbaiah, C.C.; Bush, D.S. & Sachs, M.M. (1998). Mitochondrial contribution to the anoxic Ca^{2+} signal in maize suspension-cultured cells. *Plant Physiology and Biochemistry*, Vol.118, pp. 759–771, ISSN 0981-9428.

Subbaiah, C.C.; Kollipara, K.P. & Sachs, M.M. (2000). A Ca^{2+}- dependent cysteine protease is associated with anoxia-induced root tip death in maize, *Journal of Experimental Botany*, Vol. 51, pp. 721– 730, ISSN 0022-0957.

Sun, W.; van Montagu, M. & Verbruggen, N. (2002). Small heat shock proteins and stress tolerance in plants. *Biochimica et Biophysica Acta*, Vol.1577, pp.1-9, ISSN 0006-3002.

Talanova, V.V.; Titov, A.F. & Boeva, N.P. (2000). Effect of increasing concentrations of lead and cadmium on cucumber seedlings. *Biologia Plantarum*, Vol.43, pp.441-444, ISSN 0006-3134.

Thomashow, M.F. (1999). Plant cold acclimation: freezing tolerance genes and regulatory mechanisms. *Annual Review Plant Physiology*, Vol.50, pp.571-599, ISSN 0066-4294.

Topa, M. A. & Mcleod, K. W. (1986). Aerenchyma and lenticel formation in pine seedlings: A possibleavoidance mechanism to anerobic growth conditions. *Plant Physiology*, Vol. 68, pp. 540-550, ISSN 0032-0889.

Vantoi, T.T.; Beuerlein, J.E.; Schmithenner, A.F. & Martin, S.K.St. (1994). Genetic variability for flooding tolerance in soybeans. *Crop Science*, Vol. 34, pp. 1112-1115, 0011-183X.

Vartapetian, B. B. & Jackson, M. B. (1997). Plant adaptations to anaerobic stress. *Annals of Botany*, Vol. 79, pp. 3-20, ISSN 0305-7364.

Vinit-Dunand, F.; Epron, D.; Alaoui-Sosse, B. & Badot, P.M. (2002). Effects of copper on growth and on photosynthesis of mature and expanding leaves in cucumber plant. *Plant Science*, Vol.163, pp.53-58, ISSN 0168-9452.

Viswanathan, C. & Zhu, J-K. (2002). Molecular genetic analysis of cold-regulated gene transcription. *Philosophical Transactions of the Royal Society of London. B, Biological Sciences*, Vol.357, pp.877-886, ISSN 0080-4622.

Wanner, L.A. & Junttila, O. (1999). Cold-induced freezing tolerance in Arabidopsis. *Plant Physiology*, Vol.120, pp.391-399, ISSN 0032-0889.

White, S. D. & Ganf, G. G. (2002). A comparison of the morphology, gas space anatomy and potential for internal aeration in Phragmites australis under variable and static water regimes. *Aquatic Botany*, Vol. 73, pp. 115-127, ISSN 0304-3770.

Xin, Z. & Browse, J. (2000). Cold comfort farm: the acclimation of plants to freezing temperatures. *Plant Cell Environmental*, Vol.23, pp.893-902.

Zhu, J-K. (2001a). Cell signaling under salt, water and cold stresses. *Current Opinion in Plant Biology*, Vol.4, pp. 401-406, ISSN 1369-5266.
Zhu, J-K. (2001b). Plant salt tolerance. *Trends Plant Science*, Vol.6, pp.66-71, ISSN 1360-1385.
Zhu, J-K. (2002). Salt and drought stress signal transduction in plants. *Annual Review Plant Biology*, Vol.53, pp.247-273, ISSN 1543-5008.
Zielinski, R.E. (1998). Calmodulin and calmodulin-binding proteins in plants, *Annual Review of Plant Physiology and Plant Molecular Biology*, Vol. 49, pp. 697–725, ISSN 1040-2519.

3

Oxygen Metabolism in Chloroplast

Boris Ivanov, Marina Kozuleva and Maria Mubarakshina
Institute of Basic Biological Problems Russian Academy of Sciences
Russia

1. Introduction

Oxygen was almost non-existent in the Earth's atmosphere before the oxygenic photosynthetic bacteria appeared. Since O_2 is capable of combining with most chemical elements, the stable level of O_2 in the atmosphere is the result of it being continuously regenerated by the oxygenic photosynthetic organisms, *i.e.* the cyanobacteria, algae and plants.

The molecular mechanism of water oxidation to O_2 is still unclear, although many structural details are known and some of the details of the charge accumulating cycle are well worked out (reviewed in Barber, 2008; Brudvig, 2008). The water-oxidizing complex, with a Mn_4Ca cluster as the active site, is an integral part of the Photosystem II (PSII), one of the main complexes of the photosynthetic electron transport chain (PETC). When the energy of a quantum of light absorbed by a chlorophyll molecule in this photosystem reaches the reaction center, photochemistry occurs leading to charge separation. The electron is used to reduce plastoquinone, while the electron hole is used to oxidize a Mn ion of the cluster and eventually used to oxidize water. Two sequential photochemical turnovers are required to reduce quinone to quinol, while four sequential turnovers are required to oxidize two water molecules forming O_2. It is important to note that the water oxidation/oxygen evolution process is the most easily damaged function of the PETC under stress conditions.

Sixty years ago, the first data were published indicating the light-induced reduction of O_2 in the chloroplasts (Mehler, 1951) (see 2.2). There has been much debate concerning what is the proportion of the total electron flow from water that ends up on O_2. It seems likely that there is no generally applicable answer to this question and it seems that the best answer is that it depends on the conditions. Under continuous illumination the proportion of electrons transferred to O_2 was reported to be less than 10 % in C_3-plants, up to 15 % in C_4-plants (mesophyll cells), and even 30 % in algae (Badger et al., 2000). In a recent study with leaves of *Hibiscus rosa-sinensis,* it was concluded that in this plant it was almost 40 % (Kuvykin et al., 2008). We believe that both the rate of oxygen reduction and its proportion of the total electron transport depends on i) the plant species, the genome of which determines the range of these values, ii) environmental factors (light, temperature, mineral nutrition, supply of water, and so on), and iii) the age of the plant.

The reduction of O_2 by the PETC in chloroplasts results in the formation of a series of reduced forms of O_2 that are termed Reactive Oxygen Species (ROS), namely, superoxide anion radical ($O_2^{\bullet -}$), hydrogen peroxide (H_2O_2), and hydroxyl radical ($OH^{\bullet}$). ROS also include the singlet oxygen ($^1O_2$), which is not generated by O_2 reduction but by energy transfer from other molecule, mainly from excited chlorophyll triplet state (see 2.2.2).

The above ROS-generating reactions should be distinguished from ROS-mediated reactions, in which the ROS themselves interact with components of the chloroplast. The reactions of both types have "positive" and "negative" effects on chloroplast functions. The occurrence of both types of ROS reactions and to what degree their influence is positive or negative can change as conditions change during the life of the plant, being primarily determined by the level of stress encountered.

2. Oxygen metabolism in chloroplast

2.1 The properties of O_2 molecule and reactive oxygen species

Under usual conditions in the nature, oxygen is a gas composed of diatomic molecules O_2, **dioxygen.** Triplet is the ground state of the dioxygen since the molecule has two electrons with parallel spins in two antibonding molecular orbitals. Since these electrons are unpaired, dioxygen is a biradical. However, the reaction of this biradical with cell components has quantum-mechanical constraint because these components are in the singlet state, *i.e.* they have the valence electrons with antiparallel spins. Due to the above reasons the spontaneous reactions of cell metabolites with dioxygen are highly retarded despite its high oxidizing potential, $E_0' = +0.845$ V of the full reduction of O_2 to H_2O. Such situation is saving for organisms, and the reactions of cell metabolites with O_2 proceed generally with involvement of enzymes, which activate a substrate to speed up these reactions. However the oxidation of cell components can readily proceed by ROS.

Singlet oxygen, $^1O_2$, is formed as the result of the spin flip of one of unpaired electrons. The transformation of $^1O_2$ to triplet is relatively slow; its lifetime in the cell was estimated to be appr. 3 μs (Hatz et al., 2007). This estimation is higher than the previous one for cytoplasm, 0.2 μs (Matheson et al., 1975). In the apolar media this lifetime is higher, 12 μs in ethanol and 24 μs in benzene, and in the heavy water the lifetime increases almost twentyfold and reaches 68 μs (Krasnovsky, 1998). The chloroplast is a prevailing source of $^1O_2$ in the living organisms.

Superoxide anion radical, $O_2^{\bullet -}$, can appear if one additional electron is transferred to the antibonding orbital of O_2. This transfer is possible only if a donor molecule has a redox potential close or lower than the redox potential of pair $O_2/O_2^{\bullet -}$. In the aqueous solutions E_0' ($O_2/O_2^{\bullet -}$) is equal to −0.16 V *vs.* the normal hydrogen electrode (NHE) at 1 M O_2. This value should be used in all thermodynamic consideration of the reactions in the aqueous solutions, instead of −0.33 V, which is the standard potential at 1 atm of O_2. The value of the midpoint redox-potential in aprotic media is much lower, in the region −0.55 ÷ −0.6 V *vs.* NHE (Afanas'ev, 1989). Thus in aprotic media $O_2^{\bullet -}$ is a very strong reductant.

The heavy solvation of $O_2^{\bullet -}$ in aqueous solutions evidently determines its moderate activity in deprotonation reaction in this media; pK_a value of perhydroxyl radical, $HO_2^{\bullet}$, is equal to 4.8. Thus in the aqueous solutions at physiological pH 7.7 the amount of $HO_2^{\bullet}$ is near 0.25 % from total amount of $HO_2^{\bullet} + O_2^{\bullet -}$. The basicity of superoxide ion is much stronger in aprotic media; it was estimated that 'thermodinamic' value of pK_a is close to 12. However more detailed consideration of full deprotonation process leads to a statement that in such media

$O_2^{\bullet-}$ should be considered as a deprotonating agent with pK_a of approximately 24 (Afanas'ev et al., 1987). Moreover considering deprotonation of any substrate by $O_2^{\bullet-}$ it is necessary to take into account that the basicity of proton donors can also increase in aprotic medium, and *e.g.* the rate constant of deprotonation of α-tocopherol by $O_2^{\bullet-}$ is higher in water than in aprotic solvents (Afanas'ev et al., 1987). Being the neutral free radical, $HO_2^{\bullet}$ cannot abstract a proton, but it can abstract a hydrogen atom from substrates with active C-H bonds, initiating fatty acid peroxidation (see further).

$O_2^{\bullet-}$ ion is rather stable even in aqueous solution; the half-life of $O_2^{\bullet-}$ was found to be close to 15 s at pH 11 (Fujiwara et al., 2006). The pH value is very important since the rate constant of spontaneous dismutation (Reaction 1) has maximum at pH 4.8 being equal to $10^8 M^{-1} s^{-1}$, and it sharply decreases in more alkaline media to $10^5 M^{-1} s^{-1}$ at pH 7.7.

$$O_2^{\bullet-} + O_2^{\bullet-} \rightarrow H_2O_2 + O_2 \quad (1)$$

The living cells contain the special enzyme superoxide dismutase (SOD), which catalyzes the dismutation of $O_2^{\bullet-}$ and determines a lifetime of $O_2^{\bullet-}$, and thus the possibility of its involvement in biochemical processes (see further). In the aprotic solvents the $O_2^{\bullet-}$ dismutation is prohibited, and *e.g.* in dimethylformamid $O_2^{\bullet-}$ can persist almost one month (Wei et al., 2004).

$O_2^{\bullet-}$ can interpenetrate cell membranes; the permeability coefficient of the soybean phospholipid bilayer for $O_2^{\bullet-}$ was estimated to be 20 nm s^{-1} (Takahashi & Asada, 1983). The permeability of the egg yolk phospholipid membrane for $HO_2^{\bullet}$ was estimated to be than for $O_2^{\bullet-}$ by almost three orders greater (Gus'kova et al., 1984).

Hydrogen peroxide, H_2O_2, is the most stable ROS. E_0' ($O_2^{\bullet-}/H_2O_2$) is equal to +0.94 V (Asada & Takahashi, 1987) in the aqueous solutions and in the presence of the electron donors and protons $O_2^{\bullet-}$ can react as a good oxidant producing H_2O_2. Ascorbate, quinols, glutathione, and so on can be such donors. In the absence of donors, the dismutation of $O_2^{\bullet-}$ is the main reaction of H_2O_2 production. In the cell, H_2O_2 can also be produced by two-electron oxidases such as glycolate, glucose, amino and sulfite oxidases, which oxidize these substrates by dioxygen directly (Byczkowsky & Gessener, 1988).

The lowest pK_a value of H_2O_2 is 11.8, and under physiological pHs H_2O_2 exists mostly in the neutral form. The properties of H_2O_2 in the aqueous solutions are determined mainly by hydrogen bonds between water and H_2O_2 molecules. These bonds can prevent transfer of H_2O_2 molecules from the aqueous solution to the hydrophobic solvent in spite of their neutral form. The value of E_0' (H_2O_2/H_2O) in aqueous solutions is equal to +1.3 V *vs.* NHE, and in acidic solutions H_2O_2 is one of the most powerful chemical oxidizers. The reduction of H_2O_2 to water requires the breaking of O-O bond, and under physiological conditions the main target of oxidizing action of H_2O_2 are the reduced sulfhydril groups of biomolecules.

Hydroxyl radical, $OH^{\bullet}$, the most destructive ROS, can be produced in cells in the reaction of H_2O_2 molecule decomposition, which is catalyzed by metal. The reaction in which the reductant of H_2O_2 is ferrous iron terms as the Fenton reaction (Reaction 2).

$$H_2O_2 + Fe^{2+} \rightarrow Fe^{3+} + OH^- + OH^{\bullet} \quad (2)$$

This reaction can also be catalyzed by univalent cuprous ion, which is oxidized to divalent ion. Both the oxidized iron and cuprum can be re-reduced by $O_2^{\bullet-}$, and the total reaction of H_2O_2 reduction by $O_2^{\bullet-}$ terms as the Haber-Weiss reaction. The reduction of ferric ion to ferrous can also occur by the reduced cell components, such as ascorbate.

Hydroxyl radical is the penultimate step of dioxygen reduction to water, but this ROS is the strongest oxidant with E_0' ($OH^\bullet/H_2O$) = +2.3 V. Because of high reactivity, $OH^\bullet$ is able to readily oxidize almost all biomolecules at nearly diffusion controlled rates. Therefore $OH^\bullet$ interacts with lipids, proteins and nucleic acids right in the place where it is generated. Since such generation depends on the location of H_2O_2 production, as well as the presence of both metals and reductants, all these circumstances determine the site specificity of the destructive effect of $OH^\bullet$ on biomolecules (Asada & Takahashi, 1987).

The peroxyl radical, $ROO^\bullet$, and hydroperoxide, ROOH, of organic molecule can be considered as long-lived ROS. Their generation usually occurs during the free radical chain reaction known as lipid peroxidation, where they are termed as $LOO^\bullet$ and LOOH. The lipid peroxidation is actually the oxidation of polyunsaturated fatty acid side chains of the membrane phospholipids, and it is initiated by the abstraction of hydrogen atom from the *bis*-allylic methylene of LH to produce $L^\bullet$. The abstraction can be executed by perhydroxyl radical as stated above, whereas the role of $O_2^{\bullet-}$ is usually denied (Bielski et al., 1983), as well as by hydroxyl radical, if the latter does appear in the membrane, and by other ways, *e.g.* by long-lived oxidized reaction center of PSII (see 2.2.2). Under physiological conditions the most possible reaction of $L^\bullet$ is the reaction with dioxygen, when one active electron from organic radical can occupy one of partially filled antibonding orbitals of dioxygen, producing $LOO^\bullet$. This radical is reactive enough to attack adjacent fatty acid side chain, abstracting hydrogen, producing LOOH and new $L^\bullet$; and thus propagating the chain oxidation of lipids. $^1O_2$, reacting with fatty acid can form LOOH directly. LOOH can decompose to highly cytotoxic products, among of which the aldehydes are most dangerous.

2.2 Production of ROS in chloroplasts

Mehler observed the oxygen uptake and H_2O_2 formation under illumination in broken chloroplasts, *i.e.* the chloroplasts with destroyed envelope (Mehler, 1951). Later, it was shown that the primary product of O_2 reduction in the photosynthetic electron transport chain is the $O_2^{\bullet-}$ (Allen & Hall, 1973; Asada et al., 1974). The oxygen reduction rate averages 25 μmol O_2 mg Chl^{-1} h^{-1} in isolated thylakoids under saturating light intensity (Asada & Takahashi, 1987; Khorobrykh et al., 2004). Oxygen uptake and H_2O_2 formation under illumination of thylakoids is the result of the reactions

$$2H_2O - 4e^- \rightarrow 4H^+ + O_2 \text{ (release) - water oxidation in PSII} \quad (3)$$

$$4O_2 + 4e^- \rightarrow 4\, O_2^{\bullet-} \text{ - dioxygen reduction} \quad (4)$$

and subsequent dismutation of $O_2^{\bullet-}$ (Reaction 1). Taking into account the peculiarities of this electron flow, namely that the donor and the acceptor are the forms of oxygen, and the fact that an electron does not return back to the place of its donation to PETC, this flow besides "the Mehler reaction" was termed "pseudocyclic electron transport".

2.2.1 Production of ROS in chloroplast stroma: mechanism and producers

Production of superoxide in stroma

Ferredoxin (Fd), a stromal protein and the electron carrier between PSI and $NADP^+$, has long been regarded as the main reductant of oxygen in the Mehler reaction. The addition of Fd to the suspension of isolated thylakoids led to an increase of an oxygen consumption rate (Allen, 1975*a*; Furbank & Badger, 1983; Ivanov et al., 1980). E_m for Fd/Fd^{red} is −420 mV that enables the reduced Fd (Fd^{red}) to reduce O_2 to $O_2^{\bullet-}$ in the water media. The pseudo-first order rate constant of this reacton was found to be in the region 0.07 - 0.19 s^{-1} (Golbeck & Radmer, 1984, Hosein & Palmer, 1983, Kozuleva et al., 2007). The weak capability of Fd^{red} to reduce O_2 is important for function of chloroplasts since Fd^{red} is a key metabolite that is required for many metabolic reactions in chloroplasts, first of all, the reduction of $NADP^+$.

Recently it was shown that oxygen reduction by Fd is only a part of the total oxygen reduction by PETC (Kozuleva & Ivanov, 2010). The share of oxygen reduction by Fd was measured to be 40-70 % in the absence and 1-5 % in the presence of $NADP^+$. It means that *in vivo* oxygen reduction occurs mostly by the membrane-bound components of PETC rather than by Fd^{red}, however the role of Fd can increase if the $NADP^+$ supply becomes limited.

It was shown that some stromal flavoenzymes such as ferredoxin-$NADP^+$ oxidoreductase, monodehydroascorbate reductase and glutathion reductase added to thylakoid suspension also can produce $O_2^{\bullet-}$ (Miyake et al., 1998). The authors have suggested that these enzymes are reduced by Photosystem I (PSI) directly. However *in vivo* the enzymes have to compete with Fd for electrons from terminal acceptors of PSI at the docking site that is optimized for association with Fd. So this way of oxygen reduction is unlikely under normal conditions.

Production of hydrogen peroxide in stroma

It is considered that the dismutation of $O_2^{\bullet-}$ with involvement of SOD is the main producer of H_2O_2 in chloroplasts stroma. The production of H_2O_2 in stroma through the reduction of $O_2^{\bullet-}$ by ascorbic acid or by reduced glutathione (GSH) is also possible. However the rate constants for these reactions are $3.3\times10^5\ M^{-1}s^{-1}$ (Gotoh & Niki, 1992) and 10^2-$10^3\ M^{-1}s^{-1}$ (Winterbourn & Metodiewa, 1994), respectively, *i.e.* they are considerably less than that for SOD-catalyzed dismutation, $2\times10^9\ M^{-1}s^{-1}$. Fd^{red} was also proposed to produce H_2O_2 in the reaction with $O_2^{\bullet-}$ generated in course of the Mehler reaction (Allen, 1975*b*). However *in vivo* Fd^{red} is involved in a number of reactions and its steady-state concentration is not high, and this way of H_2O_2 production in stroma should be unlikely in the case of effective operation of SOD.

Production of hydroxyl radical in stroma

The main way of $OH^\bullet$ generation is the Fenton reaction (Reaction 2). In chloroplasts stroma there are pools of iron deposited in a redox inactive form. Iron is bound with chelators such as ferritin, the iron storage protein (Theil, 2004), as well as low molecular mass chelators, *e.g.* nicotianamine (Anderegg & Ripperger, 1989). The concentration of free iron ions can be increased when the accumulation of the iron either exceeds the chelating ability of chloroplasts or the iron is released from its complex with chelators (Thomas et al., 1985). The authors have suggested that $O_2^{\bullet-}$ can cause the releasing of iron from ferritin.

The reduced ferredoxin can catalyze the Fenton reaction probably due to it has Fe in its structure (Hosein & Palmer, 1983; Snyrychova et al., 2006). However as it was noted above, the reduced ferredoxin in chloroplast is effectively used for various metabolic pathways, and its level is not high. So, this way of $OH^{•}$ generation can be significant only under stress conditions. The production of $OH^{•}$ also can occur during sulfite oxidation in chloroplasts, and both sulfite radical and hydroxyl radical can initiate oxidative damage of unsaturated lipids and chlorophyll molecules (Pieser et al., 1982).

2.2.2 Production of ROS in thylakoid membrane: mechanism and producers

Production of singlet oxygen in thylakoid membrane

The main route of $^1O_2$ generation in thylakoids is the transfer of energy from the chlorophyll in triplet state to molecular oxygen (Neverov & Krasnovsky Jr., 2004; Rutherford & Krieger-Liszkay, 2001). The main place of the chlorophyll triplet state formation in thylakoids is PSII, presumably a chlorophyll *a* molecule located on the surface of the pigment-protein complexes and a chlorophyll *a* molecule of the special pair (P680) (Neverov & Krasnovsky Jr., 2004). The chlorophyll triplet state and hence $^1O_2$ are usually formed under conditions that are favourable for the charge recombination in $P680^+Pheo^-$ when forward electron transport is very limited (for review see Krieger-Liszkay, 2005), for example when the plastoquinone pool (PQ-pool) becomes over-reduced. This leads to the full reduction of Q_A and results in a low yield of charge separation due to the electrostatic effect of Q_A^- on the $P680^+Pheo^-$ radical pair. This is known as closed PSII however still around 15 % of charge separation occurs at such conditions leading to the formation of the chlorophyll triplet state.

The chlorophyll triplet state formation can occur by a true back reaction through $P680^+Pheo^-$ or by a direct (tunneling) recombination (Keren et al., 1995). These processes can happen under normal functional conditions but with a very low rate. The distribution of these two routes is determined by the energy gap between the $P680^+Pheo^-$ radical pair and the $P680^+Q_A^-$ radical pair. It was shown that true back reactions with the electron coming back from Q_B^- leads to deactivation of some steps in water-oxidizing cycle giving rise to the chlorophyll triplet state formation and $^1O_2$ generation (Rutherford & Inoue, 1984).

It was found that the treatment of plants by some herbicides that are known to bind to Q_B site in PSII and to block photosynthetic electron transport results in formation of the chlorophyll triplet state and $^1O_2$ that finally leads to death of plants (Krieger-Liszkay & Rutherford, 1998).

Production of superoxide in thylakoid membrane

As had been repeatedly proposed $O_2^{•-}$ can be generated within thylakoid membrane (Kruk et al., 2003; Mubarakshina et al., 2006; Takahashi & Asada, 1988) and the first direct evidence was recently obtained using detectors of $O_2^{•-}$ with different lipophilicity (Kozuleva et al., 2011).

PSI. Traditionally it was supposed that the components of acceptor side of PSI, which have highly negative E_m values are the main reductants of oxygen. $O_2^{•-}$ production can possibly occur under oxidation by oxygen of the FeS centers F_A and F_B, which are located in PsaC subunit of PSI exposed to stroma. This $O_2^{•-}$ production would occur outside the thylakoid membrane. The media within thylakoid membrane has low permittivity where E_m of $O_2/O_2^{•-}$ pair could be approximately −600 mV (see 2.1). The components of PSI that are situated below the surface of the membrane, phylloquinone A_1 and the FeS cluster F_X, have E_m values −820 mV and −730 mV, respectively (Brettel & Leibl, 2001). Thus the reduction of O_2 by these centers is thermodynamically allowed.

PSII. The $O_2^{\bullet-}$ generation in PSII has been also shown (Ananyev et al., 1994). However oxygen reduction in this photosystem can achieve only about 1-1.5 μmol O_2 mg Chl^{-1} h^{-1} at physiological pHs (Khorobrykh et al., 2002). In PSII thermodynamically only $Pheo^-$ (E_m of $Pheo/Pheo^-$ is −610 mV) is able to reduce O_2 to $O_2^{\bullet-}$. However under normal functional conditions fast electron transfer from $Pheo^-$ to Q_A^- (300–500 ps (Dekker & Grondelle, 2000)) prevents the electron transfer from $Pheo^-$ to O_2. If Q_A^- is fully reduced (*e.g.* under strong stress conditions) this process likely can occur. It is discussed in the literature (Bondarava et al., 2010; Pospíšil, 2011) that other components of PSII such as Q_A^- (E_m of Q_A/Q_A^- is −80 mV (Krieger et al., 1995)) and low-potential form of cytochrome b_{559} (E_m is 0−80 mV (Stewart & Brudvig, 1998)) can reduce molecular oxygen. However these processes are less favorable thermodynamically and probably do not occur under normal functional conditions.

The plastoquinone pool. Plastoquinone (PQ) is the mobile electron carrier between PS II and cytochrome b_6/f complexes in the thylakoid lipid bilayer phase and it simultaneously transfers the protons across the thylakoid membrane. TKhorobrykh & Ivanov (2002) provided the evidences of the involvement of the PQ-pool in the process of oxygen reduction. Using the inhibitor of the plastoquinol oxidation by cytochrome b_6/f complexes, dinitrophenylether of 2-iodo-4-nitrothymol (DNP-INT), the rate of oxygen uptake was measured to be 9-10 μmol O_2 mg Chl^{-1} h^{-1} at pHs higher than 6.5. It was shown that in the course of oxygen reduction in the PQ-pool, $O_2^{\bullet-}$ was produced. Thermodynamical analysis of the data revealed that only plastosemiquinone ($PQ^{\bullet-}$) (E_m of $PQ/PQ^{\bullet-}$ is −170 mV) in the PQ-pool could reduce O_2 to $O_2^{\bullet-}$ (Reaction 5).

$$PQ^{\bullet-} + O_2 \rightarrow PQ + O_2^{\bullet-} \qquad (5)$$

It was proposed that the Q-cycle operation eliminates an appearance of long-lived $PQ^{\bullet-}$ in the plastoquinol-oxidizing site (Osyczka et al., 2004). However the free $PQ^{\bullet-}$ can be produced in the reaction of plastoquinone/plastoquinol disproportionation (Rich, 1985) and thus the $PQ^{\bullet-}$ can reduce oxygen to $O_2^{\bullet-}$ under normal functional conditions. It was estimated that the product of the free $PQ^{\bullet-}$ concentration and the rate constant of the reaction between semiquinone and O_2 for quinones with E_m values close to those of $PQ/PQ^{\bullet-}$, is very similar to the experimentally observed rates of oxygen reduction in the presence of DNP-INT (Mubarakshina & Ivanov, 2010). Moreover the detailed consideration of this process leads to a conclusion that the reaction between $PQ^{\bullet-}$ and O_2 proceeds at the membrane-water interface.

PTOX. Plastid terminal oxidase (PTOX) is the enzyme that oxidizes plastoquinol and reduces oxygen to water thus it is involved in chlororespiratory and play important role in many processes under stress conditions (for review see Nixon & Rich, 2006). Using Tobacco plants with over-expressing of PTOX it was proposed that PTOX also can reduce dioxygen to $O_2^{\bullet-}$ (Heyno et al., 2009). However under normal functional conditions this process (even if occurs) should not give the essential contribution to the overall generation of $O_2^{\bullet-}$ in PETC taking into account that the quantity of PTOX per PSII is ~1 % only (Andersson & Nordlund, 1999; Lennon et al., 2003).

Production of hydrogen peroxide in thylakoid membrane

Spontaneous dismutation of $O_2^{\bullet-}$ in the thylakoid membrane should be very low owing to a strong electrostatic repulsion in the membrane interior with low permittivity. However it has been found that H_2O_2 is produced within the membrane with significant rate and the production increases with an increase of light intensity (Mubarakshina et al., 2006). On the basic of the data presented in (Ivanov et al., 2007; Khorobrykh et al., 2004; Mubarakshina et al., 2006) it was proposed that H_2O_2 within thylakoid membrane is produced due to the reduction of $O_2^{\bullet-}$ by plastoquinol (Reaction 6) (for review see Mubarakshina & Ivanov, 2010).

$$PQH_2 + O_2^{\bullet-} \rightarrow PQ^{\bullet-} + H_2O_2 \qquad (6)$$

H_2O_2 can also be produced at PSII donor and acceptor sides. At the acceptor side, H_2O_2 can be formed outside thylakoids by the dismutation of $O_2^{\bullet-}$ produced within membrane (Arato et al., 2004; Khorobrykh et al., 2002; Klimov et al., 1993) or inside the membrane by the interaction of $O_2^{\bullet-}$ with non-heme iron of PSII (Pospíšil et al., 2004). At the donor side H_2O_2 can be formed as an intermediate during water oxidizing cycle operation if this cycle is seriously disrupted (Ananyev et al., 1992; Hillier & Wydrzynski, 1993). Thus H_2O_2 production at PSII donor and acceptor sides should be largely neglected under normal conditions.

Production of hydroxyl radical in thylakoid membrane

The various treatments of isolated PSII particles can lead to hydroxyl radical generation (Arato et al., 2004; Pospíšil et al., 2004). Production of hydroxyl radical by PSII is limited under normal functional conditions unlike under the strong stress conditions. It was suggested that in PSI the reduced F_A and F_B can catalyze the Fenton's reaction and form $OH^{\bullet}$ (Snyrychova et al., 2006). The presence of effective electron acceptors from PSI such as methylviologen (Snyrychova et al., 2006) and probably Fd and $NADP^+$, results in a decrease of $OH^{\bullet}$ generation. So *in vivo* the production of $OH^{\bullet}$ by PSI would be minor.

Production of organic peroxides (ROOH) in thylakoid membrane

It was shown that oxygen uptake in the PSII particles at pH above 8 and after the Tris treatment was not the result of oxygen reduction to $O_2^{\bullet-}$ only (Khorobrykh et al., 2002). These conditions can lead to destruction of the water-oxidizing complex and it was proposed that this can result in the formation of long-lived $P680^+$, which can oxidize the close lipids. These lipids can react with oxygen producing the lipid peroxides and thus increasing the oxygen uptake. Using the fluorescent probe Spy-HP it has been recently shown that organic peroxides (ROOH) are produced in PSII membranes when the function of the water-oxidizing complex is disrupted (Khorobrykh et al., accepted).

2.3 Negative effects of ROS in chloroplasts. ROS scavenging systems as the part of chloroplast metabolism

2.3.1 Destructive action of ROS in chloroplasts

The destructive action of ROS in chloroplasts as well as in other parts of the cell is targeted on proteins, nucleic acids and lipids, which can lose their specific functions even due to small changes in their structure after interaction with ROS. Chloroplasts contain own genome represented by the DNA with 110-120 genes, accompanied by own system of the protein biosynthesis, including RNA and ribosomes (Cui et al., 2006). It is interesting that in every chloroplast there are a few tens of genome copies, and this may be an adaptation to the existence under conditions of continuous ROS production by PETC. $OH^{\bullet}$ is considered as the main ROS injuring DNA. It preferably attacks the thimines and cytosines, and in a less extent, adenines, guanines, and the rest of desoxyribose (Cadet et al., 1999). $O_2^{\bullet -}$ has weaker effects on the DNA, and attacks preferably guanines. Since chloroplast genome contains the genes coding some components of PETC, the breakdown of the operation of such genes can affect the normal electron transfer, and the modified PETC in its turn can increase the production of $O_2^{\bullet -}$. In stroma, a toxic $O_2^{\bullet -}$ action is aimed mostly at heme-containing enzymes, such as peroxidases (Asada, 1994). In the thylakoid membrane, perhydroxyl radical, can initiate lipid peroxidation that leads to disturbing the membrane structure and its functions, such as barrier, transport, maintenance of the membrane proteins, and so on.

The damaging effect of H_2O_2 on the genome is determined by the production of $OH^{\bullet}$ in the vicinity of DNA. More specific effect of H_2O_2 in chloroplasts is the inhibition of photosynthesis. It was found that H_2O_2 inhibits the photosynthesis in intact chloroplasts with a half-inhibition at 10 μM (Kaiser, 1976). Electron transfer through PETC is rather resistant to H_2O_2, and the photosynthesis inhibition in the presence of H_2O_2 occurs due to oxidation of thiol groups of enzymes involved in carbon fixation cycle (Charles & Halliwell, 1980; Kaiser, 1979). It can be calculated that in chloroplasts 10 μM H_2O_2 can arise during less than for 1 min under usual photosynthesis rates even if only 1 % of electrons are transferred to O_2. The survival of the chloroplast is provided by the protective (antioxidant) system, which is very active in chloroplasts (see further).

$^1O_2$ being produced in PSII interacts mainly with D1 protein of the core complex of PSII reaction center (Aro et al., 1993; Trebst et al., 2002). This process possibly explains the very high rate of the replacement of D1 by newly synthesized proteins at high light intensity. It may be noted that the PSII activity can also be destroyed not only by $^1O_2$ produced in PSII but also by ROS produced in PSI (Krieger-Liszkay et al., 2011; Tjus et al., 2001).

2.3.2 Mechanisms and components of ROS scavenging reactions in stroma and in the thylakoid membrane

Chloroplasts of the leaf cells are the building sites of the plant. Since potentially harmful ROS are continuously produced in chloroplasts in the light, these organelles are supplied with an efficient system of ROS scavenging. This system may be divided into stromal and membrane parts, however, these parts are connected by common metabolites and operate jointly to maintain chloroplast function. Averaged O_2 concentration in chloroplasts under illumination does not estimably differ from the one in the dark owing to a fast equilibration of new O_2 molecules in the water phase (Ligeza et al., 1998). The quasi-stationary O_2 concentration in the thylakoid membrane in the light can be higher than in other compartments of a chloroplast, due to the production of O_2 molecules in water-oxidizing complex. Futhemore the O_2 concentration in hydrophobic media is approximately ten times higher than in water. Taking into account the primary generation of ROS by the membrane components, the thylakoid membrane requires particularly strong protection.

2.3.2.1 Stromal defense system

Superoxide dismutase. SODs are the water-soluble proteins. The main chloroplast isoform of SOD in all plants is CuZn-SOD, and some plants also contain Fe-SOD in stroma (Kurepa et al., 1997). Immunogold labeling of the chloroplastic CuZn-SOD revealed that the enzyme is mostly concentrated, almost 70 % of its total amount, in 5-nm layer in the vicinity of the thylakoid membrane surface (Ogawa et al., 1995). Authors stated its local concentration in this layer as about 1 mM. Thus SOD prevents the incoming of $O_2^{\bullet-}$ from the membrane to stroma. SOD scavenges $O_2^{\bullet-}$ also in the bulk of stroma, where $O_2^{\bullet-}$ can emerge due to oxidation of Fd^{red} or some other enzymes by oxygen.

Ascorbate and ascorbate peroxidase. The concentration of ascorbate in chloroplasts is very high, achieving 10 - 50 mM (for review see Smirnoff, 2000), and even about 300 mM in alpine plants (Streb et al., 1997). Ascorbate can act as an effective quencher of $O_2^{\bullet-}$ with a high rate constant. Moreover ascorbate is involved in regeneration of the α-tocopherol radicals formed during detoxification of lipid peroxide radicals. The scavenging of H_2O_2 in chloroplasts is performed by ascorbate peroxidase (APX), which catalyzes the reaction of H_2O_2 with ascorbate. Catalase was not found in chloroplasts, although the low catalase activity of thylakoids and some stromal components is not ruled out. Having more low value of $K_m(H_2O_2)$ as compared with catalase, 80 µM *vs.* 25 mM, APX can provide more low H_2O_2 concentration; and this is important, taking into account the inhibitory effect of H_2O_2 on the Calvin cycle enzymes (see 2.3.1). The reaction, which is catalyzed by APX has a high rate constant, 10^7 $M^{-1}s^{-1}$. Chloroplasts contain APX in two isofoms, thylakoid-bound and soluble stromal ones (Miyake & Asada, 1992). Both APXs are highly specific to ascorbate as the electron donor, and they are promptly inactivated, during 10 s, in its absence (Nakano & Asada, 1987). These peroxidases form two defending lines to protect stromal components from H_2O_2.

Glutathione and glutathione peroxidase. The reduced form of glutathione (GSH) plays an important role in the stabilization of many stromal enzymes. For the antioxidant function it is important that it serves as a substrate for dehydroascorbate reductase. GSH is able to react directly with ROS including H_2O_2 (Dalton et al., 1986), hydroxyl radical (Smirnoff & Cumbes, 1989) and even $^1O_2$ (Devasagayam et al., 1991). Chloroplasts also contain phospholipid hydroperoxide-scavenging glutathione peroxidase (Eshdat et al., 1997) that may be involved in the reduction of lipid peroxide of thylakoid membranes to its alcohol, suppressing the chain oxidation of thylakoid phospholipids. This glutathione peroxidase may be considered as the part of the membrane defense system.

Osmolytes. Osmolytes are the group of metabolites that decrease water potential inside the cell and prevent intracellular water loss. This group includes soluble sugars, glycine, betaine, proline etc. The antioxidant capacity of proline is the result of its ability to quench $^1O_2$ and scavenge $OH^\bullet$ (Matysik et al., 2002). Recently it was shown that synthesis of proline occurs, at least partly, in chloroplasts (Székely et al., 2008) where proline can execute the antioxidant function and protect both the membranes against lipid peroxidation and the stromal enzymes against desactivation.

Some soluble sugars were recently recognized as antioxidants (for review see Bolouri-Moghaddam et al., 2010). Addition of mannitol to thylakoid suspension resulted in decrease of $OH^\bullet$ production, and the transgenic tobacco plants with enhanced mannitol production targeted to the chloroplast had the increased $OH^{\bullet}$ scavenging capacity (Shen et al., 1997). The mutants of *Arabidopsis* with overexpressed enzymes providing the elevated concentration of galactinol and raffinose in leaves were more resistant to oxidative stress caused by methylviologen treatment than wild-type plants (Nishizawa et al., 2008). The authors concluded that antioxidant capacity of these sugars could be explained by their reaction with $OH^{\bullet}$ (the rate constants were measured as $7.8\times10^9\,M^{-1}s^{-1}$ and $8.4\times10^9\,M^{-1}s^{-1}$ for galactinol and raffinose, respectively).

Peroxiredoxins. Peroxiredoxins (PRXs) are identified as antioxidant enzymes for detoxification of H_2O_2 (for review see Dietz et al., 2006). Furthermore it was found that PRXs can also detoxify alkyl hydroperoxides and peroxinitrite, and probably can modulate oxolipid-dependent and NO-related signalling (Baier and Dietz, 2005; Rhee et al., 2005; Sakamoto et al., 2003). PRXs are 17-22 kDa enzymes that possess N-terminal cysteine residue(s) responsible for peroxidase activity. Four PRXs that are targeted to chloroplasts were identified in *Arabidopsis*: 2-cysteine (2-Cys) PRXs dimeric and oligomeric forms, PRX Q and PRX II E (Dietz et al., 2006). 2-Cys PRXs and PRX Q are associated with thylakoid membrane components while PRX II E has been identified as stromal enzyme. PRXs become oxidized after reaction with H_2O_2. Re-activation of oxidized PRXs in chloroplasts occurs via action of thioredoxin and thioredoxin-like proteins (Broin et al., 2002).

Flavonoids. Flavonoids were found to perform an antioxidant function in tissues exposed to a wide range of environmental stressors (Babu et al., 2003; Reuber et al., 1996). It has been recently assumed that antioxidant activity of flavonoids outperforms that of well-known antioxidants, such as ascorbate and α-tocopherol (Hernández et al., 2008). Flavonoids effectively scavenge the free radicals (for review see Rice-Evans et al., 1996). This can occur due to their ability to quench unpaired electrons of radicals, *e.g.* $O_2^{\bullet -}$ (Sichel et al., 1991). It was also shown that flavonoids situated in chloroplasts can scavenge $^1O_2$ (Agati et al., 2007). Flavonoids include the substances with different lipophilicity, and thus perform their antioxidant functions in stroma as well as in the membrane.

2.3.2.2 Membrane defense system

Vitamin E. Vitamin E is the class of lipophilic compounds (α-, β-, γ- and δ-tocopherols (Tocs); α-, β-, γ- and δ- tocotrienols and their derivatives). Vitamin E is synthesized in the plastid envelope and is stored in plastoglobuli (for review see Lichtenthaler, 2007). The greatest amount of Tocs is found in the membranes of chloroplasts (including thylakoid membranes) where they execute the antioxidant function.

Vitamin E can react with almost all ROS. It can reduce $O_2^{\bullet -}$ with the rate constant of $10^6\,M^{-1}s^{-1}$ (Polle & Rennenberg, 1994). It was shown that vitamin E has scavenging activity against $OH^{\bullet}$ (Wang & Jiao, 2000) and can decompose H_2O_2 (Srivastava et al., 1983). α-Toc protects PSII from oxidative damage. Trebst et al. (2002) showed that inhibition of Toc biosynthesis in *Chlamydomonas* resulted in a stimulation of light-induced loss of PSII activity and D1 protein degradation. This implies that Toc can come close to the site of $^1O_2$ generation in the reaction center of PSII. The rate constants for $^1O_2$ quenching by different Tocs are appr. 0.13 - $3.13\times10^8\,M^{-1}\,s^{-1}$ in organic solvents (Gruszka et al., 2008), so Tocs can be the effective scavengers of $^1O_2$ within the membrane. It is also possible that α-Tocs can protect the β-carotene molecules in PSII, thereby preventing the PSII damage (Havaux et al., 2005). Tocs can reduce fatty acyl peroxy radicals, thus terminating lipid peroxidation chain reactions (Polle & Rennenberg, 1994). The regeneration of Tocs occurs with involving of water-soluble antioxidants. For example, formation of α-Toc from α-Toc quinone has been reported to take place *in vitro* in the presence of ascorbate (Gruszka et al., 2008).

Carotenoids. There are two major types of carotenoids: the hydrocarbon class, or carotenes, and the oxygenated (alcoholic) class, or xanthophylls. Carotenoids can efficiently quench the dangerous triplet state of chlorophylls that is the origin of the $^1O_2$ (Cogdell et al., 2000). This mostly occurs in the antenna system (Mozzo et al., 2008) but not in the reaction center. It is also known, that carotenoids, namely β-carotene, can quench $^1O_2$ directly (Foote and Denny, 1968). It was shown that a lack of such carotenoids as zeaxanthin and lutein leads to $^1O_2$ accumulation in thylakoids (Alboresi et al., 2011).

Plastoquinone. Plastoquinone (PQ-9), which as the chemical substance is the isoprenoid prenyllipid, is present in thylakoid membranes, chloroplast envelope and osmiophilic plastoglobuli of the stroma (Lichtenthaler, 2007). Plastoglobuli represent the storage compartments for plastoquinone, mainly in its reduced state. In the thylakoid membrane, PQ-pool maintained in the reduced state can execute antioxidant function, preventing membrane lipid peroxidation and pigment bleaching (Hundal et al., 1995). Furthermore it was shown that *in vitro* plastoquinol has an antioxidant activity similar or even higher than that of tocopherols (Kruk et al., 1994, 1997). It was also found that the added quinones can quench the excited states of chlorophyll molecules (Rajagopal et al., 2003), thus inhibiting the $^1O_2$ generation. Moreover plastoquinone can also directly scavenge $^1O_2$ that is produced by the reaction center triplet chlorophyll of PSII (Kruk & Trebst, 2008; Yadav et al., 2010). It is very possible that plastoquinol effectively scavenges $O_2^{\bullet -}$ and perhydroxyls in thylakoid membrane (Reaction 6) (for review see Mubarakshina & Ivanov, 2010). These reactions are the mechanisms by which the PQ-pool can prevent membrane lipid peroxidation. It is known that even in the dark the PQ-pool can be in the reduced state owing to operation of the Ndh complex. This can provide the protective function of the PQ-pool in the membrane if ROS are produced under stress in the dark. It was found that the extent of the PQ-pool reduction in the dark increased upon heat stress, and this was considered as involvement of the Ndh complex in the defense system (Sazanov et al., 1998). Thus it is possible to assume that the higher amount of plastoquinone than of other components of PETC is needed in order to execute the antioxidant function rather than the electron carrier function.

2.4 Role of reaction of oxygen with chloroplast components in the constructive metabolism

2.4.1 Photorespiration

Photorespiration is a pathway of oxidative carbon metabolism which resulted from oxygenase activity of ribulose-1,5-bisphosphate carboxylase/oxygenase (RubisCO) (for review see Maurino & Peterhansel, 2010). Photorespiration cycle starts in chloroplasts from the reaction of ribulose-1,5-bisphosphate with O_2 molecule. As the result 3-phosphoglycerate and 2-phosphoglycolate are produced. The latter is dephosphorylated to glycolate, a toxic molecule. The following reactions of glycolate metabolism lead to a recover of 3-phosphoglycerate and occur in peroxisomes, mitochondria, cytosol and, finally, in chloroplasts again. At current atmospheric levels of CO_2 and O_2, photorespiration in C_3-plants dissipates 25 % of the carbon fixed during CO_2 assimilation (Sharkey, 1988).

Photorespiration is required for all photosynthetic organisms, including cyanobacteria (Eisenhut et al., 2008) and even higher plants with C_4-type of photosynthesis (Zelitch et al., 2008). As a part of chloroplast metabolism, photorespiration can be considered as one of the sinks for the excess of the energetic equivalents such as ATP and NAD(P)H under different stresses (for review see Wingler et al., 2000). Photorespiration operates also as a safety valve protecting PETC from over-reduction, and thus preventing the generation of ROS by components of PETC. Furthermore it can prevent the ROS formation in stroma by using NAD(P)H itself, since NAD(P)H can to some extent spontaneously reduce oxygen to $O_2^{\bullet-}$ giving H_2O_2.

2.4.2 The electron transport to oxygen in PETC and ATP production for chloroplast metabolism

Oxygen reduction in PETC was mainly considered as an important part of the chloroplasts metabolism due to possibility of additional ATP synthesis coupled with this process (Badger, 1985; Heber, 1973). There were calculations of ATP synthesis coupled with O_2 reduction from the total ATP synthesis during simultaneous reductions of both $NADP^+$ and O_2 (Furbank & Badger, 1983; Ivanov et al., 1980; Robinson & Gibbs, 1982). Observed in some studies lower efficiency for ATP production of the pseudocyclic electron transport as compare with the efficiency of the non-cyclic electron transport in C_3- (Woo, 1983) as well as in C_4-plants (Ivanov & Edwards, 2000) can be probably explained now as a result of the use of the protons bound by PQH_2 not only for the ΔpH formation across the thylakoid membrane but also for the reduction of $O_2^{\bullet-}$ to H_2O_2 in the membrane (see 2.2.2).

The Mehler reaction is now considered as a part of so called water-water cycle (WWC). The last term was proposed by Asada, who described the reactions involved in WWC in the fullest detail (Asada, 1999). Briefly, WWC includes the transport of the electrons from water to oxygen, the $O_2^{\bullet-}$ dismutation catalyzed by SOD, the reduction of H_2O_2 by ascorbate with involvement of APX (see 2.3.2.1), followed by the reduction of monodehydroascorbate (MDHA) appearing in the latter reaction to ascorbate by the electrons from PETC (*i.e.* ultimately from water). The reduction of O_2 in PETC is the slowest reaction in WWC, while other reactions proceed with almost diffusion-controlled rates. The latter ensures the minimal accumulation of $O_2^{\bullet-}$ and H_2O_2, preventing their interaction with the target molecules in stroma. The equal amounts of electrons reduce O_2 and MDHA in WWC. Thus WWC constitutes in the total photosynthetic electron flow the double of what is the electron transport to O_2 itself. The reduction of MDHA can deprive the Calvin cycle of electrons, *i.e.* can cease the photosynthesis. The photosynthesis of the intact chloroplasts was completely suppressed at H_2O_2 addition in the light and the CO_2 fixation restarted after the H_2O_2 was exhausted (Nakano & Asada, 1980). This result demonstrated that the system of H_2O_2 scavenging has priority over the system of CO_2 fixation in receiving electrons from PETC; this conclusion was supported in (Backhausen et al., 2000).

It was shown that WWC does play the important role in chloroplasts, and the breakdown of its normal operation negatively influences on the metabolism of these organelles (Rizhsky et al., 2003). This may mainly result from the destructive effects of ROS, but possibly also from the decrease of ATP/NADPH ratio. It is accepted now, that the electron flow to $NADP^+$ provides, the ATP/NADPH of 1.5 owing to Q-cycle operation (Ivanov, 1993). Exactly such ratio is required in order the Calvin cycle reactions to proceed, but ATP is necessary in chloroplasts not only for the cycle, but also for biosynthesis of protein and numerous transport processes. The cyclic electron transport around PSI, which can be essential producer of ATP also cannot operate under anaerobic conditions when PETC is overreduced, and the necessary redox poising for this transport is provided by the electron transfer to O_2 (Ziem-Hanck & Heber, 1980).

2.4.3 Role of electron transfer to oxygen in the protection of PETC from photoinhibiton

The electron flows involved in WWC besides the ATP production play an important role in the protection of photosynthetic apparatus from photoinhibition under illumination. The excess of photon energy beyond the necessary one to fulfil CO_2 assimilation can arise not only in strong light, but also in moderate and even low light when the environmental conditions (the improper temperature, insufficient water supply, high salt concentration, the presence of pollutants, and so on) lead to suppression of the capacity of the photosynthetic apparatus for CO_2 assimilation.

WWC and cyclic electron transport are primarily coupled with the proton pumping into the thylakoid lumen, and these protons in the lumen initiate conformational and biochemical changes, which accelerate thermal energy dissipation in antenna and reaction centers (Horton et al., 1996). The pigment apparatus state providing the dissipation of the photon energy into a heat originates due to violaxanthin de-epoxidation to antheraxanthin and zeaxanthin. This reaction, which requires ascorbate, is catalyzed by violaxanthin deepoxidase (VDE) situated at the lumen side of thylakoid membrane; VDE is activated by a decrease of lumen pH to 5-6 (for review see Demmig-Adams, 1990). Since WWC does not consume ATP, the accumulation of protons in lumen is very fast in the absence of inorganic carbon fixation in C_3- and in C_4-plants (Ivanov et al., 1998; Ivanov & Edwards, 2000). This allows WWC to respond to such changes in light intensity as sunflecks.

The overreduction of the acceptor side of PSII is one of the pre-conditions of photoinhibition initiation, and the electron withdrawal from PETC to O_2 and MDHA within WWC also can effectively use the excess of light energy for the electron transfer. This is especially important under limiting supply of CO_2. It is known that photoinhibition of PSII owing to overreduction occurs under anaerobic conditions even at rather low light intensity (Park et al., 1996) and is prevented even by low dioxygen concentrations when WWC begins to function. The drain of electrons directly from the PQ-pool to oxygen (see 2.2.2) can be a very important mechanism in protection of PETC from photoinhibiton since the PQ-pool operation is known to be a limiting step of photosynthesis.

2.5 Signalling pathways under stress conditions.

The system of signal transfer in plants during stress conditions is a very complicated regulatory mechanism (for recent reviews see Foyer & Noctor, 2009; Li et al., 2009; Mullineaux, 2009; Suzuki et al., 2011). The biosynthesis of chloroplast proteins are coordinated by both chloroplast and nuclear genomes. Thus the tight cooperation between these two systems is obligatory for the assembly of functionally active chloroplasts. This can only happen due to well-co-ordinated work of regulatory signals coming from nucleus to chloroplasts and from chloroplasts to nucleus. The latter is called the retrograde signalling (Beck, 2005; Chan et al., 2010; Mullineaux & Karpinski, 2002).

2.5.1 ROS as important signalling agents

Enhanced ROS production under stress conditions is considered to be a signal in order to regulate the cell redox homeostasis. ROS play a key role in the regulation of plant development, programmed cell death and also in biotic and abiotic stress responses (Apel & Hirt, 2004; Desikan et al., 2001; Mittler et al., 2004). Many of the $^1O_2$-responsive genes are different from those activated by $O_2^{\bullet-}$ or H_2O_2, assuming that the signalling by different ROS occurs via distinct pathways (Laloi et al., 2006; op den Camp et al., 2003) however some $^1O_2$-responsive nuclear genes can be activated by other ROS (Anthony et al., 2005; Gadjev et al., 2006). Moreover it has been recently shown that H_2O_2 may also either directly or indirectly antagonize $^1O_2$-mediated signalling (Baruah et al., 2009; Ledford et al., 2007).

Production of $^1O_2$ in *Arabidopsis thaliana* under stress conditions was shown using a specific fluorescent dye (Hideg et al., 2001; op den Camp et al., 2003). Generation of $^1O_2$ leads to a rapid change in nuclear gene expression that reveals the transfer of $^1O_2$-derived signals from the plastid to the nucleus (Laloi et al. 2006; op den Camp et al. 2003). Using the *flu* mutant with disturbed chlorophyll biosynthesis it was shown that $^1O_2$ forms an integral part of a signalling network that is important not only for stress responses but also for the plant development (Baruah et al., 2009). It was proposed that $^1O_2$ accumulation in thylakoids represents a signalling pathway in the early stages of stress acclimation (Alboresi et al., 2011).

A role for $O_2^{\bullet-}$ in retrograde signalling was suggested using gene expression arrays (Scarpeci et al., 2008) and mutations in chloroplastic CuZn-SOD (Rizhsky et al., 2003). The generation of $O_2^{\bullet-}$ in the absence of H_2O_2 accumulation revealed a subset of nuclear encoded genes that are likely to be specific for an $O_2^{\bullet-}$-mediated signalling pathway (Scarpeci et al., 2008). In CuZn-SOD mutants, the accumulation of $O_2^{\bullet-}$ results in activation of chloroplast-encoded genes that is not stimulated by other ROS (Rizhsky et al., 2003).

H_2O_2 has been recognized as the ROS causing the largest changes in the levels of gene expression in plants in retrograde signalling (Bechtold et al., 2008; Dat et al., 2000; Fahnenstich et al., 2008; Foyer and Noctor, 2009; Li et al., 2009). H_2O_2 possibly induces protein phosphorylation by mitogen-activated protein kinases (MAPKs) (Desikan et al., 1999) which are involved in signalling pathways regulating gene expression (Grant et al., 2000). The reversible redox modulation of Cys residues of the proteins is perhaps the most obvious mechanism for the H_2O_2-mediated activation of MAPK pathways. Transcriptomic analyses of *Arabidopsis* plants have revealed hundreds of H_2O_2-responsive genes (Ding et al., 2010; Yun et al., 2010). It was shown that H_2O_2 produced inside the chloroplasts can leave the chloroplasts thus escaping the effective antioxidant systems located inside the chloroplast (Mubarakshina et al., 2010). It is possible that the appearance of H_2O_2 in cytoplasm can be important for the executing of retrograde signalling mediated by H_2O_2.

2.5.2 Other signalling agents and crosstalk between different signalling pathways and ROS

Antioxidants can also be the sensors of the stress conditions by the regulation of the level of ROS. It was proposed that the ascorbate content could influence the expression of antioxidative genes in *Arabidopsis* (Noctor et al., 2000), and APXs located in chloroplasts are crucial for photoprotection and signalling (Danna et al., 2003; Kangasjärvi et al., 2008). It was shown that the level of Tocs is also elevated in response to a variety of abiotic stresses, including high-intensity light, drought, toxic metals, and high and low temperatures (Maeda & DellaPenna, 2007). Furthermore it was suggested that α-Toc may affect intracellular signalling in plant cells by interacting with key components of the signalling cascade (Munné-Bosch et al., 2007).

Thioredoxin, glutathione, peroxiredoxins and glutaredoxins , which contain sulfhydryl groups (-SH), represent a part of the reduction-oxidation signalling network (for more detailed review see Scheibe & Dietz, 2011, see also Coupe et al., 2006; Fey et al., 2005; Geigenberger et al., 2005; Lindahl & Kieselbach, 2009). The oxidation of (-SH) groups can occur not only by the interaction between the components having -SH groups in their structure but also with involvement of H_2O_2 since H_2O_2 can directly oxidizes -SH groups (Quesada et al., 1996). In plants two chloroplastic thioredoxins, named thioredoxin *f* and thioredoxin *m*, were originally identified as light dependent regulators of several carbon metabolism enzymes including Calvin cycle enzymes (Lemaire et al., 2007). It was found that glutathione is involved in the control of gene expression (Dron et al., 1988; Wingate et al., 1988). H_2O_2 can be sensed by glutathione peroxidase 3 of *Arabidopsis*, which modulates activities of phosphatases, protein kinases transcription factors and ion channels involved in abscisic acid signalling pathways (Wang & Song, 2008).

The PQ-pool can also be considered as both the component of PETC, the redox state of which is an important factor for the redox signalling and the antioxidant, the level of which increases under stress conditions. The treatment of plants by the pathogen-derived elicitor resulted in elevated ROS production, lipid peroxidation and lipoxygenase followed by a significant increase in total plastoquinone level (Maciejewska et al., 2002). High light conditions also lead to a massive accumulation of plastoquinone, preferentially plastoquinol in leaves (Lichtenthaler, 2007; Szymańska & Kruk, 2010). The changing of the antenna size is one of the main mechanisms of plants acclimate to changes in environmental light conditions. It was revealed that the redox state of the PQ-pool regulates the antenna size of PSII under different light conditions (Escoubas et al., 1995; Fey et al., 2005; Lindahl et al., 1995; Pfannschmidt et al., 1999; Yang et al., 2001). In the works (Frigerio et al., 2007; Morosinotto et al., 2006) it was found that the regulation of the antenna size is carried out by the changing of the antenna proteins quantity at the post-transcriptional level. However the signal from the PQ-pool for the light acclimation has still remained largely unsolved. It was shown that H_2O_2 production in the PQ-pool increases with an increase of light intensity (Khorobrykh et al., 2004; Mubarakshina et al., 2006) that is correlated with an increase of the redox state of the PQ-pool. Possibly, H_2O_2 can be the best candidate in order to be a signal of the PQ-pool over reduction.

Phytohormones are also the components of cell signalling. The first steps of synthesis of such stress hormones as abscisic, salicylic and jasmonic asids occur in plastids (Vernooij et al., 1994; Wasilewska et al., 2008; Wasternack, 2007) and their influences on the photosynthesis have been demonstrated (Filella et al., 2006; Mateo et al., 2006). However the involvement of phytohormones in chloroplast-to-nucleus signalling is still under debate.

Another component that was shown to be important for the regulation of nuclear genes by chloroplasts is Mg-protoporphyrin (Mg-Proto), the precursor of chlorophyll (Gray, 2003). It was suggested that after illumination, Mg-protoporphyrins can move from chloroplast to cytoplasm where they can interact with a signal conductor (von Gromoff et al., 2006). Another possibility is that Mg-protoporphyrins can change the activity of regulatory proteins directly in chloroplasts (for review see Yurina & Odintsova, 2007). There are no clear evidences about the crosstalk between Mg-protoporphyrin-derived signalling and ROS. It was found that it is unlikely that the accumulation of Mg-Proto is linked to the production of $^1O_2$ (Mullineaux, 2009). However it was suggested that light is needed to promote the export of Mg-Proto to the cytosol from the chloroplast (Kropat et al., 1997). Probably this export may involve the co-operation with ROS.

Photosynthesis-derived metabolites can also act as a signal. One of these metabolites is sugar. The regulatory effect of sugars on the expression of nuclear photosynthetic genes and on plant metabolism was established (for review see Gupta & Kaur, 2005). Both glucose and sucrose were shown to be a signal molecule, and other sugars produced in a number of plant metabolic pathways are also potential signalling agents. The most regulatory effects can be ascribed to glucose (Ramon et al., 2009). Hexokinase, the first enzyme in glucose catabolism can reduce the intracellular ROS levels (Sun et al., 2008). It was shown that mitohondrial hexokinases play a key role as a regulator of ROS levels (Camacho-Pereira et al., 2009). In the work (Bolouri-Moghaddam et al., 2010) it was postulated that chloroplastic hexokinases also can carry out this function. It was shown that the plant with higher levels of hexokinase was more resistant to oxidative stress induced by methylviologen (Sarowar et al., 2008).

3. Conclusion

The ways of ROS production and ROS scavenging in chloroplasts are very important for the metabolisms of cell and whole organism. These ways can differ under normal functional conditions and under stress conditions as shown in Figure 1.

As stated here, ROS production can be divided into the stromal part and the membrane part. Probably, the ways of ROS production in different parts of chloroplasts initiate different signalling pathways. The signalling network in chloroplast stroma is widely studied, and many of its characteristics become clear. We also pay the attention to the signalling pathways that may begin in the thylakoid membrane. H_2O_2 that is produced in the thylakoid membrane by the reaction of superoxide with plastoquinol can diffuse out of the thylakoid membrane not only to stroma but also to lumen, thus avoiding the antioxidant systems of chloroplast stroma. Some of signal sensors such as thylakoid protein kinases are associated with the thylakoid membrane and the signalling pathway initiated by such kinases may involve H_2O_2 produced in the membrane.

The interesting fact is that during stress conditions even such components as osmolytes, flavonoids etc. that have their own functions, also are involved in ROS scavenging. Since the control of ROS level is intrinsically involved in executing their signal roles, the cooperation of the whole system of chloroplast metabolites obviously provides sustaining the chloroplast function under stress conditions.

Although the role of ROS in signalling is commonly accepted now and a lot of studies are focused on the changes in both ROS production under stress conditions and the abundance of the antioxidant and other stress enzymes, the molecular mechanisms of the signalling pathways remain unclear. For example, it is not totally clear, how ROS fulfill signalling functions, being inside chloroplasts and how after leaving for cytoplasm. ROS produced inside chloroplast can be the signal for the inside-organelle signalling to regulate the chloroplast genes. If ROS leave the chloroplast and appear in cytoplasm, it can affect the integrated signalling network of the whole cell.

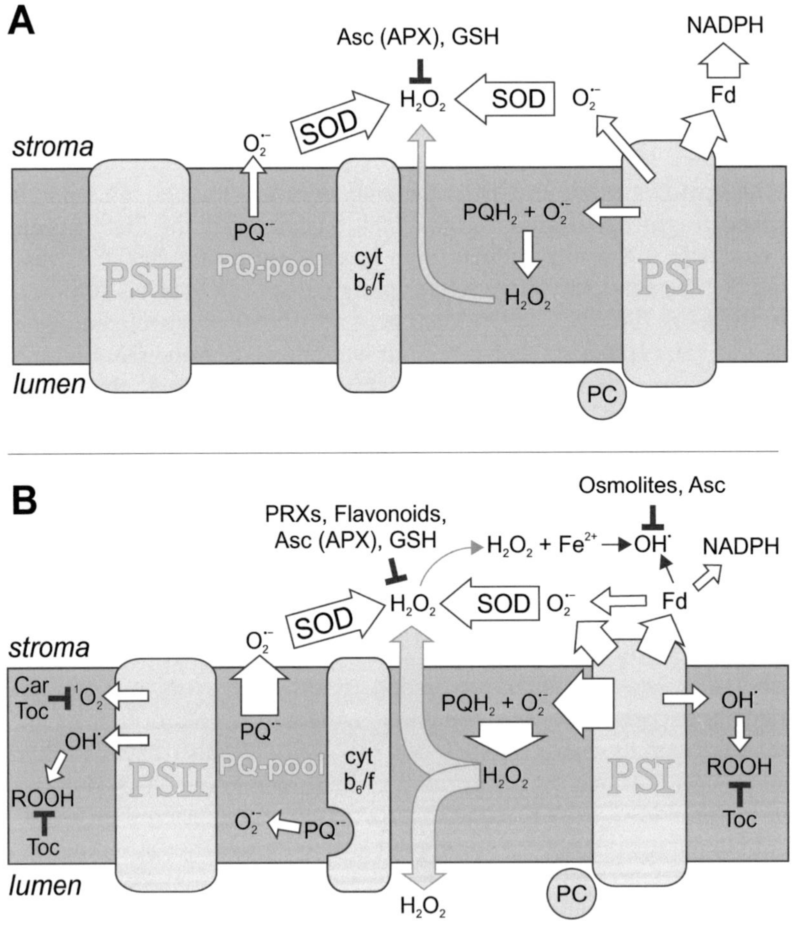

Fig. 1. The ways of ROS production and ROS scavenging in chloroplasts under normal functional conditions (A) and stress conditions (B). The scheme represents the putative ways of ROS production and scavenging that play the most important role in oxygen metabolism. APX, ascorbate peroxidase; Asc, ascorbate; Car, carotenoids; cyt b_6/f, cytochrome b_6/f complex; Fd, ferredoxin; GSH, reduced glutathione; PC, plastocyanin; PQ, plastoquinone; $PQ^{\bullet -}$, plastosemiquinone; PQH_2, plastoquinol; PRXs, peroxiredoxins; PSI and PSII, photosystem I and II, respectively; SOD, superoxide dismutase; Toc, tocopherols. For details see text.

4. References

Afanas'ev, I.B., Grabovetskii, V.V. & Kuprianova, N.S. (1987). Kinetics and Mechanism of the Reactions of Superoxide Ion in Solution. Part 5. Kinetics and Mechanism of the Interaction of Superoxide Ion with Vitamin E and Ascorbic Acid. *Journal of the Chemical Society, Perkin Transactions, 2,* No. 3, (n.d.), pp. 281-285, ISSN 0300-9580

Afanas'ev, I.B. (1989). *Superoxide Ion: Chemistry and Biological Implications*, CRC Press, ISBN 978-084-9354-51-9, Boca Raton, Florida, USA

Agati, G., Matteini, P., Goti, A. & Tattini, M. (2007). Chloroplast-Located Flavonoids Can Scavenge Singlet Oxygen. *The New Phytologist*, Vol. 174, No. 1, (April 2007), pp. 77–89, ISSN 0028-646X

Alboresi, A., Dall'Osto, L., Aprile, A., Carillo, P., Roncaglia, E., Cattivelli, L. & Bassi, R. (2011). Reactive Oxygen Species and Transcript Analysis upon Excess Light Treatment in Wild-Type *Arabidopsis thaliana vs* a Photosensitive Mutant Lacking Zeaxanthin and Lutein. *BMC Plant Biology*, Vol. 11:62 (April 2011) ISSN 1471-2229

Allen, J.F. & Hall, D.O. (1973). Superoxide Reduction as a Mechanism of Ascorbate-Stimulated Oxygen-Uptake by Isolated Chloroplasts. *Biochemical and Biophysical Research Communications*, Vol. 52, No. 3, (June 1973), pp. 856–862, ISSN 0006-291X

Allen, J.F. (1975). Oxygen Reduction and Optimum Production of ATP in Photosynthesis. *Nature*, Vol. 256, No. 5518, (August 1975), pp. 599–600, ISSN 0028-0836

Allen, J.F. (1975). A Two-Step Mechanism for Photosynthetic Reduction of Oxygen by Ferredoxin. *Biochemical and Biophysical Research Communications*, Vol. 66, No. 1, (September 1975), pp. 36–43, ISSN 0006-291X

Ananyev, G., Wydrzynski, T., Renger, G. & Klimov, V. (1992). Transient Peroxide Formation by the Managense-Containing, Redox-Active Donor Side of Photosystem II upon Inhibition of O_2 Evolution with Lauroylcholine Chloride. *Biochimica et Biophysica Acta*, Vol. 1100, No. 3, (June 1992), pp. 303–311, ISSN 0005-2728

Ananyev, G., Renger, G., Wacker U. & Klimov V. (1994). The Photoproduction of Superoxide Radicals and the Superoxide Dismutase Activity of Photosystem II. The Possible Involvement of Cytochrome b559. *Photosynthesis Research*, Vol. 41, No. 2 , (August 1994), pp. 327-338, ISSN 0166-8595

Anderegg, G. & Ripperger, H. (1989). Correlation between Metal Complex Formation and Biological Activity of Nicotianamine Analogues. *Journal of the Chemical Society, Chemical Communications*, Vol. 10, (n.d.), pp. 647-650, ISSN 0022-4936

Andersson, M.E. & Nordlund, P. (1999). A Revised Model of the Active Site of Alternative Oxidase. *FEBS Letters*, Vol. 449, No. 1, (April 1999), pp. 17–22, ISSN 0014-5793

Anthony, J.R., Warczak, K.L. & Donohue, T.J. (2005). A Ttranscriptional Response to Singlet Oxygen, a Toxic Byproduct of Photosynthesis. *Proceedings of the National Academy of Sciences of the USA*, Vol. 102, No. 18, (May 2005), pp. 6502–6507, ISSN 0027-8424

Apel, K. & Hirt, H. (2004). Reactive Oxygen Species: Metabolism, Oxidative Stress, and Signal Transduction. *Annual Review of Plant Biology*, Vol. 55, (June 2004), pp. 373–399, ISSN 1543-5008

Arato, A., Bondarava, N. & Krieger-Liszkay, A. (2004). Production of Reactive Oxygen Species in Chloride- and Calcium-Depleted Photosystem II and their Involvement in Photoinhibition. *Biochimica et Biophysica Acta*, Vol. 1608, No. 2-3, (February 2004), pp. 171–180, ISSN 0005-2728

Aro, E.M., Virgin, I. & Andersson, B. (1993). Photoinhibition of Photosystem II. Inactivation, Protein Damage and Turnover. *Biochimica et Biophysica Acta*, Vol. 1143, No. 2, (July 1993), pp. 113–134, ISSN 0005-2728

Asada, K., Kiso, K. & Yoshikawa K. (1974). Univalent Reduction of Molecular Oxygen by Spinach Chloroplasts on Illumination. *The Journal of Biological Chemistry*, Vol. 249, No. 7, (April 1974), pp. 2175-2181, ISSN 0021-9258

Asada, K. & Takahashi, M. (1987). Production and Scavenging of Active Oxygen in Photosynthesis, In: *Photoinhibition*, D.J. Kyle, C.B. Osmond & C.J. Arntzen, (Eds), pp. 227-287, Elsevier, ISBN 978-044-4415-96-7, Amsterdam, The Netherlands

Asada, K. (1994). Production and Action of Active Oxygen Species in Photosynthesis Tissues, In: *Causes of Photooxidative Stress and Amelioration of Defense Systems in Plants*, C.H. Foyer & P.M. Mullineaux, (Eds), pp. 78-104, CRC Press, ISBN 978-084-9354-43-4, Boca Raton, Florida, USA

Asada, K. (1999). The Water-Water Cycle in Chloroplasts: Scavenging of Active Oxygens and Dissipation of Excess Photons. *Annual Review of Plant Physiology & Plant Molecular Biology*, Vol. 50, (June 1999), pp. 601–639, ISSN 1040-2519

Babu, T.S., Akhtar, T.A., Lampi, M.A., Tripuranthakam, S., Dixon, R. & Greenber, B.M. (2003). Similar Stress Responses Are Elicited by Copper and Ultraviolet Radiation in the Aquatic Plant *Lemma gibba*: Implication of Reactive Oxygen Species as Common Signals. *Plant & Cell Physiology*, Vol. 44, No. 12, (December 2003), pp. 1320–1329, ISSN 0032-0781

Backhausen, J.E., Kitzmann, C., Horton, P. & Scheibe, R. (2000). Electron Acceptors in Isolated Intact Spinach Chloroplasts Act Hierarchically to Prevent Over-Reduction and Competition for Electrons. *Photosynthesis Research*, Vol. 64, No. 1, (April 2000), pp. 1-13, ISSN 0166-8595

Badger, M.R. (1985). Photosynthetic Oxygen Exchange. *Annual Review of Plant Physiology*, Vol. 36, (June 1985), pp. 27-53, ISSN 0066-4294

Badger, M.R., von Caemmerer, S., Ruuska, S. & Nakano, H. (2000). Electron Flow to Oxygen in Higher Plants and Algae: Rates and Control of Direct Photoreduction (Mehler Reaction) and Rubisco Oxygenase. *Philosophical Transactions of the Royal Society of London. Series B, Biological Sciences*, Vol. 355, No. 1402, (October 2000), pp. 1433 - 1446, ISSN 0962-8436

Baier, M. & Dietz, K.J. (2005). Chloroplasts as Source and Target of Cellular Redox Regulation: a Discussion on Chloroplast Redox Signals in the Context of Plant Physiology. *Journal of Experimental Botany*, Vol. 56, No. 416, (June 2005), pp. 1449–1462, ISSN 0022-0957

Barber, J. (2008). Crystal Structure of the Oxygen-Evolving Complex of Photosystem II. *Inorganic Chemistry*, Vol. 47, No. 6, (March 2008), pp. 1700-1710, ISSN 0020-1669

Baruah, A., Simkova, K., Apel, K. & Laloi, C. (2009). *Arabidopsis* Mutants Reveal Multiple Singlet Oxygen Signaling Pathways Involved in Stress Response and Development. *Plant Molecular Biology*, Vol. 70, No. 5, (July 2009), pp. 547–563, ISSN 0167-4412

Bechtold, U., Richard, O., Zamboni, A., Gapper, C., Geisler, M., Pogson, B., Karpinski, S. & Mullineaux, P. (2008). Impact of Chloroplastic- and Extracellular-Sourced ROS on High Light-Responsive Gene Expression in *Arabidopsis*. *Journal of Experimental Botany*, Vol. 59, No. 2, (January 2008), pp. 121–133, ISSN 0022-0957

Beck, C.F. (2005). Signaling Pathways from the Chloroplast to the Nucleus. *Planta*, Vol. 222, No. 5, (November 2005), pp. 743–756, ISSN 0032-0935

Bielski, B.H.J., Arudi, R.L. & Sutherland, M.W. (1983). A Study of the Reactivity of $HO_2/O_2^{\bullet -}$ with Unsaturated Fatty Acids. *The Journal of Biological Chemistry*, Vol. 258, No. 8, (April 1983), pp. 4759-4761, ISSN 0021-9258

Bolouri-Moghaddam, M.R., Le Roy, K., Xiang, L., Rolland, F. & van den Ende, W. (2010). Sugar Signalling and Antioxidant Network Connections in Plant Cells. *FEBS Journal*, Vol. 277, No. 9, (May 2010), pp. 2022-2037, ISSN 1742-464X

Bondarava, N., Gross, C.M., Mubarakshina, M.M., Golecki, J.R., Johnson, G.N. & Krieger-Liszkay, A. (2010). Putative Function of Cytochrome b559 as a Plastoquinol Oxidase. *Physiologia Plantarum*, Vol. 138, No. 4, (April 2010), pp. 463–473., ISSN 0031-9317

Brettel, K. & Leibl, W. (2001). Electron Transfer in Photosystem I. *Biochimica et Biophysica Acta*, Vol. 1507, No. 1-3, (October 2001), pp. 100–114, ISSN 0005-2728

Broin, M., Cuiné, S., Eymery, F. & Rey, P. (2002). The Plastidic 2-Cysteine Peroxiredoxin Is a Target for a Thioredoxin Involved in the Protection of the Photosynthetic Apparatus against Oxidative Damage. *The Plant Cell*, Vol. 14, No. 6, (June 2002), pp. 1417-1432, ISSN 1040-4651

Brudvig, G.W. (2008). Water Oxidation Chemistry of Photosystem II. *Philosophical Transactions of the Royal Society of London. Series B, Biological Sciences*, Vol. 363, No. 1494, (March 2008), pp. 1211-1218, ISSN 0962-8436

Byczkowsky, J.Z. & Gessener, T. (1988). Biological Role of Superoxide Ion-Radical. *The International Journal of Biochemistry*, Vol. 20, No. 6, (February 1988), pp. 569-580, ISSN 0020-711X

Cadet, J., Delatour, T., Douki, T., Gasparutto, D., Pouget, J., Ravanat, J. & Sauvaigo, S. (1999). Hydroxyl Radicals and DNA Base Damage. *Mutation Research*, Vol. 424, No. 1–2, (March 1999), pp. 9–21, ISSN 0027-5107

Camacho-Pereira, J., Meyer, L.E., Machado, L.B., Oliveira, M.F. & Galina, A. (2009). Reactive Oxygen Species Production by Potato Tuber Mitochondria is Modulated by Mitochondrially Bound Hexokinase Activity. *Plant Physiology*, Vol. 149, No. 2, (February 2009), pp. 1099-1110, ISSN 0032-0889

Chan, K.X., Crisp, P.A., Estavillo, G.M. & Pogson, B.J. (2010). Chloroplast-to-Nucleus Communication: Current Knowledge, Experimental Strategies and Relationship to Drought Stress Signaling. *Plant Signaling & Behavior*, Vol. 5, No. 12, (December 2010), pp. 1575–1582, ISSN 1559-2316

Charles, S.A. & Halliwell, B. (1980). Effect of Hydrogen Peroxide on Spinach (*Spinacia oleracea*) Chloroplast Fructose Bisphosphatase. *The Biochemical Journal*, Vol. 189, No. 2, (August 1980), pp. 373-376, ISSN 0264-6021

Cogdell, R.J., Howard, T.D., Bittl, R., Schlodder, E., Geisenheimer, I. & Lubitz, W. (2000). How Carotenoids Protect Bacterial Photosynthesis. *Philosophical Transactions of the Royal Society of London. Series B, Biological Sciences*, Vol. 355, No. 1402, (October 2000), pp. 1345-1349, ISSN 0962-8436

Coupe, S.A., Palmer, B.G. & Lake, J.A. (2006). Systemic Signaling of Environmental Cues in *Arabidopsis* Leaves. *Journal of Experimental Botany*, Vol. 57, No. 2, (January 2006) pp. 329-341, ISSN 0022-0957

Cui, L., Veeraraghavan, N., Richter, A., Wall, K., Jansen, R.K., Leebens-Mack, J., Makalowska, I. & de Pamphilis, C.W. (2006). ChloroplastDB: the Chloroplast Genome Database. *Nucleic Acids Research,* Vol. 34, (January 2006), pp. D692-696, ISSN 0305-1048

Dalton, D.A., Russell, S.A., Hanus, F.J., Pascoe, G.A. & Evans, H.J. (1986). Enzymatic Reactions of Ascorbate and Glutathione That Prevent Peroxide Damage in Soybean Root Nodules. *Proceedings of the National Academy of Sciences of the USA,* Vol. 83, No. 11, (June 1986), pp. 3811-3815, ISSN 0027-8424

Danna, C.H., Bartoli, C.G., Sacco, F., Ingala, L.R., Santa-Maria, G.E., Guiamet, J.J. & Ugalde, R.A. (2003). Thylakoid-Bound Ascorbate Peroxidase Mutant Exhibits Impaired Electron Transport and Photosynthetic Activity. *Plant Physiology,* Vol. 132, No. 4, (August 2003), pp. 2116–2125, ISSN 0032-0889

Dat, J., Vandenabeele, S., Vranova, E., van Montagu, M., Inze, D. & van Breusegem, F. (2000). Dual Action of the Active Oxygen Species During Plant Stress Responses. *Cellular & Molecular Life Sciences,* Vol. 57, No. 5, (May 2000), pp. 779–795, ISSN 1420-682X

Dekker, J.P. & van Grondelle, R. (2000). Primary Charge Separation in Photosystem II. *Photosynthesis research,* Vol. 63, No. 3, (March 2000), pp. 195–208, ISSN 0166-8595

Demmig-Adams, B. (1990). Carotenoids and Photoprotection in Plants: A Role for the Xanthophyll Zeaxanthin. *Biochimica et Biophysica Acta,* Vol. 1020, No. 1, (October 1990), pp. 1-24, ISSN 0005-2728

Desikan, R., Clarke, A., Hancock, J.T. & Neill, S.J. (1999). H_2O_2 Activates a MAP Kinase-Like Enzyme in *Arabidopsis thaliana* Suspension Cultures. *Journal of Experimental Botany,* Vol. 50, No. 341, (December 1999), pp. 1863–1866, ISSN 0022-0957

Desikan, R., Mackerness, S., Hancock, J.T. & Neill, S.J. (2001). Regulation of the *Arabidopsis* Transcriptome by Oxidative Stress. *Plant Physiology,* Vol. 127, No. 1, (September 2001), pp. 159–172, ISSN 0032-0889

Devasagayam, T.P., Sundquist, A.R., Di Mascio, P., Kaiser, S. & Sies, H. (1991). Activity of Thiols as Singlet Molecular Oxygen Quenchers. *Journal of Photochemistry and Photobiology. B, Biology,* Vol. 9, No. 1, (April 1991), pp. 105-116, ISSN 1011-1344

Dietz, K.J., Jacob, S., Oelze, M.L., Laxa, M., Tognetti, V., de Miranda, S.M.N., Baier, M. & Finkemeier, I. (2006). The Function of Peroxiredoxins in Plant Organelle Redox Metabolism. *Journal of Experimental Botany,* Vol. 57, No. 8, (May 2006), pp. 1697–1709, ISSN 0022-0957

Ding, M.Q., Hou, P.C., Shen, X., Wang, M.J., Deng, S.R., Sun, J., Xiao, F., Wang, R., Zhou, X., Lu, C., Zhang, D., Zheng, X., Hu, Z. & Chen, S. (2010). Salt-Induced Expression of Genes Related to Na+/K+ and ROS Homeostasis in Leaves of Salt-Resistant and Salt-Sensitive Poplar Species. *Plant Molecular Biology,* Vol. 73, No. 3, (June 2010), pp. 251–269, ISSN 0167-4412

Dron, M., Clouse, S.D., Dixon, R.A., Lawton, M.A. & Lamb, C.J. (1988). Glutathione and Fungal Elicitor Regulation of a Plant Defense Gene Promoter in Electroporated Protoplasts. *Proceedings of the National Academy of Sciences of the USA,* Vol. 85, No. 18, (September 1988), pp. 6738-6742, ISSN 0027-8424

Eisenhut, M., Ruth, W., Haimovich, M., Bauwe, H., Kaplan, A. & Hagemann, M. (2008). The Photorespiratory Glycolate Metabolism Is Essential for Cyanobacteria and Might Have Been Conveyed Endosymbiontically to Plants. *Proceedings of the National*

Academy of Sciences of the USA, Vol. 105, No. 44, (November 2008), pp. 17199-17204, ISSN 0027-8424

Escoubas, J.M., Lomas, M., LaRoche, J. & Falkowski P.G. (1995). Light Intensity Regulation of cab Gene Transcription is Signaled by the Redox State of the Plastoquinone Pool. *Proceedings of the National Academy of Sciences of the USA,* Vol. 92, No. 22, (October 1995), pp. 10237-10241, ISSN 0027-8424

Eshdat, Y., Holland, D., Faltin, Z. & Ben-Hayyim, G. (1997). Plant Glutathione Peroxidases. *Physiologia Plantarum,* Vol. 100, No. 2, (June 1997), pp. 234-240, ISSN 0031-9317

Fahnenstich, H., Scarpeci, T.E., Valle, E.M., Fliigge, U.I. & Maurino, V.G. (2008). Generation of Hydrogen Peroxide in Chloroplasts of *Arabidopsis* Overexpressing Glycolate Oxidase as an Inducible System to Study Oxidative Stress. *Plant Physiology,* Vol. 148, No. 2, (October 2008), pp. 719- 729, ISSN 0032-0889

Fey, V., Wagner, R., Brautigam, K., Wirtz, M., Hell, R., Dietzmann, A., Leister, D., Oelmüller, R. & Pfannschmidt, T. (2005). Retrograde Plastid Redox Signals in the Expression of Nuclear Genes for Chloroplast Proteins of *Arabidopsis thaliana. The Journal of biological chemistry,* Vol. 280, No. 7, (February 2005), pp. 5318–5328, ISSN 0021-9258

Filella, I., Peñuelas, J. & Llusià, J. (2006). Dynamics of the Enhanced Emissions of Monoterpenes and Methyl Salicylate, and Decreased Uptake of Formaldehyde, by *Quercus ilex* Leaves after Application of Jasmonic Acid. *The New phytologist,* Vol. 169, No. 1, (January 2006), pp.135-144, ISSN 0028-646X

Foote, C.S. & Denny, R.W. (1968). Chemistry of Singlet Oxygen. VII. Quenching by β-Carotene. *Journal of the American Chemical Society,* Vol. 90, No. 22, (October 1968), pp. 6233–6235, ISSN 0002-7863

Foyer CH, & Noctor G. (2009) Redox Regulation in Photosynthetic Organisms: Signaling, Acclimation, and Practical Implications. *Antioxidants & Redox Signaling,* Vol. 11, No. 4, (April 2009), pp. 861-905, ISSN 1523-0864

Frigerio, S., Campoli, C., Zorzan, S., Fantoni, L.I., Crosatti, C., Drepper, F., Haehnel, W., Cattivelli, L., Morosinotto, T. & Bassi, R. (2007). Photosynthetic Antenna Size in Higher Plants Is Controlled by the Plastoquinone Redox State at the Post-transcriptional Rather than Transcriptional Level. *The Journal of Biological Chemistry,* Vol. 282, No. 40, (October 2007), pp. 29457-29469, ISSN 0021-9258

Fujiwara, K., Kumata, H., Kando, N., Sakuma, E., Aihara, M., Morita, Y. & Miyakawa, T. (2006). Flow Injection Analysis to Measure the Production Ability of Superoxide with Chemiluminescence Detection in Natural Waters. *International Journal of Environmental Analytical Chemistry,* Vol. 86, No. 5, (April 2006), pp. 337-346, ISSN 0306-7319

Furbank, R. & Badger, M. (1983). Oxygen Exchange Associated with Electron Transport and Photophosphorilation in Spinach Thylakoids. *Biochimica et Biophysica Acta,* Vol. 723, No. 3, (June 1983), pp. 400–409, ISSN 0005-2728

Gadjev, I., Vanderauwera, S., Gechev, T.S., Laloi, C., Minkov, I.N., Shulaev, V., Apel, K., Inze, D., Mittler, R. & van Breusegem, F. (2006). Transcriptomic Footprints Disclose Specificity of Reactive Oxygen Species Signaling in *Arabidopsis. Plant Physiology,* Vol. 141, No. 2, (June 2006), pp. 436–445, ISSN 0032-0889

Geigenberger, P., Kolbe, A. & Tiessn, A. (2005). Redox Regulation of Carbon Storage and Partitioning in Response to Light and Sugars. *Journal of Experimental Botany,* Vol. 56, No. 416, (June 2006), pp. 1469-1479, ISSN 0022-0957

Golbeck, J. & Radmer, R. (1984). Is the Rate of Oxygen Uptake by Reduced Ferredoxin Sufficient to Account for Photosystem I - Mediated O_2 Reduction?, In: *Advances in Photosynthesis Research,* C. Sybesma (Ed), pp. 1.4.561 - 1.4.564, Martinus Nijhoff/Dr W. Junk, ISBN 978-902-4729-42-5, Hague, The Netherlands

Gotoh, N. & Niki E. (1992). Rates of Interactions of Superoxide with Vitamin E, Vitamin C and Related Compounds as Measured by Chemiluminescence. *Biochimica et Biophysica Acta,* Vol. 1115, No. 3, (January 1992), pp. 201-207, ISSN 0005-2728

Grant, J.J., Yun, B.W. & Loake, G.J. (2000). Oxidative Burst and Cognate Redox Signalling Reported by Luciferase Imaging: Identification of a Signal Network that Functions Independently of Ethylene, SA and Me-JA but Is Dependent on MAPKK Activity. *The Plant Journal,* Vol. 24, No. 5, (December 2000), pp. 569–582, ISSN 0960-7412

Gray, J.C. (2003). Chloroplast-to-Nucleus Signalling: a Role for Mg-Protoporphyrin. *Trends in Genetics,* Vol. 19, No. 10, (October 2003), pp. 526-529, ISSN 0168-9525

Gruszka, J., Pawlak, A. & Kruk, J. (2008). Tocochromanols, Plastoquinol, and Other biological Prenyllipids as Singlet Oxygen Quenchers-Determination of Singlet Oxygen Quenching Rate Constants and Oxidation Products. *Free Radical Biology & Medicine,* Vol. 45, No. 6, (September 2008), pp. 920-928, ISSN 0891-5849

Gupta, A.K. & Kaur, N. (2005). Sugar Signalling and Gene Expression in Relation to Carbohydrate Metabolism under Abiotic Stresses in Plants. *Journal of Biosciences,* Vol. 30, No. 5, (December 2005), pp. 761-776, ISSN 0250-5991

Gus'kova, R.A., Ivanov, I.I., Kol'tover, V.K., Akhobadze, V.V. & Rubin, A.B. (1984). Permeability of Bilayer Lipid Membranes for Superoxide ($O_2^{\bullet-}$) Radicals. *Biochimica et Biophysica Acta,* Vol. 778, No. 3, (December 1984), pp. 579–585, ISSN 0005-2728

Hatz, S., Lambert, J.D.C. & Ogilby, P.R. (2007). Measuring the Lifetime of Singlet Oxygen in a Single Cell: Addressing the Issue of Cell Viability. *Photochemical & Photobiological Sciences,* Vol. 6, No. 10, (October 2007), pp. 1106-1116, ISSN 1474-905X

Havaux, M., Eymery, F., Porfirova, S., Rey, P. & Dörmann, P. (2005). Vitamin E Protects against Photoinhibition and Photooxidative Stress in *Arabidopsis thaliana. The Plant Cell,* Vol. 17, No. 12, (December 2005), pp. 3451-69, ISSN 1040-4651

Heber, U. (1973). Stoichiometry of Reduction and Phosphorylation during Illumination of Intact Chloroplasts. *Biochimica et Biophysica Acta,* Vol. 305, No. 1, (April 1973), pp. 140-152, ISSN 0005-2728

Hernández, I., Alegre, L., van Breusegem, F. & Munné-Bosch, S. (2009). How Relevant are Flavonoids as Antioxidants in Plants? *Trends in Plant Science,* Vol. 14, No.3, (March 2009), pp. 125-132, ISSN 1360-1385

Heyno, E., Gross, C.M., Laureau, C., Culcasi, M., Pietri, S. & Krieger-Liszkay, A. (2009). Plastid Alternative Oxidase (PTOX) Promotes Oxidative Stress When Overexpressed in Tobacco. *The Journal of Biological Chemistry,* Vol. 284, No. 45, (November 2009), pp. 31174-31180, ISSN 0021-9258

Hideg, E.E., Ogawa, K., Kalai, T. & Hideg, K. (2001). Singlet Oxygen Imaging in *Arabidopsis thaliana* Leaves under Photoinhibition by Excess Photosynthetically Active Radiation. *Physiologia Plantarum,* Vol. 112, No. 1, (May 2001), pp. 10–14, ISSN 0031-9317

Hillier, W. & Wydrzynski, T. (1993). Increases in Peroxide Formation by the O_2-Evolving Catalytic Site Upon the Removal of the Extrinsic 16-, 23-, and 33-kDa Proteins Are

Reversed by $CaCl_2$ Addition. *Photosynthesis Research,* Vol. 38, No. 3, (January 1993), pp. 417–423, ISSN 0166-8595

Horton, P., Ruban, A.V. & Walters, R.G. (1996) Regulation of Light Harvesting in Green Plants. *Annual Review of Plant Physiology & Plant Molecular Biology,* Vol. 47, (June 1996), pp. 655-684, ISSN 1040-2519

Hosein, B. & Palmer, G. (1983). The Kinetics and Mechanism of Oxidation of Reduced Spinach Ferredoxin by Molecular Oxygen and its Reduced Products. *Biochimica et Biophysica Acta,* Vol. 723, No. 3, (June 1983), pp. 383–390, ISSN 0005-2728

Hundal, T., Forsmark-Andrée, P., Ernster, L. & Andersson, B. (1995). Antioxidant Activity of Reduced Plastoquinone in Chloroplast Thylakoid Membranes. *Archives of Biochemistry and Biophysics,* Vol. 324, No. 1, (December 1995), pp. 117–122, ISSN 0003-9861

Ivanov, B.N., Red'ko, T.P., Shmeleva, V.L. & Mukhin, E.N. (1980). The Role of Ferredoxin in Pseudocyclic Electron Transport in Isolated Pea Chloroplasts. *Biochemistry (Moscow),* Vol. 45, No. 8, (August 1980), pp. 1425-1432, ISSN 0320-9725

Ivanov, B.N. (1993). Stoichiometry of Proton Uptake by Thylakoids during Electron Transport in Chloroplasts. In: *Photosynthesis. Photoreactions to Plant Productivity,* Y. P. Abrol, P. Mohanty & Govindjee (Eds.), pp. 109-132, Kluwer Academic Publishers, ISBN 978-079-2319-43-6, New Delhi: Oxford

Ivanov, B.N., Kobayashi, Y., Bukhov, N.K. & Heber, U. (1998). Photosystem I-Dependent Cyclic ElectronF in Intact Spinach Chloroplasts: Occurrence, Dependence on Redox Conditions and Electron Acceptors and Inhibition by Antimycin A. *Photosynthesis Research,* Vol. 57, No., (), pp. 61-70, ISSN 0166-8595

Ivanov, B. & Edwards, G. (2000). Influence of Ascorbate and the Mehler Peroxidase Reaction on Non-Photochemical Quenching of Chlorophyll Fluorescence in Maize Mesophyll Chloroplasts. *Planta,* Vol. 210, No. 5, (April 2000), pp. 765-774, ISSN 0032-0935

Ivanov, B.N., Mubarakshina, M.M. & Khorobrykh, S.A. (2007). Kinetics of the Plastoquinone Pool Oxidation Following Illumination. Oxygen Incorporation into Photosynthetic Electron Transport Chain. *FEBS Letters,* Vol. 581, No. 7, (April 2007), pp. 1342–1346., ISSN 0014-5793

Kaiser, W. (1976). The Effect of Hydrogen Peroxide on CO_2-Fixation of Isolated Chloroplasts. *Biochimica et Biophysica Acta,* Vol. 440, No. 3, (September 1976), pp. 476-482, ISSN 0005-2728

Kaiser, W. (1979). Reversible Inhibition of the Calvin Cycle and Activation of Oxidative Pentose Phosphate Cycle in Isolated Intact Chloroplasts by Hydrogen Peroxide. *Planta,* Vol. 145, No. 4, (January 1979), pp. 377-382, ISSN 0032-0935

Kangasjärvi, S., Lepistö, A., Hännikäinen, K., Piippo, M., Luomala, E.M., Aro, E.M. & Rintamäki, E. (2008). Diverse Roles for Chloroplast Stromal and Thylakoid-Bound Ascorbate Peroxidases in Plant Stress Responses. *Biochemical Journal,* Vol. 412, No. 2, (June 2008), pp 275-285, ISSN 0264-6021

Keren, N., Gong, H. & Ohad, I. (1995). Oscillations of Reaction Centre II D1 Protein Degradation *in vivo* Induced by Repetitive Flashes. *The Journal of Biological Chemistry,* Vol. 270, No. 2, (January 1995), pp. 806–814, ISSN 0021-9258

Khorobrykh, S.A. & Ivanov, B.N. (2002). Oxygen Reduction in a Plastoquinone Pool of Isolated Pea Thylakoids. *Photosynthesis Research,* Vol. 71, No. 3, (March 2002), pp. 209-219, ISSN 0166-8595

Khorobrykh, S.A., Khorobrykh, A.A., Klimov, V.V. & Ivanov, BN. (2002). Photoconsumption of Oxygen in Photosystem II Preparations under Impairment of Water-Oxidation Complex. *Biochemistry (Moscow),* Vol. 67, No. 6, (June 2002), pp. 683–688, ISSN 0006-2979

Khorobrykh, S., Mubarakshina, M. & Ivanov, B. (2004). Photosystem I is not Solely Responsible for Oxygen Reduction in Isolated Thylakoids. *Biochimica et Biophysica Acta,* Vol. 1657, No. 2-3, (July 2004), pp. 164-167, ISSN 0005-2728

Khorobrykh, S.A., Khorobrykh, A.A., Yanykin, D.V., Ivanov, B.N., Klimov, B.N. & Mano, J., submitted. Photoproduction of Catalase-Insensitive Peroxides on the Donor Side of Manganese-Depleted Photosystem II: Evidence with a Specific Fluorescent Probe. *Biochemistry (accepted)*

Klimov, V.V., Ananyev, G.M., Zastrizhnaya, O.M., Wydrzynski, T. & Renger, G. (1993). Photoproduction of Hydrogen Peroxide in Photosystem II Membrane Fragments. A Comparison of Four Signals. *Photosynthesis Research,* Vol. 38, No. 3, (January 1993), pp. 409-416, ISSN 0166-8595

Kozuleva, M.A., Naidov, I.A., Mubarakshina, M.M. & Ivanov, B.N. (2007). Participation of Ferredoxin in Oxygen Reduction by the Photosynthetic Electron Transport Chain. *Biofizika,* Vol. 52, No. 4, (July-August 2007), pp. 650-655, ISSN 0006-3029

Kozuleva, M.A. & Ivanov, B.N. (2010). Evaluation of the Participation of Ferredoxin in Oxygen Reduction in the Photosynthetic Electron Transport Chain of Isolated Pea Thylakoids. *Photosynthesis Research,* Vol. 105, No. 1, (July 2010), pp. 51-61, ISSN 0166-8595

Kozuleva, M., Klenina, I., Proskuryakov, I., Kirilyuk, I. & Ivanov, B. (2011). Production of Superoxide in Chloroplast Thylakoid Membranes. ESR Study with Cyclic Hydroxylamines of Different Lipophilicity. *FEBS Letters,* Vol. 585, No. 7, (April 2011), pp. 1067-1071, ISSN 0014-5793

Krasnovsky, A.A., Jr. (1998). Singlet Molecular Oxygen in Photobiochemical Systems: IR Phosphorescence Studies. *Membrane & Cell Biology,* Vol. 12, No. 5, (n.d.), pp. 665-690, ISSN 1023-6597

Krieger, A., Rutherford, A.W. & Johnson, G.N. (1995). On the Determination of Redox Midpoint Potential of the Primary Quinone Electron Acceptor, Q_A, in Photosystem II. *Biochimica et Biophysica Acta,* Vol. 1229, No. 2, (April 1995), pp. 193-201, ISSN 0005-2728

Krieger-Liszkay, A. & Rutherford, A.W. (1998). Influence of Herbicide Binding on the Redox Potential of the Quinone Acceptor in Photosystem II: Relevance to Photodamage and Phytotoxicity. *Biochemistry,* Vol. 37, No. 50, (December 1998), pp. 17339-17344, ISSN 0006-2960

Krieger-Liszkay, A. (2005). Singlet Oxygen Production in Photosynthesis. *Journal of Experimental Botany,* Vol. 56, No. 411, (January 2005), pp. 337–346, ISSN 0022-0957

Krieger-Liszkay, A., Kós, P.B., Hideg, E. (2011). Superoxide Anion Radicals Generated by Methylviologen in Photosystem I Damage Photosystem II. *Physiologia Plantarum,* Vol. 142, No. 1, (May 2011), pp. 17-25, 0031-9317

Kropat, J., Oster, U., Rüdiger, W. & Beck, C.F. (1997). Chlorophyll Precursors are Signals of Chloroplast Origin Involved in Light Induction of Nuclear Heat Shock Genes. *Proceedings of the National Academy of Sciences of the USA,* Vol. 94, No. 25, (December 1997), pp. 14168 - 14172, ISSN 0027-8424

Kruk, J., Schmid, G.H. & Strzałka, K. (1994). Antioxidant Properties of Plastoquinol and Other Biological Prenylquinols in Liposomes and Solution. *Free Radical Research,* Vol. 21, No. 6, (November-December 1994), pp. 409-416, ISSN 1029-2470

Kruk, J., Jemioła-Rzemińska, M. & Strzałka, K. (1997). Plastoquinol and α-Tocopherol Quinol Are More Active than Ubiquinol and α-Tocopherol in Inhibition of Lipid Peroxidation. *Chemistry and Physics of Lipids,* Vol. 87, No. 1, (May 1997), pp. 73-80, ISSN 0009-3084

Kruk, J., Jemiola-Rzeminska, M., Burda, K., Schmid, G. & Strzalka, K. (2003). Scavenging of Superoxide Generated in Photosystem I by Plastoquinol and Other Prenyllipids in Thylakoid Membranes. *Biochemistry,* Vol. 42, No. 1, (October 2003), pp. 8501-8505, ISSN 0264-6021

Kruk, J. & Trebst, A. (2008). Plastoquinol as a Singlet Oxygen Scavenger in Photosystem II. *Biochimica et Biophysica Acta,* Vol. 1777, No. 2, (February 2008), pp. 154-162, ISSN 0005-2728

Kurepa, J., Hérouart, D., van Montagu, M. & Inzé D. (1997). Differential Expression of CuZn-and Fe-Superoxide Dismutase Genes of Tobacco during Development, Oxidative Stress, and Hormonal Treatments. *Plant & Cell Physiology,* Vol. 38, No. 4, (April 1997), pp. 463-470, ISSN 0032-0781

Kuvykin, I.V., Vershubskii, A.V., Ptushenko, V.V. & Tikhonov, A.N. (2008). Oxygen as an Alternative Electron Acceptor in the Photosynthetic Electron Transport Chain of C3 Plants. *Biochemistry (Moscow),* Vol.73, No. 10, (October 2008), pp. 1063-1075, ISSN 0006-2979

Laloi, C., Przybyla, D. & Apel, K. (2006). A Genetic Approach Towards Elucidating the Biological Activity of Different Reactive Oxygen Species in *Arabidopsis thaliana. Journal of Experimental Botany,* Vol. 57, No. 8, (May 2006), pp. 1719–1724, ISSN 0022-0957

Ledford, H.K., Chin, B.L. & Niyogi, K.K. (2007). Acclimation to Singlet Oxygen Stress in *Chlamydomonas reinhardtii. Eukaryotic Cell,* Vol. 6, No. 6, (June 2007), pp. 919–930, ISSN 1535-9786

Lemaire, D.S., Michelet, L., Zaffagnini, M., Massot, V. & Issakidis-Bourguet, E. (2007). Thioredoxins in Chloroplasts. *Current Genetics,* Vol. 51, No. 6, (June 2007), pp. 343-365, ISSN 0172-8083

Lennon, A.M., Prommeenate, P. & Nixon, P.J. (2003). Location, Expression and Orientation of the Putative Chlororespiratory Enzymes, Ndh and IMMUTANS, in Higher-Plant Plastids. *Planta,* Vol. 218, No. 2, (December 2003), pp. 254–260, ISSN 0032-0935

Li, Z., Wakao, S., Fischer, B.B. & Niyogi, K.K. (2009). Sensing and Responding to Excess Light. *Annual Review of Plant Biology,* Vol. 60, (June 2009), pp. 239-260, 1543-5008

Lichtenthaler, H.K. (2007). Biosynthesis, Accumulation and Emission of Carotenoids, α-Tocopherol, Plastoquinone, and Isoprene in Leaves under High Photosynthetic Irradiance. *Photosynthesis Research,* Vol. 92, No. 2, (May 2007), pp. 163–179, ISSN 0166-8595

Ligeza, A., Tikhonov, A.N., Hyde, J.S. & Subczynski, W.K. (1998). Oxygen Permeability of Thylakoid Membranes: Electron Paramagnetic Resonance Spin Labeling Study. *Biochimica et Biophysica Acta,* Vol. 1365, No. 3, (July 1998), pp. 453-463, ISSN 0005-2728

Lindahl, M. & Kieselbach, T. (2009). Disulphide Proteomes and Interactions with Thioredoxin on the Track Towards Understanding Redox Regulation in Chloroplasts and Cyanobacteria. *Journal of Proteomics,* Vol. 72, No. 3, (April 2009), pp. 416–438, ISSN 1876-7737

Maciejewska, U., Polkowska-Kowalczyk, L., Swiezewska, E. & Szkopinska, A. (2002). Plastoquinone: Possible Involvement in Plant Disease Resistance. *Acta Biochimica Polonica,* Vol. 49, No. 3, (n.d.), pp. 775–780, ISSN 0001-527X

Maeda, H. & DellaPenna, D. (2007). Tocopherol functions in photosynthetic organisms. *Current Opinion in Plant Biology,* Vol. 10, No.3, (June 2007), pp. 260-265, ISSN 1369-5266

Mateo, A., Funck, D., Mühlenbock, P., Kular, B., Mullineaux, P.M. & Karpinski, S. (2006). Controlled Levels of Salicylic Acid Are Required for Optimal Photosynthesis and Redox Homeostasis. *Journal of Experimental Botany,* Vol. 57, No. 8, (May 2006), pp. 1795-1807, ISSN 0022-0957

Matheson, I.B.C., Etheridge, R.D., Kratowich, N.R. & Lee, J. (1975). The Quenching of Singlet Oxygen by Amino Acids and Proteins. *Photochemistry & Photobiology,* Vol. 21, No. 3, (March 1975), pp. 165-171, ISSN 0031-8655

Matysik, J., Alia, Bhalu, B. & Mohanty, P. (2002). Molecular Mechanisms of Quenching of Reactive Oxygen Species by Proline under Stress in Plants. *Current Science,* Vol. 82, No. 5, (March 2002), pp.525–532, ISSN 0011-3891

Maurino, V.G.,& Peterhansel, C. (2010). Photorespiration: Current Status and Approaches for Metabolic Engineering. *Current Opinion in Plant Biology,* Vol. 13, No. 3, (June 2010), pp. 249-256, ISSN 1369-5266

Mehler, A.C. (1951). Studies on Reactions of Illuminated Chloroplasts. I. Mechanism of the Reduction of Oxygen and Other Hill Reagents. *Archives of Biochemistry and Biophysics,* Vol. 33, No. 1, (August 1951), pp. 65-77, ISSN 0003-9861

Mittler, R., Vanderauwera, S., Gollery, M. & Van Breusegem, F. (2004). The Reactive Oxygen Gene Network in Plants. *Trends in Plant Science,* Vol. 9, No. 10, (October 2004), pp. 490–498, ISSN 1360-1385

Miyake, C. & Asada, K. (1992). Thylakoid-Bound Ascorbate Peroxidase in Spinach Chloroplasts and Photoreduction of Its Primary Oxidation Product Monodehydroascorbate Radicals in Thylakoids. *Plant & Cell Physiology,* Vol. 33, No. 5, (July 1992), pp. 541-553, ISSN 0032-0781

Miyake, C., Schreiber, U., Hormann, H., Sano, S. & Asada, K. (1998). The FAD-Enzyme Monodehydroascorbate Radical Reductase Mediates Photoproduction of Superoxide Radicals in Spinach Thylakoid Membranes. *Plant & Cell Physiology,* Vol. 39, No. 8, (August 1998), pp. 821 – 829, ISSN 0032-0781

Morosinotto, T., Bassi, R., Frigerio, S., Finazzi, G., Morris, E. & Barber, J. (2006). Biochemical and Structural Analyses of a Higher Plant Photosystem II Supercomplex of a Photosystem I-Less Mutant of Barley. Consequences of a Chronic Over-Reduction of the Plastoquinone Pool. *The FEBS Journal,* Vol. 273, No. 20, (October 2006), pp. 4616-4630, ISSN 1742-464X

Mozzo, M., Passarini, F., Bassi, R., van Amerongen, H. & Croce, R. (2008). Photoprotection in Higher Plants: the Putative Quenching Site is Conserved in All Outer Light-Harvesting Complexes of Photosystem II. *Biochimica et Biophysica Acta,* Vol. 1777, No. 10, (October 2008), pp. 1263-1267, ISSN 0005-2728

Mubarakshina, M., Khorobrykh, S. & Ivanov, B. (2006). Oxygen Reduction in Chloroplast Thylakoids Results in Production of Hydrogen Peroxide Inside the Membrane. *Biochimica et Biophysica Acta,* Vol. 1757, No. 11, (November 2006), pp. 1496-1503, ISSN 0005-2728

Mubarakshina, M.M., Ivanov B.N., Naydov I.A., Hillier W., Badger M.R., Krieger-Liszkay A. (2010). Production and diffusion of chloroplastic H_2O_2 and its implication to signalling. *Journal of Experimental Botany*, Vol. 61, No. 13, pp. 3577–3587, ISSN 0022-0957

Mubarakshina, M.M. & Ivanov, B.N. (2010). The Production and Scavenging of Reactive Oxygen Species in the Plastoquinone Pool of Chloroplast Thylakoid Membranes. *Physiologia Plantarum,* Vol. 140, No. 2, (October 2010), pp. 103-110, ISSN 0031-9317

Mullineaux, P. & Karpinski, S. (2002). Signal Transduction in Response to Excess Light: Getting Out of the Chloroplast. *Current Opinion in Plant Biology,* Vol. 5, No. 1, (February 2002), pp. 43–48, ISSN 1369-5266

Mullineaux, P.M. (2009). ROS in Retrograde Signalling from the Chloroplast to the Nucleus, In: *Reactive Oxygen Species in Plant Signaling. Signaling and Communication in Plants,* L.A. Rio & A. Puppo (Eds), pp. 221-240, Springer, ISBN 978-364-2003-90-5, Berlin Heidelberg, Germany

Munné-Bosch, S., Weiler, E.W., Alegre, L., Müller, M., Düchting, P. & Falk, J. (2007). α-Tocopherol May Influence Cellular Signaling by Modulating Jasmonic Acid Levels in Plants. *Planta,* Vol. 225, No. 3, (February 2007), pp. 681-691, ISSN 0032-0935

Nakano, Y. & Asada, K. (1980). Spinach Chloroplasts Scavenge Hydrogen Peroxide on Illumination. *Plant & Cell Physiology,* Vol. 21, No. 7, (November 1980), pp. 1295–1307, ISSN 0032-0781

Nakano, Y. & Asada, K. (1987). Purification of Ascorbate Peroxidase in Spinach Chloroplasts: Its Inactivation in Ascorbate-Depleted Medium and Reactivation by Monodehydroascorbate Radical. *Plant & Cell Physiology,* Vol. 28, No., (January 1987), pp. 131-140, ISSN 0032-0781

Neverov, K.V. & Krasnovsky Jr., A.A. (2004). Phosphorescence Analysis of the Chlorophyll Triplet States in Preparations of Photosystem II. *Biofizika,* Vol. 49, No. 3, (May-June 2004), pp. 493-498, ISSN 0006-3029

Nishizawa, A., Yukinori, Y. & Shigeoka, S. (2008). Galactinol and Raffinose as a Novel Function to Protect Plants from Oxidative Damage. *Plant Physiology,* Vol. 147, No. 3, (July 2008), pp. 1251–1263, ISSN 0032-0889

Nixon, P.J. & Rich, P.R. (2006). Chlororespiratory Pathways and Their Physiological Significance. In: *The Structure and Function of Plastids. Advances in Photosynthesis and Respiration,* R.R. Wise & J.K. Hoober (Eds), pp 237-251,Springer, ISBN 978-140-2040-61-0, Dordrecht, The Netherlands

Noctor, G., Veljovic-Jovanovic, S. & Foyer, C.H. (2000). Peroxide Processing in Photosynthesis: Antioxidant Coupling and Redox Signaling. *Philosophical Transactions of the Royal Society of London. Series B, Biological Sciences,* Vol. 355, No. 1402, (October 2000), pp. 1465-1475, ISSN 0962-8436

Ogawa, K., Kanematsu, S., Takabe, K. & Asada, K. (1995). Attachment of CuZn-Superoxide Dismutase to Thylakoid Membranes at the Site of Superoxide Generation (PSI) in Spinach Chloroplasts: Detection by Immuno-Gold Labelling after Rapid Freezing and Substitution Method. *Plant & Cell Physiology,* Vol. 36, No. 4, (June 1995), pp. 565-573, ISSN 0032-0781

Op den Camp, R.G.L., Przybyla, D., Ochsenbein, C., Laloi, C., Kim, C., Danon, A., Wagner, D., Hideg, E., Göbel, C., Feussner, I., Nater, M. & Apel, K. (2003). Rapid Induction of Distinct Stress Responses after the Release of Singlet Oxygen in *Arabidopsis*. *The Plant Cell,* Vol. 15, No. 10, (October 2003), pp. 2320–2332, ISSN 1040-4651

Osyczka, A., Moser, C.C., Daldal, F. & Dutton, P.L. (2004). Reversible Redox Energy Coupling in Electron Transfer Chains. *Nature,* Vol. 427, No. 6975, (February 2004), pp. 607–612, ISSN 0028-0836

Park, Y.-I., Chow, W.S., Osmond, C.B. & Anderson, J.M. (1996). Electron Transport to Oxygen Mitigates against the Photoinactivationof Photosystem II *in vivo*. *Photosynthesis Research,*, Vol. 50, No. 1, (October 1996), pp. 23-32, ISSN 0166-8595

Peiser, G.D., Lizada, M.C. & Yang, S.F. (1982). Sulfite-Induced Lipid Peroxidation in Chloroplasts as Determined by Ethane Production. *Plant Physiology,* Vol. 70, No. 4, (October 1982), pp. 994-998, ISSN 0032-0889

Pfannschmidt, T., Nillson A. & Allen, J.F. (1999). Photosynthetic Control of Chloroplast Gene Expression. *Nature,* Vol. 397, No. 6720, (February 1999), pp. 625-628, ISSN 0028-0836

Polle, A.R. & Rennenberg, H. (1994). Photooxidative Stress in Trees. In: *Causes of Photooxidative Stress and Amelioration of Defense Systems in Plants,* C.H. Foyer & P.M. Mullineaux (Eds), pp. 199-209, CRC Press, ISBN 978-084-9354-43-4, Boca Raton, Florida, USA

Pospíšil, P., Arató, A., Krieger-Liszkay, A. & Rutherford, A.W. (2004). Hydroxyl Radical Generation by Photosystem II. *Biochemistry,* Vol. 43, No. 21, (June 2004), pp. 6783–6792, ISSN 0006-2960

Pospíšil, P. (2011) Enzymatic Function of Cytochrome b559 in Photosystem II. *Journal of Photochemistry and Photobiology,* Vol. 104, No. 1-2, (July-August 2011), pp 341-347, ISSN 1011-1344

Quesada, A.R., Byrnes, R.W., Krezoski, S.O. & Petering, D.H. (1996). Direct Reaction of H_2O_2 with Sulfhydryl Groups in HL-60 Cells: Zinc-Metallothionein and Other Sites. *Archives of Biochemistry & Biophysics,* Vol. 334, No. 2, (October 1996), pp. 241-250, ISSN 0003-9861

Rajagopal, S., Egorova, E.A., Bukhov, N.G. & Carpentier, R. (2003). Quenching of excited states of chlorophyll molecules in submembrane fractions of Photosystem I by exogenous quinones. *Biochimica et Biophysica Acta,* Vol. 1606, No. 1-3, (September 2003), pp. 147-152, ISSN 0005-2728

Ramon, M., Rolland, F. & Sheen, J. (2009). Sugar Sensing and Signaling, In: *The Arabidopsis Book,*06.11.2008, Available from http://www.bioone.org/doi/full/10.1199/tab.0117

Reuber, S., Bornman, J.F. & Weissenbock, G. (1996). Phenylpropanoid Compounds in Primary Leaf Tissues of Rye (*Secale cereale*). Light Response of Their Metabolism and the Possible Role in UV-B Protection. *Physiologia Plantarum,* Vol. 97, No. 1, (May 1996), pp. 160–168, ISSN 0031-9317

Rhee, S.G., Chae, H.Z. & Kim, K. (2005). Peroxiredoxins: a Historical Overview and Speculative Preview of Novel Mechanisms and Emerging Concepts in Cell Signalling. *Free Radical Biology & Medicine,* Vol. 38, No. 12, (June 2005), pp. 1543–1552, ISSN 0891-5849

Rice-Evans, C.A., Miller, N. & Paganga, G. (1996). Structure-Antioxidant Relationships of Flavonoids and Phenolic Acids. *Free Radical Biology & Medicine,* Vol. 20, No. 7, (n.d.) pp. 933–956, ISSN 0891-5849

Rich, P.R. (1985). Mechanisms of Quinol Oxidation in Photosynthesis. *Photosynthesis Research,* Vol. 6, No. 4, (December 1985), pp. 335-348, ISSN 0166-8595

Rizhsky, L., Liang, H. & Mittler, R. (2003). The Water-Water Cycle Is Essential for Chloroplast Protection in the Absence of Stress. *The Journal of Biological Chemistry,* Vol. 278, No. 40, (October 2003), pp. 38921–38925, ISSN 0021-9258

Robinson, J.M. & Gibbs, M. (1982). Hydrogen Peroxide Synthesis in Isolated Spinach Chloroplast Lamellae: An Analysis of the Mehler Reaction in the Presence of NADP Reduction and ATP Formation. *Plant Physiology,* Vol. 70, No. 5, (November 1982), pp. 1249-1254, ISSN 0032-0889

Rutherford, A.W. & Inoue, Y. (1984). Oscillation of Delayed Luminescence from PSII: Recombination of S2QB and S3QB. *FEBS Letters,* Vol. 165, No. 2, (January 1984), 163–170, ISSN 0014-5793

Rutherford, A.W. & Krieger-Liszkay, A. (2001). Herbicide-Induced Oxidative Stress in Photosystem II. *Trends in Biochemical Sciences,* Vol. 26, No. 11, (November 2001), pp. 648-653, ISSN 0968-0004

Sakamoto, A., Tsukamoto, S., Yamamoto, H., Ueda-Hashimoto, M., Takahashi, M., Suzuki, H. & Morikawa, H. (2003). Functional Complementation in Yeast Reveals a Protective Role of Chloroplast 2-Cys Peroxiredoxin against Reactive Nitrogen Species. *The Plant Journal,* Vol. 33, No. 5, (March 2003), pp. 841–851, ISSN 0960-7412

Sarowar, S., Lee, J.-Y., Ahn, E.-R. & Pai, H.-S. (2008). A Role of Hexokinases in Plant Resistance to Oxidative Stress and Pathogen Infection. *Journal of Plant Biology,* Vol. 51, No. 5, (September 2008), pp. 341-346, ISSN 1867-0725

Sazanov, L.A., Burrows, P.A. & Nixon, P.J. (1998). The Chloroplast Ndh Complex Mediates the Dark Reduction of the Plastoquinone Pool in Response to Heat Stress in Tobacco Leaves. *FEBS Letters,* Vol. 429, No. 1, (June 1998), pp. 115-118, ISSN 0014-5793

Scarpeci, T.E., Zanor, M.I., Carrillo, N., Mueller-Roeber, B. & Valle, E.M. (2008). Generation of Superoxide Anion in Chloroplasts of *Arabidopsis thaliana* during Active Photosynthesis: A Focus on Rapidly Induced Genes. *Plant Molecular Biology,* Vol. 66, No. 4, (March 2008), pp. 361–378, ISSN 0167-4412

Scheibe, R. & Dietz, K.J. (2011). Reduction–Oxidation Network for Flexible Adjustment of Cellular Metabolism in Photoautotrophic Cells. *Plant, Cell & Environment,* doi: 10.1111/j.1365-3040.2011.02319.x

Sharkey, T.D. (1988). Estimating the Rate of Photorespiration in Leaves. *Physiologia Plantarum,* Vol. 73, No. 1, (May 1988), pp. 147–152, ISSN 0031-9317

Shen, B., Jensen, R.G. & Bohnert, H.J. (1997). Increased Resistance to Oxidative Stress in Transgenic Plants by Targeting Mannitol Biosynthesis to Chloroplasts. *Plant Physiology,* Vol. 113, No. 4, (April 1997), pp. 1177-1183, ISSN 0032-0889

Sichel, G., Corsaro, C., Scalia, M., di Bilio, A. & Bonomo, R.P. (1991). *In vitro* Scavenger Activity of Some Flavonoids and Melanins against $O_2^{\bullet-}$. *Free Radical Biology & Medicine,* Vol. 11, No. 1, (n.d.), pp. 1-8, ISSN 0891-5849

Smirnoff, N. & Cumbes, Q.J. (1989). Hydroxyl Radical Scavenging Activity of Compatible Solutes. *Phytochemistry,* Vol. 28, No. 4, (n.d.), pp. 1057-1060, ISSN 0031-9422

Smirnoff, N. (2000). Ascorbate Biosynthesis and Function in Photoprotection. *Philosophical Transactions of the Royal Society of London. Series B, Biological Sciences,* Vol. 355, No. 1402, (October 2000), pp. 1455–1464, ISSN 0962-8436

Snyrychova, I., Pospısil, P. & Naus, J. (2006). Reaction Pathways Involved in the Production of Hydroxyl Radicals in Thylakoid Membrane: EPR Spin-Trapping Study. *Photochemical & Photobiological Sciences,* Vol. 5, No. 5, (May 2006), pp 472–476, ISSN 1474-905X

Srivastava, S., Phadke, R.S., Govil, G. & Rao, C.N.R. (1983). Fluidity, Permeability and Antioxidant Behaviour of Model Membranes Incorporated with α-Tocopherol and Vitamin E Acetate. *Biochimica et Biophysica Acta,* Vol. 734, No. 2, (October 1983), pp. 353-362, ISSN 0005-2728

Stewart, D. & Brudvig, G. (1998). Cytochrome b559 of Photosystem II. *Biochimica et Biophysica Acta,* Vol. 1367, No. 1-3, (October 1998), pp. 63–87, ISSN 0005-2728

Streb, P., Feierabend, J. & Bligny, R. (1997). Resistance to Photoinhibition of Photosystem II and Catalase and Antioxidative Protection in High Mountain Plants. *Plant, Cell & Environment,* Vol. 20, No. 8, (August 1997), pp. 1030 - 1040, ISSN 0140-7791

Sun, L., Shukair, S., Naik, T.J., Moazed, F. & Ardehali, H. (2008). Glucose Phosphorylation and Mitochondrial Binding Are Required for the Protective Effects of Hexokinases I and II. *Molecular & Cellular Biology,* Vol. 28, No. 3, (February 2008), pp. 1007-1017, ISSN0270-7306

Suzuki, N., Koussevitzky, S., Mittler, R. & Miller, G. (2011). ROS and redox signalling in the response of plants to abiotic stress. *Plant, Cell & Environment,* doi: 10.1111/j.1365-3040.2011.02336.x

Székely, G., Abrahám, E., Cséplo, A., Rigó, G., Zsigmond, L., Csiszár, J., Ayaydin, F., Strizhov, N., Jásik, J., Schmelzer, E., Koncz, C. & Szabados, L. (2008). Duplicated P5CS Genes of *Arabidopsis* Play Distinct Roles in Stress Regulation and Developmental Control of Proline Biosynthesis. *The Plant Journal,* Vol. 53, No. 1, (January 2008), pp. 11-28, ISSN 0960-7412

Szymańska, R. & Kruk, J. (2010). Plastoquinol is the Main Prenyllipid Synthesized During Acclimation to High Light Conditions in Arabidopsis and is Converted to Plastochromanol by Tocopherol Cyclase. *Plant & Cell Physiology,* Vol. 51, No. 4, (April 2010), pp. 537-545, ISSN 0032-0781

Takahashi, M.A. & Asada, K. (1983). Superoxide Anion Permeability of Phospholipids Membranes and Chloroplast Thylakoids. *Archives of Biochemistry and Biophysics,* Vol. 266, No. 2, (October 1983), pp. 558–566, ISSN 0003-9861

Takahashi, M. & Asada, K. (1988). Superoxide Production in Aprotic Interior of Chloroplast Thylakoids. *Archives of biochemistry & biophysics,* Vol. 267, No. 2, (December 1988), pp. 714-722, ISSN 0003-9861

Theil, E.C. (2004) Iron, Ferritin, and Nutrition. *Annual Review of Nutrition,* Vol. 24, (July 2004), pp. 327–343, ISSN 0199-9885

Thomas, C.E., Morehouse, L.A. & Aust, S.D. (1985). Ferritin and Superoxide-Dependent Lipid Peroxidation. *The Journal of Biological Chemistry,* Vol. 260, No. 6, (March 1985), pp. 3275-3280, ISSN 0021-9258

Tjus, S.E., Scheller, H.V., Andersson, B. & Møller, B.L. (2001). Active Oxygen Produced during Selective Excitation of Photosystem I Is Damaging not Only to Photosystem I, but Also to Photosystem II. *Plant Physiology,* Vol. 125, No. 4, (April 2001), pp. 2007-2015, ISSN 0032-0889

Trebst A., Depka, B. & Holländer-Czytko, H. (2002). A Specific Role for Tocopherol and of Chemical Singlet Oxygen Quenchers in the Maintenance of Photosystem II Structure and Function in *Chlamydomonas reinhardii. FEBS Letters,* Vol. 516, No. 1-3, (April 2002), pp. 156–160, ISSN 0014-5793

Vernooij, B., Friedrich, L., Morse, A., Reist, R., Kolditz-Jawhar, R., Ward, E., Uknes, S., Kessmann, H. & Ryals, J. (1994). Salicylic Acid Is Not the Translocated Signal Responsible for Inducing Systemic Acquired Resistance but Is Required in Signal Transduction. *The Plant Cell,* Vol. 6, No. 7, (July 1994), pp. 959-965, ISSN 1040-4651

von Gromoff, E.D., Schroda, M., Oster, U. & Beck, C.F. (2006). Identification of a Plastid Response Element that Acts as an Enhancer within the *Chlamydomonas* HSP70A Promoter. *Nucleic Acids Research,* Vol. 34, No. 17, (October 2006), pp. 4767-4779, ISSN 0305-1048

Wang, P.T. & Song, C.P. (2008). Guard-Cell Signalling for Hydrogen Peroxide and Abscisic Acid. *The New Phytologist,* Vol. 178, No. 4, (June 2008), pp. 703–718, ISSN 0028-646X

Wang, S.Y. & Jiao, H. (2000). Scavenging Capacity of Berry Crops on Superoxide Radicals, Hydrogen Peroxide, Hydroxyl Radicals, and Singlet Oxygen. *Journal of Agricultural and Food Chemistry,* Vol. 48, No. 11, (November 2000), 5677-5684, ISSN 0021-8561

Wasilewska, A., Vlad, F., Sirichandra, C., Redko, Y., Jammes, F., Valon, C., Frei dit Frey, N. & Leung, J. (2008). An Update on Abscisic Acid Signaling in Plants and More... *Molecular Pant,* Vol. 1, No. 2, (March 2008), pp. 198-217, ISSN 1674-2052

Wasternack, C. (2007). Jasmonates: an Update on Biosynthesis, Signal Transduction and Action in Plant Stress Response, Growth and Development. *Annals of Botany,* Vol. 100, No. 4, (October 2007), pp. 681–697, ISSN 0305-7364

Wei, Y., Dang, X. & Hu, S. (2004). Electrochemical Properties of Superoxide Ion in Aprotic Media. *Russian Journal of Electrochemistry,* Vol. 40, No. 4, (April 2004), pp. 400-403, ISSN 1023-1935

Wingler, A., Lea, P.J., Quick, W.P. & Leegood, R,C. (2000). Photorespiration: Metabolic Pathways and Their Role in Stress Protection. *Philosophical Transactions of the Royal Society of London. Series B, Biological Sciences,* Vol. 355, No. 1402, (October 2000), pp. 1517-1529, ISSN 0962-8436

Winterbourn, C.C. & Metodiewa, D. (1994). The Reaction of Superoxide with Reduced Glutathione. *Archives of Biochemistry and Biophysics,* Vol. 314, No. 2, (November 1994), pp. 284-290, ISSN 0003-9861

Woo, K.C. (1983). Evidence for Cyclic Photophosphorylation during CO(2) Fixation in Intact Chloroplasts: Studies with Antimycin A, Nitrite, and Oxaloacetate. *Plant Physiology,* Vol. 72, No. 2, (June 1983), pp. 313-320, ISSN 0032-0889

Yadav, D.K., Kruk, J., Sinha, R.K. & Pospíšil, P. (2010). Singlet Oxygen Scavenging Activity of Plastoquinol in Photosystem II of Higher Plants: Electron Paramagnetic

Resonance Spin-Trapping Study. *Biochimica et Biophysica Acta,* Vol. 1797, No. 11, (November 2010), pp. 1807-1811, ISSN 0005-2728

Yang, D.H., Andersson, B., Aro, E.M. & Ohad, I. (2001). The Redox State of the Plastoquinone Pool Controls the Level of the Light-Harvesting Chlorophyll a/b Binding Protein Complex II (LHC II) during Photoacclimation. *Photosynthesis Research,* Vol. 68, No. 2, (May 2001), pp. 163-174, ISSN 0166-8595

Yun, K.Y., Park, M.R., Mohanty, B., Herath, V., Xu, F.Y., Mauleon, R., Wijaya, E., Bajic, V.B., Bruskiewich, R. & de Los Reyes, B.G. (2010). Transcriptional Regulatory Network Triggered by Oxidative Signals Configures the Early Response Mechanisms of Japonica Rice to Chilling Stress. *BMC Plant Biology,* Vol. 10:16, (January 2010), pp., ISSN 1471-2229

Yurina, N.P. & Odintsova, M.S. (2007). Plant Signaling Systems. Plastid-Generated Signals and Their Role in Nuclear Gene Expression. *Russian Journal of Plant Physiology,* Vol. 54, No. 4, (July 2007), pp. 427-438, ISSN 1021-4437

Zelitch, I., Schultes, N.P., Peterson, R.B., Brown, P. & Brutnell, T.P. (2008). High Glycolate Oxidase Activity Is Required for Survival of Maize in Normal Air. *Plant Physiology,* Vol. 149, No. 1, (January 2008), pp. 195-204, ISSN 0032-0889

Ziem-Hanck, U. & Heber, U. (1980). Oxygen Requirement of Photosynthetic CO_2 Assimilation. *Biochimica et Biophysica Acta,* Vol. 591, No. 2, (July 1980), pp. 266-274, ISSN 0005-2728

4

Stress and Cell Death in Yeast Induced by Acetic Acid

M. J. Sousa[1], P. Ludovico[2,3], F. Rodrigues[2,3], C. Leão[2,3] and M. Côrte-Real[1]
[1]*Molecular and Environmental Research Centre (CBMA)/Department of Biology, University of Minho, Braga*
[2]*Life and Health Sciences Research Institute (ICVS), School of Health Sciences, University of Minho, Braga*
[3]*ICVS/3B's - PT Government Associate Laboratory, Braga/Guimarães*
Portugal

1. Introduction

Yeasts are nowadays relevant microorganisms in both biotechnology, with important economic impact in several fields, and fundamental research where *Saccharomyces cerevisiae* appears as one of the most used and versatile eukaryotic cell models. In industrial fermentations, yeasts are subjected to different stress conditions, such as those imposed by low water activity and by the presence of cytotoxic compounds. Yeast cells react to adverse conditions by triggering a stress response, enabling them to adapt to the new environment. However, upon a severe cell cue the elicited stress responses may be insufficient to guarantee cell survival and cell death may occur. The simplicity of yeast and its amenability to manipulation and genetic tractability make this unicellular eukaryotic microorganism a powerful tool in deciphering the mechanisms of eukaryotic cellular processes and their modes of regulation. Despite the differences in signalling pathways between yeast and higher eukaryotes current knowledge on cellular stress responses and programmed cell death confirms that several steps are phylogenetically conserved and therefore yeasts are ideal model systems to study the molecular pathways underlying these processes.

In this chapter we focus on the molecular mechanisms associated with stress response and cell death in yeast triggered by acetic acid. We start with a general introduction devoted to the physiological responses to acetic acid, and to the high resistance of the food spoilage yeast *Zygosaccharomyces bailii* to this acid in comparison with *S. cerevisiae* and other yeast species. Basic aspects of programmed cell death are also covered. The subsequent sections are dedicated to an overview of ours and other authors' studies highlighting the kinetics, components and pathways already identified in acetic acid-induced cell death.

1.1 Acetic acid physiological responses

Acetic acid is a normal by-product of the alcoholic fermentation carried out by *S. cerevisiae* and of contaminating lactic and acetic acid bacteria (Du Toit & Lambrechts, 2002; Pinto et al., 1989; Vilela-Moura et al., 2011) or it can be originated from acid-catalyzed hydrolysis of lignocelluloses (Lee et al., 1999; Maiorella et al., 1983). Above certain concentrations accepted as normal (0.2 to 0.6 g/l), acetic acid has a negative impact on the organoleptic qualities of wine and may affect the course of fermentation, leading to sluggish or arrested fermentations (Alexandre & Charpentier, 1998; Bely et al., 2003; Santos et al., 2008). In bioethanol production from lignocellulosic acid hydrolysates, acetic acid may also be associated with the inhibition of alcoholic fermentation, limiting the productivity of the process (Lee et al., 1999; Maiorella et al., 1983; Palmqvist & Hahn-Hägerdal, 2000). Therefore, acetic acid has a negative impact on yeast performance, restraining the production efficiency of wine, bioethanol or of products obtained by heterologous expression with engineered yeast cells under fermentative conditions. On the other hand, the cytotoxic effect of acetic acid is exploited in food industry, where it is used as a preservative. Some non-*Saccharomyces* species such as *Z. bailli* are highly resistant to acetic acid. Understanding the molecular determinants underlying such acid resistance phenotype is relevant for the design of strategies aiming at the genetic improvement of industrial *S. cerevisiae* strains, and the prevention of food and beverage spoilage by resistant species.

In most strains of *S. cerevisiae*, acetic acid is not metabolized by glucose-repressed yeast cells and enters the cell in the non-dissociated form by simple diffusion. Inside the cell, the acid dissociates and, if the extracellular pH is lower than the intracellular pH, this will lead to an intracellular acidification and to the accumulation of its dissociated form (which depends on the pH gradient), affecting cellular metabolism at various levels (Casal et al., 1996; Guldfeldt & Arneborg, 1998; Leão & van Uden, 1986; Pampulha & Loureiro, 1989;). Intracellular acidification caused by acetic acid leads to trafficking defects, hampering vesicle exit from the endosome to the vacuole (Brett et al., 2005). Though acetic acid induces plasma membrane ATPase activation (50 mM, pH 3.5), this enhanced activity is not enough to counteract cytosolic and vacuolar acidification (Carmelo et al., 1997). The toxic effects of the undissociated form of the acid also translate into an exponential inhibition of growth and fermentation rates (Pampulha & Loureiro, 1989; Phowchinda et al., 1995). Studies on glucose transport and enzymatic activities showed that the sugar uptake is not inhibited and that enolase is the glycolitic enzyme most affected by acetic acid, presumably resulting in a limitation of glycolytic flux (Pampulha & Loureiro-Dias, 1990). As revealed by the proteomic analysis of acetic acid-treated cells, carbohydrate metabolism is strongly affected, in agreement with a decreased glycolytic rate. Levels of the glycolytic proteins phosphofructokinase (Pfk2p) and fructose 1,6-bisphosphate aldolase (Fba1p) were decreased whereas the pyruvate decarboxylase isoenzyme (Pdc1p) suffered several post-translational modifications (Almeida et al., 2009). Growth in batch cultures following cellular adaptation to acetic acid is associated not only with a decrease in the maximum specific growth rate and in the ATP yield, but also with a recovery in intracellular pH and an increase in the specific glucose consumption rate, indicating that metabolic energy was diverted from metabolism (Pampulha & Loureiro-Dias, 2000). Using anaerobic chemostat cultures, it was shown that higher trehalose contents induced by lower growth rates or by the presence of ethanol are related to higher tolerance of *S. cerevisiae* to acetic acid (Arneborg et al., 1995, 1997). However, internal acidification caused by the acid can lead to the activation of trehalase (Valle et al., 1986). Hypersensitivity to acetic acid was observed in auxotrophic mutants with requirements for aromatic amino acids. Consistently, prototrophic *S. cerevisiae* strains are more resistant to acetic acid treatment (Gomes et al., 2007). Though there is no direct evidence, these phenotypes are probably explained by an inhibition of the amino acid uptake, since sensitivity is suppressed by supplementing the medium with high levels of tryptophan (Bauer et al., 2003). Accordingly, it was recently shown that acetic acid causes severe intracellular amino-acid starvation (Almeida et al., 2009), as referred below (section 4.4.). In another study, it was found that deletion of *FPS1*, coding for an aquaglyceroporin channel, abolishes acetic acid accumulation at low pH (Mollapour & Piper, 2007). This observation was explored to improve acetic acid resistance and fermentation performance of an ethanologenic industrial strain of *S. cerevisiae* through the disruption of *FPS1* (Zhang et al., 2011). The acetic acid-tolerance phenotype of the disrupted mutant was mainly explained by the preservation of plasma membrane integrity, higher *in vivo* activity of the H^+-ATPase, and lower oxidative damage after acetic acid treatment.

1.2 The high resistance of *Zygosaccharomyces bailii* to acetic acid

Acetic acid, due to its toxic effects, is used in food industry as a preservative against microbial spoilage. As a weak monocarboxylic acid with a pK_a of 4.76, its toxicity is strongly dependent on the pH of the medium, exerting an antimicrobial effect mainly at low pH values (below pK), where the protonated form predominates. However, there are some yeast species that are able to spoil foods and beverages due to their capacity to survive and grow under these stress conditions where other microorganisms are not competitive. *Z. bailli* is one of the most widely represented spoilage yeast species, particularly resistant to organic acids in acidic media with sugar (Thomas & Davenport, 1985). Another interesting feature of *Z. bailii* is its ability to grow under strictly anaerobic conditions (with trace amounts of oxygen) in complex medium, whereas in synthetic medium under strictly anaerobic conditions *Z. bailii* displays an extremely slow and linear growth compatible with oxygen-limitation (Rodrigues et al., 2001). These differential requirements for anaerobic growth, different from those associated with Tween 80 and ergosterol, are still a matter of debate (Rodrigues et al., 2005). This species is much more tolerant to acetic acid than *S. cerevisiae* and is able to grow in medium with acetic acid concentrations well above those tolerated by the later yeast, a phenotype that seems to be related to the metabolism of the acid. Glucose respiration and fermentation in *Z. bailii* and *S. cerevisiae* express different sensitivity patterns to ethanol and acetic acid. Inhibition of fermentation is much less pronounced in *Z. bailii* than in *S. cerevisiae*, and the inhibitory effects of acetic acid on *Z. bailii* are not significantly potentiated by ethanol (Fernandes et al., 1997).

One of the peculiar traits of *Z. bailii* is the mechanism underlying the transport of acetic acid into the cell and its regulation, the first step of acid metabolism. Either glucose or acetic acid grown cells display activity of mediated transport systems for acetic acid (Sousa et al., 1996). This is in contrast with what has been described so far in other yeast species, namely *S. cerevisiae, Candida utilis,* and *Torulaspora delbrueckii* where active transport of acetate by a H^+-symport is inducible and subject to glucose repression (Casal & Leão, 1995; Cássio et al., 1987, 1993; Leão & van Uden, 1986). Additionally, in the presence of glucose, *Z. bailii* displays a reduced passive permeability to the acid when compared with *S. cerevisiae* (Sousa et al., 1996). Unlike most strains of *S. cerevisiae,* which are unable to metabolize acetic acid in the presence of glucose, *Z. bailii* is able to simultaneous use the two substrates due to the high activity of the enzyme acetyl-CoA synthetase (Sousa et al., 1998). Thus, it appears that in *Z. bailii* both membrane transport and acetyl-CoA synthetase could assume particular physiological relevance in regards to the high resistance of this yeast species to environments containing mixtures of sugars and acetic acid, such as those often present during wine fermentation. Under these conditions, both the membrane transport flux and the intracellular metabolic flux of the acid seem to be regulated in such a way that cell can cope with the cytotoxic effects of the acid. These physiological traits have been related to the high resistance of Z. *bailii* to acidic media containing ethanol since this alcohol inhibits the mediated transport of the acid (Sousa et al., 1998).

1.3 A brief overview of programmed cell death: From multicellular organisms to yeast

The designation "Programmed Cell Death" (PCD) was first introduced by Lockshin (Gewies, 2003). Though PCD was initially related to the physiological cell death during organism's development, it has been generalised to alternative suicide processes that cells activate in response to various environmental aggressions. The term active cell death means that the process is genetically regulated in opposition to passive or accidental death, which is an uncontrolled death that occurs after exposure to an excessive dose of the lethal agent. These processes play an important role in the normal development, homeostasis mechanisms and disease control of multicellular organisms. Among the different forms of PCD (Kromer et al., 2009), namely apoptosis, autophagic cell death and programmed necrosis, apoptosis is the most common morphological expression of PCD. The main morphological features of an apoptotic cell, since the initial description by Kerr, Wyllie and Currie (1972), are the reduction of cellular volume (pyknosis), chromatin condensation and nuclear fragmentation (karyorrhexis) and engulfment by resident phagocytes (*in vivo*). All these changes take place in cells which display little or no ultrastructural modifications of cytoplasmic organelles and maintain plasma membrane integrity until the final stages of the process. Exposure of posphatidylserine on the outer leaflet of the plasma membrane of apoptosing cells, which promote phagocytosis by scavenging macrophages, is often an early event of apoptosis. Though not exclusive of apoptosis, other biochemical and functional changes such as oligonucleosomal DNA fragmentation, and the presence of proteolytically active caspases (cysteine-dependent aspartate-specific proteases) or of cleavage products of their substrates, may accompany the dismantling of the apoptotic cell. While in some settings apoptosis occurs independently of caspases, in others these proteases are key regulators of the death process and responsible for morphological and biochemical alterations typical of apoptosis (e.g., cellular blebbing and shrinkage, DNA fragmentation, and plasma membrane changes), as well as for the rapid clearance of the dying cell (Hengartner, 2000).

At least two major apoptotic pathways have been described in mammalian cells. One requiring the participation of mitochondria, called "intrinsic pathway," and another one in which mitochondria are bypassed and caspases are activated directly, called "extrinsic pathway" (Hengartner, 2000; Matsuyama et al., 2000). Regarding the mitochondrial pathway, two main events have been proposed as integral control elements in the cell's decision to die, namely, the permeabilization of the mitochondrial membrane and the release of several apoptogenic factors like cytochrome *c* (cyt *c*), apoptosis inducing factor (AIF), endonuclease G (Endo G), HtrA2/OMI and Smac/DIABLO (Hengartner, 2000; Matsuyama et al., 2000). Release of cyt *c* to the cytosol drives the assembly of a high-molecular-weight complex, the apoptosome, that activates caspases (Adrian & Martin,

2001). Translocation of cyt *c* to the cytosol is, therefore, a pivotal event in apoptosis. Cyt *c* is a soluble protein loosely bound to the outer face of the inner mitochondrial membrane, and its release is associated with an interruption of the normal electron flow at the complex III site, that could divert electron transfer to the generation of superoxide (Cai & Jones, 1998).

Beside caspases, members of the Bcl-2 protein family are key regulators of apoptosis, playing a crucial role in the regulation of the mitochondrial apoptosis pathway in vertebrates (Roset et al., 2007). The Bcl-2 family members have been identified and classified accordingly to their structure and function. At first, this family was usually divided in anti- and pro-apoptotic members. Currently, with new results obtained for a sub-group of this family, the BH-3 only proteins, they are divided into four categories (Chipuk et al., 2010). The anti-apoptotic Bcl-2 proteins (A1, Bcl-2, Bcl-w, Bcl-xL and Mcl-1), Bcl-2 effector proteins, (Bak and Bax), direct activator BH3-only proteins (Bid, Bim and Puma) and sensitizer/de-repressor proteins (Bad, Bik, Bmf, Hrk and Noxa). Complex interactions between members of this family control the integrity of the mitochondrial outer membrane (Green et al., 2002). The pro-apoptotic members of this family (Bax and Bak) are critical for mitochondrial membrane permeabilization, since deletion of both proteins impairs this event (Wei et al., 2001). Multicellular organisms have developed different regulatory complex mechanisms that coordinate cell death and cell proliferation and guarantee tissue homeostasis and normal development. Dysfunction of apoptosis is associated with severe human pathologies such as cancer and neurodegenerative diseases. Therefore, the identification of components of the different apoptotic pathways and the understanding of mechanisms underlying their regulation is critical for the development of new strategies for prevention and treatment against those diseases.

For several years, *Caenorhabditis elegans* and *Drosophila melanogaster* have been chosen as core models for cell death research, and until a decade ago it was not conceivable that unicellular organisms including yeast could possess a PCD process. This assumption was supported by the absence of key regulators of mammalian PCD in yeast, as indicated by plain homology searches and by the difficulty to explain the sense of cell suicide and its evolutionary advantage in a unicellular organism. However, in the late 1990s early 2000s, evidence indicating the presence of some basic features characteristic of an apoptotic phenotype in *S. cerevisiae* was reported (Madeo et al., 1997). This study showed that the expression of a point-mutated *CDC48* gene (cdc48^{S565G}), essential in the endoplasmic reticulum (ER)-associated protein degradation pathway, leads to a characteristic apoptotic phenotype. Later on it was shown in *S. cerevisiae* that depletion of glutathione or exposure to low external doses of H_2O_2 triggers the cell into apoptosis, whereas depletion of reactive oxygen species (ROS) or hypoxia prevents apoptosis (Madeo et al., 1999). In addition, an intracellular accumulation of ROS was detected in the cell cycle mutant cdc48^{S565G} of *S. cerevisiae* and in yeast cells expressing mammalian Bax (Ligr et al., 1998). These results allowed the identification of ROS production as a key cellular event common to the known scenarios of apoptosis in yeast and animal cells (Madeo et al., 1999). Subsequent studies revealed that acetic induces apoptosis in *S. cerevisiae* through the involvement of mitochondria, indicating the conservation in yeast of an intrinsic death pathway (Ludovico et al., 2001, 2002). These former studies led to the emergence of a new research field that profited from the recognized advantages of yeast for the study of biological processes. Currently, there is increasing evidence that apoptotic-like cell death pathways exist in unicellular organisms such as yeast and that this ability confers selective advantage in adapting to adverse environmental conditions and thus ensuring survival of the clone (Herker et al., 2004). Therefore, it is consensual that yeast can undergo cell death with typical markers of mammalian apoptosis in response to different stimuli and possess orthologs of mammalian apoptosis regulators, supporting the existence of a primordial apoptotic machinery similar to that present in higher eukaryotic cells (for a revision see Carmona-Gutierrez et al., 2010; Pereira et al., 2008).

2. Stress response pathways and key components

Under unfavourable environmental conditions, the yeast cell induces a common set of functional changes as a broad response to stress. These changes include, on one hand, reduction in activities linked with cell proliferation and protein synthesis, anabolic pathways and other processes associated with high energy expense, and, on the other, increase in activities related to protection and repair of damage of different molecules (DNA, proteins and lipids) and cellular structures. Gasch et al. (2000) reported that there are changes in a common set of about 900 genes (termed "environmental stress response" - ESR) in response to 12 different adverse environmental transitions, which mainly depend on the transcription factors Yap1, Msn2 and Msn4. From the 900 genes of the ESR, ≈600 are repressed and ≈ 300 are induced. The last set incorporates approximately 50 genes previously described as part of general stress response and which bear the stress response element (STRE) promoter sequence recognized by Msn2p and Msn4p. The different environmental conditions not only produce a set of common changes, probably accounting for the cross-resistances to different unrelated stress, but also generate specific responses reflecting the particular cell targets for each stress. Genome-wide functional analyses using the yeast disruptome, as well as gene expression profiling, have been exploited to identify key components of stress response induced by different weak carboxylic acids, namely sorbic, citric, benzoic, propionate, lactic and acetic acids (Abbot et al., 2008; Kawahata et al., 2006; Mira et al., 2009, 2010; Mollapour et al., 2004; Schuller et al., 2004). The first studies combining the two approaches were performed with sorbic acid. They were used to identify key players in biological response to the acid and to differentiate the essential genes from those displaying expression changes but which were not critical, or even relevant, for the ability of the cell to cope with a particular stress (Schuller et al., 2004). In this line, it was observed that although most of the genes induced by sorbic acid stress were dependent on the Msn2/4 transcription factors, the double knockout mutant was not more sensitive to sorbate stress. Resistance to sorbic acid, on the other hand, is predominantly associated with the activities of the previously described efflux pump Pdr12 (Piper et al., 1998) and of its dedicated transcription factor War1p. Oxidative stress-sensitive mutants, as well as mutants defective in mitochondrial function, vacuolar acidification and protein sorting (vps), ergosterol biosynthesis (erg mutants) and in actin and microtubule organization were also identified as sorbate-sensitive by genome-wide screening (Mollapour et al., 2004). Sorbate resistance increased with deletion of 34 genes categorize in several different functions, including *TPK2*, coding for one of the protein kinase A (PKA) isoforms, and the genes coding for the Yap5 transcription factor, two B-type cyclins (Clb3p, Clb5p), and a plasma membrane calcium channel activated by endoplasmic reticulum stress (Cch1p/Mid1p).

Lawrence et al. (2004) combined genome-wide phenotypic studies, expression profiling and proteome analysis to investigate citric acid stress. These authors described for the first time the involvement of mitogen-activated protein kinase (MAPK) high-osmolarity glycerol (HOG) pathway in the regulation of stress induced by a weak acid. Sixty nine mutants displaying sensitivity to 400 mM citric acid (pH 3.5) were detected in the screening, but no resistant strains were found. Citric acid up-regulated many stress response genes. However, in accordance with the results from the sorbic acid study, little correlation is observed between gene deletions associated with the citric acid-sensitive phenotype and those with measurable changes in the levels of transcript or protein expressed, although they belong to the same gene ontology families. Also, as found for sorbic acid, vacuolar acidification seems to be crucial for adaptation to citric acid. Transcription factors mediating glucose derepression, enzymes involved in amino acid biosynthesis and a plasma membrane calcium channel seem essential for adaptation to citric acid as well.

Applying genome-wide functional analysis and gene expression profiling to the study of acidic stress caused by lactic and acetic acid revealed a connection between Aft1p-regulated intracellular metal metabolism and resistance (Kawahata et al., 2006). As for sorbic and citric acids, vacuolar acidification and the Hog1p pathway seem to be important for resistance to lactic and acetic acid at low pH. In accordance, a sub-lethal growth inhibitory concentration of acetic acid was shown to promote the phosphorylation of Hog1p and Slt2p, two MAP kinases in *Saccharomyces cerevisiae* (Mollapour & Piper, 2006). However, from the 101 viable kinase mutants of the Euroscarf collection, only *hog1Δ*, *pbs2Δ*, *ssk1Δ* and *ctk2Δ* exhibited deficient growth in the presence of acetic acid. Activation of Hog1p by acetic acid was shown to depend on the presence of *SSK1* and *PBS2*, but not of *SHO1* or *STE11*. In the same screening, loss of the cell integrity MAP kinase (Slt2p/Mpk1p) was found to slightly increase acetate resistance. In what concerns the known plasma membrane sensors of MAPK pathways, acetate-induced Hog1p activation appears to involve the Sln1p, as also found for citric acid (Lawrence et al., 2004), whereas Slt2p activation was dependent on Wsc1p (Mollapour et al., 2009). It was also shown that the activation of Hog1p by acetic acid causes the removal of protein-channel Fps1p from the plasma membrane and limits the accumulation of the acid (Mollapour & Piper, 2007). The transcription factor Haa1p was also associated with resistance to acetic acid in glucose medium, where the knockout mutant displayed an increased lag phase (A. R. Fernandes et al., 2005). This effect was mainly attributed to the downregulation of genes coding for the plasma membrane multidrug transporters, *TPO2* and *TPO3*, and for the cell wall glycoprotein, *YGP1*. Genome-wide screening of the *S. cerevisiae* Euroscarf mutant collection identified 650 determinants of acetic acid tolerance, clustering essentially in the functional categories of carbohydrate metabolism, transcription, intracellular trafficking, ion transport, biogenesis of mitochondria, ribosome and vacuole, and nutrient sensing and response to external stimulus (Mira et al., 2010). Accordingly, a proteomic analysis of *S. cerevisiae* cells treated with acetic acid revealed that proteins from amino-acid biosynthesis, transcription/translation machinery, carbohydrate metabolism, nucleotide biosynthesis, stress response, protein turnover and cell cycle are affected (Almeida et al., 2009). Twenty eight transcription factors were identified as required for acetic acid resistance, from which Msn2p, Skn7p and Stb5p were found to have the highest percentage of targets among the genes required for acetic acid tolerance. The transcription factor Rim101p, previously described to counteract propionic acid-induced toxicity (Mira et al., 2009), was also found to be necessary for acetic acid resistance. Differential transcriptome profiling in response to acetic acid revealed changes in the expression of 227 genes (Li & Yuan, 2010). The downregulated genes are associated with mitochondrial ribosomal proteins and with carbohydrate metabolism and regulation, whereas those related to arginine, histidine, and tryptophan metabolism were upregulated. Data indicated that acetic acid disturbs mitochondrial functions at translation, electron transport chain and ATP production levels, interrupts reserve metabolism (glycogen and trehalose metabolism and glucan synthesis), and regulates the central carbon metabolism and amino acid biosynthesis in yeast.

3. Cell death induced by acetic acid and its dependence on the temperature

Temperature profiles are an expression of the temperature dependence of growth and death in batch culture. Metabolites that accumulate in the medium and added drugs of industrial, medical or general scientific interest may profoundly change the temperature profile of yeast. Analysis of such modified profiles may shed light on the nature and localization of the targeted sites. Moreover, this analysis allows for predictions of the temperature-dependence of yeast performance in industrial fermentations and the effects of the temperature on the cytoxicity of preservatives on yeast in food, wine and other beverages (van Uden, 1984). It was shown that in *S. cerevisiae*, under certain conditions, acetic acid compromises cell viability and ultimately results in two types of cell death, high (HED) and low enthalpy (LED) cell death (Pinto et al., 1989). At concentrations similar to those that may occur during vinification and other alcoholic yeast fermentations, acetic acid and other weak acids enhance thermal death, causing a shift of the lethal temperatures of glucose-grown cell populations of *S. cerevisiae* to lower values. This type of cell death (HED) represents a thermal death enhanced exponentially by the acid which predominates at lower acetic acid concentrations (<0.5%, w/v) and higher temperatures. The knowledge acquired by the study of HED is of practical importance since the HED contributes to the so called "heat-sticking" of alcoholic yeast fermentations, particularly of red wine and fuel ethanol fermentations in warm countries in the absence of efficient temperature control. The second type of death (LED) induced by acetic acid occurred at intermediate and lower temperatures at which thermal death is not detectable, and could be considered a consequence of the cytoplasm acidification. Ethanol and other alkanols also induced these types of cell death, but acetic acid is over 30-times more toxic than ethanol. Cell death induced by acetic acid alone or with ethanol is strongly dependent not only on growth phase and pre-culture conditions of the cells before exposure to acetic acid, but also on the experimental conditions. Culture parameters such as temperature, pH, oxygen and nutrient availability, and, in particular, glucose concentration which determines the proportion between fermentative and respiratory metabolism, influence the percentage of dead cells in response to a given dose of acetic acid.

It was also observed that acetic acid and other weak acids enhanced death in glucose-grown cells of the spoilage yeast *Z. bailii*, but the effects were much lower than those described for *S. cerevisiae*, and only detectable at higher acid concentrations (Fernandes et al., 1999). *Z. bailii* is more resistant than *S. cerevisiae* to short-term intracellular pH changes caused by acetic acid (Arneborg et al., 2000). Furthermore, while in *S. cerevisiae* the enhancement of death by weak acids at intermediate temperatures could be considered a consequence of the acidification of the cytoplasm, in *Z. bailii* the intracellular acidification induced by weak acids is less pronounced and appears not to have a significant role in death at such temperature range. This reinforces the idea that in *Z. bailii*, as opposed to *S. cerevisiae*, weak acids in general and acetic acid in particular only enhance thermal death and not LED, which may occur at lower temperatures. Furthermore, significant HED at these lower temperatures requires rather high concentrations of the toxic compounds, which, at least in the case of acetic acid, are much less realistic for alcoholic fermentations than the ones that induced significant death of this type in *S. cerevisiae*. As a consequence, it could be postulated that cell viability of *Z. bailii* will not be significantly affected, even at the end of the normal alcoholic fermentation processes, where the concentration of ethanol is high. Specifically in wine, this property of *Z. bailii* may be associated with its presence at the end of the process where the environmental conditions are too severe to allow survival of *S. cerevisiae*. As referred above the ability of *Z. bailii* to use acetic acid simultaneously with glucose, even at low pH values such as 3.5, in contrast to *S. cerevisiae*, which is often unable to metabolize the acid under these conditions, probably contributes to those different patterns of behavior between the two species (Sousa et al., 1998). The responses of the yeast to stress conditions could be considered using both non- and adapted cells. Adaption of cells to acetic acid in the growth medium modifies the cell death sensitivity pattern to acid environments. In *Z. bailii*, the negative effects induced by acetic acid in cell viability were only slightly lower in adapted than in non-adapted cells, which is consistent with the fact that in *Z. bailii* transport and intracellular acetic acid metabolism operate independently of the presence of glucose in the growth medium.

4. Acetic acid as an inducer of programmed cell death

In previous sections we focused on the cytotoxic effects of acetic acid and on the cellular responses triggered by the acid. This section encompasses the characterization of the cell death process induced by acetic acid, and covers the main molecular components/pathways involved and their regulation.

4.1 Molecular components and pathways

The first studies regarding the assessment of cell structural and functional changes associated with acetic acid-induced cell death in populations of *S. cerevisiae* were performed by flow cytometry multiparametric analysis combining different viability dyes (Prudêncio et al., 1998). Kinetic changes in esterase activity, intracellular dye processing, and membrane integrity were monitored, and to detect those changes three assays involving fluorescein diacetate hydrolysis, FUN-1 processing, and propidium iodide exclusion, were used, respectively. This approach allowed establishing the temporal order of appearance of the cell changes that pointed to the decrease in the ability to process FUN-1, which preceded the decrease in esterase activity, and was followed by the loss of cell membrane integrity after incubation with acetic acid. Together, these results suggested an intracellular localization of the acetic acid cellular target(s) in an early phase of cell death, rather than on the cellular membrane which occurred much later. Nevertheless, the flow cytometric analysis of mitochondrial membrane potential, $\Delta\Psi m$, (determined by rhodamine 123 staining) and plasma membrane integrity (determined by PI staining) showed that in *S. cerevisiae* acetic acid treatment (1.0% and 1.8% v/v, pH 3.0 for 130 min.) affects the proliferative capacity that is followed by the loss of plasma membrane integrity, and later by the loss of ability of mitochondria to specifically stain with Rh123. In contrast, acetic acid treatment (1.8% and 3% v/v, pH 3.0 for 130 min.) of *Z. bailii* cells affects much less the ability of mitochondria to specifically stain with Rh123, and the loss of plasma membrane integrity observed for higher acetic acid concentrations is correlated with the loss of proliferative capacity (Ludovico 1999). Altogether these results clearly indicate that plasma membrane and mitochondria are targeted by acetic acid in *S. cerevisiae* at lower concentrations than in *Z. bailii* cells in accordance with the higher resistance phenotypes of the latter species. As mentioned above, acetic acid induces a PCD process in *S. cerevisiae* which shares common features with an apoptotic phenotype (Ludovico et al., 2001). It was found that acetic acid in concentrations between 20 and 120 mM induces a cycloheximide-inhibitable PCD process in exponentially growing *S. cerevisiae* cells that displays the most common apoptotic hallmarks, such as: (i) chromatin condensation along the nuclear envelope verified by transmission electron microscopy and DAPI staining; (ii) exposure of phosphatidylserine on the surface of the cytoplasmic membrane revealed by the FITC–annexin V reaction; and (iii) occurrence of internucleosomal DNA fragmentation demonstrated by the terminal deoxynucleotidyl transferase-mediated dUTP nick end labeling (TUNEL) assay. Exposure of cells to a higher acetic acid concentration (200 mM) resulted in cell death that was not inhibited by cycloheximide and was accompanied by ultrastructural alterations typical of necrosis. Pulsed field gel electrophoresis of chromosomal DNA from stationary phase cells dying by apoptosis after exposure to acetic acid (175 mM) revealed DNA breakdown into fragments of several hundred kilobases, consistent with the highly order chromatin degradation preceding DNA laddering in apoptotic mammalian cells (Ribeiro et al., 2006).

Caspases (cysteine aspartic proteases), key components of the mammalian apoptotic machinery, have a crucial role in cell dismantling. The metacaspase Yca1p, the only yeast ortholog of mammalian caspases identified so far, is activated in cells undergoing acetic acid-induced apoptosis in a manner strongly dependent on the cell growth phase (Pereira et al., 2007). Kex1p has been characterized as a serine carboxypeptidase B-like protease responsible for processing of prepro-factor (mating pheromone) as well as K1 and K2 killer toxin precursors (Bussey 1988; Fuller et al., 1988) while traversing the secretory pathway. This protease, besides being involved in PCD caused by defective N-glycosylation (Hauptmann et al., 2006), also contributes to the active cell death program induced by acetic acid stress. In fact, during cell death induced by acetic acid, the deletion of *KEX1* led to increased survival of cells and also a reduced production of ROS (Hauptmann et al., 2008). Though, as referred above, Yca1p was shown to be involved in acetic acid induced cell death, a caspase-independent route was also postulated (Guaragnella et al., 2006). Kex1p may be engaged in and contribute to this pathway, albeit its precise function in the cascade remains to be determined. The role of the proteasome in PCD is rather controversial. While proteosome inhibition has been found to induce PCD in certain mammal and plant cell contexts (Shinohara et al., 1996; Kim et al., 2003), transient proteasome activation is necessary for protein degradation during acetic acid-induced apoptosis in yeast (Valenti et al., 2008).

Similar to *S. cerevisiae*, acetic acid also induces in *Z. bailii* either an apoptotic or a necrotic death process, depending on the acid concentration. However, in *Z. bailii* the PCD process was found to occur at higher acetic acid concentrations (320-800 mM), described to be necrotic for *S. cerevisiae*. This is consistent with the higher resistance of *Z. bailii* compared to that of *S. cerevisiae* as discussed above. The observation that acetic acid-induced PCD can occur not only in *S. cerevisiae* but also in *Z. bailii* (Ludovico et al., 2003) and *Candida albicans* (Phillips et al., 2003) reinforces the concept of a physiological role of the PCD in the normal yeast life cycle and raises the possibility that this mode of cell death is generalized in yeast.

4.2 The involvement of mitochondria

Like in mammalian cells, the PCD process triggered by acetic acid in yeast can be mediated by mitochondria (Ludovico et al., 2002). Biochemical and molecular evidence provided by such studies included the accumulation of mitochondrial reactive oxygen species (ROS), transient hyperpolarization followed by depolarization, decrease in cytochrome oxidase activity (COX) linked to a specific decrease in the amounts of COX II subunit, affecting mitochondrial respiration, and release of lethal factors like cytochrome *c* (cyt *c*). Though in apoptosis induced by hyperosmotic shock (Silva et al., 2005) a causal relationship between cyt *c* release and caspase activation was established, this has not been shown for most of the apoptotic scenarios in yeast. Moreover, components downstream release of cyt *c* have not yet been identified in yeast. Therefore, the formation of a mammalian apoptosome-like structure in yeast and the precise role of cyt *c* release in cells undergoing apoptosis need further research. The acetic acid-induced PCD process was found to be independent of oxidative phosphorylation because it was not inhibited by oligomycin treatment. The inability of *S. cerevisiae* mutant strains (lacking mitochondrial DNA, heme lyase, or ATPase) to undergo acetic acid-induced PCD and the absence of cyt *c* release in the ATPase mutant (knockout in *ATP10*) provides further evidence that the process is mediated by a mitochondria-dependent apoptotic pathway (Ludovico et al., 2002). Accordingly, disruption of the genes *CYC1* and *CYC7* rescued cells against acetic acid-induced PCD (Guaragnella et al., 2010a). ROS, in particular hydrogen peroxide, are mediators rather than by-products in *S. cerevisiae* cells committed to apoptosis triggered by acetic acid (Guaragnella et al., 2007). This interpretation was further confirmed by the protection against acetic acid-induced PCD afforded by the overexpression of cytosolic catalase (Guaragnella et al., 2008). As aforementioned, mitochondrial outer membrane permeabilization (MOMP) is a crucial step in the apoptotic pathway. This triggers the release of proteins from the mitochondrial intermembrane space into the cytosol, where they ensure propagation of the apoptotic cascade and execution of cell death. Opening of a mitochondrial pore called the permeability transition pore complex (PTPC), which leads to the swelling of mitochondria and rupture of the mitochondrial outer membrane, has been put forward as one of the mechanisms underlying mammalian MOMP. Although the molecular composition of the pore is not completely defined, it has been proposed that its major components are the adenine nucleotide transporter (ANT), the voltage dependent anion channel (VDAC) and cyclophilin D (for a review see Kinnally et al., 2011). Yeast genetic approaches revealed that while deletion of *POR1* (yeast VDAC) enhances apoptosis triggered by acetic acid, absence of ADP/ATP carrier (AAC) proteins (yeast orthologs of ANT) protects cells exposed to acetic acid (Pereira et al., 2007). Absence of AAC proteins and the consequent impairment of cyt *c* release do not completely prevent acetic acid-induced apoptosis, suggesting that alternative cyt *c*-independent pathways are involved. One such pathway may be the translocation of Aif1p, the yeast apoptosis inducing factor, from the mitochondria to the nucleus in response to acetic acid (Wissing et al., 2004). Mammalian AIF is a bifunctional NADH oxidase which has a pro-survival role when localized in the mitochondrial intermembrane space through its involvement in mitochondrial respiration, and a lethal function upon translocation to the nucleus through a caspase-independent apoptotic process (Vahsen et al., 2004 ; Susin et al., 1999). Yeast Aif1p shares the same localization and death executing pathways as mammalian AIF, and dependence on cyclophilin A (CypA) but is partially dependent on caspase action (Wissing et al., 2004). It was recently shown that mammalian AIF mediates a programmed necrosis pathway independent of apoptosis and involving other molecules such as poly(ADP-ribose) polymerase PARP-1, calpains, Bax, Bcl-2, histone H2AX, and cyclophilin A (Delavallée et al., 2011). The yeast Yca1p is required for cyt *c* release (Guaragnella et al., 2010b) and the antioxidant N-Acetyl-L-cysteine (NAC) prevents acetic acid induced cell death by scavenging hydrogen peroxide, and inhibiting cyt *c* release and caspase activation (Guaragnella et al., 2010a). This supports the occurrence of a ROS-dependent acetic acid induced PCD. Because NAC does not prevent cells lacking Yca1p and cyt *c* to undergo PCD induced by acetic acid, it was proposed that a ROS-independent PCD can also be induced by the acid.

It was observed that during a variety of apoptotic scenarios in mammalian cells the interconnected mitochondrial network converts into punctiform morphology at early times, a process known as thread–grain transition (see reviews Parone et al., 2006; Scorrano et al.; 2005; Youle et al., 2005). This apoptotic fragmentation was proposed to be due to the activation of the physiological fission machinery, which proved also to influence the path of apoptosis (Frank et al., 2001). Inhibition of the mitochondrial fission machinery in mammals, including Drp1, ortholog of yeast Dnm1p, and Fis1p, impairs not only apoptotic fragmentation of the mitochondrial network but also cyt *c* release and the process of death itself (Frank et al., 2001; James et al., 2003). In yeast apoptosis induced by different compounds (Fannjiang et al., 2004; Kitagaki et al., 2007; Pozniakovsky et al., 2005) or upon heterologous Bax expression (Kissova et al., 2006) mitochondria also undergo extensive fragmentation suggesting it to be a general feature of yeast PCD. The mitochondrial fission protein Dnm1p and its two interactors, Mdv1p and Fis1p, have been implicated in the execution of the yeast apoptotic program induced by acetic acid (Fannjiang et al., 2004). Deletion of *DNM1* or *MDV1/NET2* inhibits cell death induced by acetic acid. Interestingly, Dnm1p-deficiency protected cells from death more efficiently than from mitochondrial fragmentation. This suggests that the absence of Dnm1p in yeast might confer protection against cell death also by mechanisms other than those related to fission of the mitochondria. In *S. cerevisiae*, Fis1p is evenly distributed along the mitochondrial surface, where it functions as a receptor to recruit Dnm1p from the cytosol to mitochondria (Modzi et al., 2000). The fact that the double mutant *dnm1Δfis1Δ* behaved like *dnm1Δ* led Fannjiang et al. (2004) to propose that Fis1p inhibits the action of Dnm1p, thus promoting cell survival. The pro-survival role of Fis1p in yeast was a surprise since in mammalian cells overexpression of *FIS1* not only triggers mitochondrial fission but also cyt *c* release and apoptosis (James et al., 2003). Nevertheless, a recent study revealed that deletion of *FIS1* essentially selects for a secondary mutation in the stress-response gene *WHI2* that confers sensitivity to cell death (Cheng et al., 2006).

Nuc1p, the yeast ortholog of the mammalian endonuclease G, also mediates apoptosis induced by acetic acid (Buttner et al., 2007). This protein shares with other mitochondrially located yeast cell-death regulators, like cyt *c* and Aif1p, and with their mammalian counterparts, a role in cell proliferation and in cell death. Furthermore, it was found that Nuc1p-mediated death is independent of Yca1p and Aif1p. Instead, the Aac2p, karyopherin Kap123p, and histone H2B interact with Nuc1p and are required for cell death upon Nuc1p overexpression, suggesting a pathway in which mitochondrial pore opening, nuclear import, and chromatin association are successively involved in EndoG-mediated death (Buttner et al., 2007). Ysp2p is another mitochondrial protein with a direct function in mitochondria-mediated PCD, since its absence hinders mitochondrial thread-to-grain transition and confers resistance to acetic acid-induced PCD (Sokolov et al., 2006).

4.3 The involvement of the vacuole

In the last decade, it has been demonstrated that organelles other than mitochondria are also engaged in the regulation of mammalian cell death processes (for a revision see Boya et al., 2008; Johansson et al., 2010; Repnik et al., 2010). Mammalian cells in response to different death stimuli may entail mitochondrial membrane permeabilization associated with the release of pro-apoptotic factors into the cytosol, as well as lysosomal membrane permeabilization (LMP) coupled with the release of cathepsins. Cathepsin D (CatD) has emerged as a central player in the apoptotic response. It was found that, following LMP, CatD is released into the cytosol and triggers a mitochondrial apoptotic cascade. However, CatD can have anti-apoptotic effects in some cellular types and specific contexts. Indeed, it is generally accepted that CatD is overexpressed and plays an important role in cancer cells (Masson et al., 2010). Pro-CatD outside the cells induces proliferation, angiogenesis, invasion and metastasis (Benes et al., 2008). Additionally, it was demonstrated that inhibition of CatD with pepstatin A induces caspase-dependent apoptosis in neuroblastoma cell lines (Kirkegaard & Jäättelä, 2008) and that overexpression of intracellular CatD in mouse xenografs using rat derived cell lines inhibits apoptosis (Masson et al., 2010). It was also reported that CatD downregulation sensitizes neuroblastoma cells to doxorubicin-induced apoptosis, while the opposite effect is observed for CatD overexpression (Sagulenko et al., 2008). In contrast, CatD mediates cyt *c* release and caspase activation in staurosporin-induced apoptosis in human fibroblasts (Johansson et al., 2003). It is therefore apparent that CatD can have opposite roles in apoptosis and that the lysosome is intrinsically connected to apoptosis through LMP.

In yeast, the vacuole seems to play a similar role to lysosomes in the regulation of apoptosis. The first study on the involvement of the vacuole in yeast apoptosis, concerns the translocation of the vacuolar protease Pep4p, the ortholog of the human CatD, into the cytosol during H_2O_2-induced apoptosis (Mason et al., 2005). It was found that, in an early phase of cell death, ROS levels and nuclear permeability increase while cell viability drops. In a later phase, the vacuolar membrane becomes permeable and provides access of the protease to nucleoporin substrates. Similar to the partial lysosomal membrane permeabilization observed during mammalian apoptosis, the release of Pep4p-EGFP from the vacuole is not linked to a rupture of the vacuolar membrane, as evidenced by a vacuolar lumen morphologically distinct from the cytosol. However, *PEP4* deleted cells are not protected from H_2O_2-induced cell death. This may be explained by the fact that migration of Pep4p out of vacuoles and nucleoporin degradation occurs only after the cells are nonviable. Release of Pep4p from the vacuolar compartment is also observed in an End3p deficient mutant displaying actin cytoskeleton stabilization-induced apoptosis (Gourlay et al., 2006). However, a role for this protease in actin-stabilized dying cells was not ascertained by the authors. Another study also documented the involvement of the vacuole in yeast apoptosis.

Deletion of class C vacuolar protein sorting genes results in drastically enhanced sensitivity of yeast to treatment with acetic acid and leads to a necrotic death, whereas death is mainly apoptotic in the wild type strain. These data indicate that a functional vacuole is required for a regulated cell death process through apoptosis (Schauer et al., 2009).

The occurrence of mitochondrial degradation following apoptosis induction is a common feature of mammalian cells and is often considered critical for its progression (reviewed in Tolkovsky et al., 2002). This event is generally mediated by lysosomes and usually occurs through an autophagic process that shows selectivity for mitochondria, termed mitophagy (Lemasters, 2005). Recent evidence supports the view that the PTP could be the trigger for mitochondrial degradation (Kim et al., 2007; Rodriguez-Enriquez et al., 2006). However, removal of mitochondria is not always dependent on the autophagic machinery (Matsui et al., 2006). In yeast cells undergoing apoptosis, mitochondrial degradation has been reported (Fannjiang et al., 2004). Heterologous expression of Bax (Kissova et al., 2007), mitochondrial dysfunction (Priault et al., 2005), osmotic swelling (Nowikovsky et al., 2007) and homeostasis of stationary phase cells (Tal et al., 2007) were also associated to selective removal of mitochondria. It was recently found that autophagy is not active during acetic acid-induced apoptosis (Pereira et al., 2010). Indeed, no increase in the amount of Atg8p, an essential autophagosome component, or in the activity of a truncated form of alkaline phosphatase, which activation is dependent on the induction of autophagy, is detected during acetic acid-induced apoptosis. Accordingly, deletion of *ATG5*, another component of the yeast autophagic machinery, did not affect cell survival. Alternatively to autophagy, the vacuolar protease Pep4p, ortholog of the human Cat D, was translocated to the cytosol and played an important role in mitochondrial degradation. Transmission electron microscopy analysis of the dying cells showed that vacuolar membrane integrity was preserved and plasma membrane integrity was maintained. Hence, Pep4p release seems to involve partial permeabilization of the vacuolar membrane rather than an extensive permeabilization typical of necrotic death. Taken together, these results suggested that Pep4p could have a role in apoptotic cell death similar to that of mammalian CatD. Instead, deletion of Pep4p confers higher susceptibility to acetic acid (Pereira et al., 2010) and leads to combined apoptotic and necrotic cell death during chronological aging (Carmona-Gutierrez et al., 2011) pointing to a function in cell protection rather than in the execution of cell death. Sustaining this hypothesis, cells overexpressing Pep4p displayed a higher resistance to acetic acid (Pereira et al., 2010) and an extension of chronological aging particularly through the anti-necrotic function of this protease rather than through its anti-apoptotic role (Carmona-Gutierrez et al., 2011). Pep4p deficient cells, like the wild type strain, exhibit mitochondrial dysfunction but are delayed in mitochondrial degradation during acid-induced apoptosis. On the other hand, Pep4p overexpression slightly enhanced mitochondrial degradation under the same conditions. Therefore, the process of removing damaged mitochondria apparently has a protective role in acetic acid-treated cells, although it is likely not the only factor affecting cell viability. Though the involvement of the vacuole and Pep4p in mitochondrial degradation is autophagy-independent, the precise mechanism is unknown. It is however apparent that this process also involves non-vacuolar proteins. Indeed, AAC-deficient cells show a decrease in mitochondrial degradation in response to acetic acid as well, and are not defective in Pep4p release. Therefore, AAC proteins seem to affect mitochondrial degradation at a step downstream to Pep4p release, possibly triggering degradation through their involvement in mitochondrial permeabilization. Accordingly, the sensitization of cells to acetic acid by deletion of *PEP4* was dependent on AAC proteins, again suggesting these proteins act downstream of Pep4p in the apoptotic cascade (unpublished results). Moreover, it was proposed that the AAC proteins relay a signal of mitochondrial dysfunction, targeting their destruction. Taken together, the aforementioned observations suggest that vacuole and mitochondria destabilization, as measured by Pep4p and cyt *c* release, respectively, are events in the cell death cascade. Even though CatD, the mammalian ortholog of Pep4p, was shown to have a role in cell death by triggering mitochondrial dysfunction and subsequent release of mitochondrial proteins, some studies have shown an inhibitory role for CatD in apoptosis. Since autophagy is not active in cells undergoing acetic acid-induced apoptosis, vacuolar membrane permeabilization associated with the release of Pep4p may act as an alternative mitochondrial degradation process. The cytosolic acidification induced by acetic acid, associated with inhibition of autophagy may favor the activity of Pep4p after its release from the vacuole. These results unveil a complex regulation and interplay between mitochondria and the vacuole in yeast PCD.

Acetic acid has also been shown to induce apoptosis in a mammalian cell model. It has been demonstrated that the short chain fatty acids (SCFA) acetic and propionic acids produced by dietary propionibacteria in the human intestine induce cell death in colorectal carcinoma cell lines (CRC) by a mitochondrial-dependent apoptotic pathway (Jan et al., 2002; Lan et al., 2007) as described for yeast cells (Ludovico et al., 2001, 2002). SCFA induced nuclei shrinkage, chromatin condensation, nuclei fragmentation into apoptotic bodies and activation of pro-caspase 3. Moreover, it was shown that the mitochondrial dysfunctions induced by SCFA in CRC cells are similar to those observed in yeast, and can also be partially inhibited by expression of anti-apoptotic members of the Bcl-2 protein family (Jan et al., 2002; Saraiva et al., 2006). Jan and co-workers determined that the adenine nucleotide transporter (ANT), a putative component of the mammalian PTPC, was a potential SCFA target. Likewise, AAC proteins, the yeast orthologs of ANT, are targets in the acetic acid-induced apoptosis pathway (Pereira et al., 2007). The observation that acetic acid triggers a mitochondrial apoptotic pathway in both yeast and CRC cells further supports the use of the yeast model system to provide insights towards enhanced understanding of the function of mitochondria in cell death. It would be interesting to assess whether acetic acid induces LMP and release of CatD in CRC cell lines. This may provide valuable insights into the enhanced understanding of the function of lysosomes in cell death and their crosstalk with mitochondria.

The peroxisome is another organelle that is involved in cell death in yeast (Jungwirth et al., 2008). Indeed, cells lacking *PEX6*, encoding a peroxisomal membrane protein involved in a key step of peroxisomal protein import, display an increased accumulation of reactive oxygen species and an enhanced loss of viability upon acetic acid treatment associated with markers of necrosis. Nevertheless, it remains to be elucidated whether this necrotic death is a regulated process.

4.4 Apoptotic signaling pathways

To date, few reports on the apoptotic-related signal transduction pathways have been published in yeast. Indeed, most studies regarding yeast apoptosis have encompassed mainly the identification of different apoptotic triggers and the components/regulators of apoptotic death. Several functional genetic analysis and pharmacological-based approaches allowed identifying components of yeast signalling cascades that, similarly to their mammalian counterparts, are engaged in conveying the information to the apoptotic apparatus. Not surprisingly, cell death signalling pathways in high eukaryotes are conserved and involved in the modulation of apoptosis in yeast.

The first evidence linking RAS/cAMP/cAMP-dependent protein kinase (PKA) pathway signalling to acetic acid-induced apoptosis in yeast came from the study by Phillips et al. (2006) with *Candida albicans*. Mutations that block Ras–cAMP–PKA signalling (*ras1*Δ, *cdc35*Δ, *tpk1*Δ, and *tpk2*Δ) suppress or delay the apoptotic response, whereas mutations that stimulate signalling (*RAS1*^{val13} and *pde2*Δ) accelerate the rate of entry into apoptosis. Consistently, pharmacological inhibition or stimulation of Ras signalling delay or promote apoptosis. Similar to the *C. albicans ras1* mutant, *RAS2* deletion is able to decrease cell death induced by acetic acid in *S. cerevisiae* (Ramsdale, 2006; Burtner et al., 2009). Transient mitochondrial hyperpolarization and ROS production (Ludovico et al., 2002) together with an increase in intracellular Ca^{2+} concentration (Pereira, C., Sousa, M.J. & Côrte-Real M., unpublished data) was found in response to acetic acid, pheromone and amiodarone. This indicates that these three stimuli converge into common death pathways. Moreover, since both pheromone- and acetic acid-induced apoptosis are inhibited by cycloheximide, it presumes transcriptional activation of target genes by upstream signalling cascades. Indeed, pheromone activates a calmodulin/calcineurin-controlled MAPK pathway (Severin et al., 2002) and the Ras pathway signals yeast apoptosis induced by harsh environmental stress, such as acetic acid (Phillips et al., 2003). Intracellular Ca^{2+} increase and ROS production are common features in these three scenarios, whereas intracellular acidification was only reported when death is induced by weak carboxylic acids, including acetic acid (Cardoso & Leão, 1992; Sokolov et al., 2006). Notably, death during chronological aging, measured as the loss of cell viability over time of stationary phase cells, is associated with acidification of the medium, and its neutralization by pH buffering compounds abolishes death. Sod2p and signalling through Ras–cAMP–PKA, including the transcription factors Msn2 and Msn4, were also shown to play a crucial role in the regulation of yeast chronological aging and death program (Fabrizio et al., 2004). Recently, a model for acetic acid as a cause of chronological aging has been proposed based on the observation that many modifiers of chronological life span also modulate acetic acid resistance. Most relevant, and besides Ras2p referred above, deletion of Sch9p kinase increases both CLS and acetic acid resistance in a Rim15p and Gis1p dependent manner (Burtner et al., 2009).

More recently, the combination of a proteomic approach using 2-DE and MS for the analysis of total cellular extracts, together with functional studies of *S. cerevisiae* cells treated with acetic acid, indicated that acetic acid causes severe intracellular amino-acid starvation, involving the general amino-acid control system as well as the TOR pathway (Almeida et al., 2009). Indeed, cells lacking Gcn4p/Gcn2p and Tor1p displayed a higher resistance to acetic acid, which in the latter case was associated with a TUNEL negative phenotype and lower ROS levels. In addition, cells lacking downstream mediators of the TOR pathway revealed that apoptotic signaling involves the phosphatases Pph21p and Pph22p but not Sit4p.

5. Conclusion

Acetic acid was early recognised as a common toxic agent present in different biotechnological processes associated with negative effects on the fermentative yeast. As mentioned above, knowledge on yeast stress mechanisms in response to acetic acid has already had an impact on the construction of industrial strains with improved performance in biotechnological processes (Zhang et al., 2011). However, most studies on stress response, including the latest on the evaluation of resistance to acetic acid at a genome-wide scale, have only been focused on the ability of yeast to divide and grow in the presence of toxic agents. Studying the effect of gene deletions on cell growth, however, does not provide the full picture of the determinants for stress resistance and survival under stress. In the last decade, acetic acid has been identified as an inducer of a programmed cell death (PCD) process. The recognition that different lethal stimuli, such as acetic acid, trigger a regulated death in yeast provides a new basis for future breeding strategies of industrial strains with improved cell survival. Thus far, the occurrence in yeast of a regulated cell death has not been exploited to control yeast performance in industrial processes. Moreover, no genome-wide studies have been performed until now regarding the elucidation of the mechanisms underlying acetic acid-induced PCD and of its regulatory pathways. Therefore, a complete picture of the main executors and regulators involved in acetic acid-induced PCD in yeast is still missing. A new genome-wide analysis could allow identifying genes involved in the execution of acetic acid induced-PCD and also of those involved in its regulation. This high-throughput approach will likely provide information on new putative targets for the control of acetic acid-induced PCD and ultimately will allow improving the performance of industrial yeast strains, and to design new strategies for food preservation by inhibiting or activating the PCD process, respectively. An overview of the current knowledge on targets and pathways underlying PCD induced by acetic acid is shown in Fig. 1.

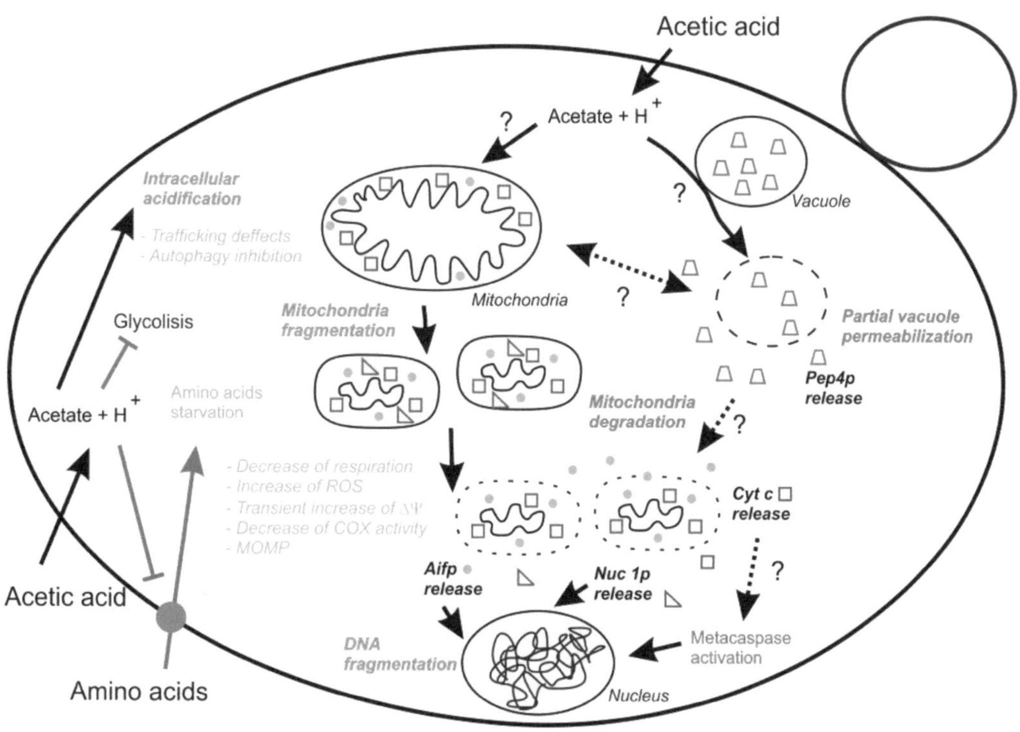

Fig. 1. Current knowledge on targets and pathways underlying PCD induced by acetic acid in *Saccharomyces cerevisiae*.

A major challenge in the future regarding stress and PCD in response to acetic acid will be to get a clearer picture on the different activated signalling pathways and of the crosstalk between them, and to comprehend how multiple cell targets and different types of damage lead the yeast cell to commit to a regulated form of suicide. As discussed above, different signalling pathways share several components, allowing the cell to convey the signal through a complex communication network. Regarding PCD induced by acetic acid, it was already shown that it targets mitochondria and the vacuole and probably other organelles, and induces different types of damage. It would be important to clarify how these organelles communicate and the different types of damage interact to ultimately trigger cell death, and whether they act in an independent, additive or synergistic manner. Understanding how the distinct stress signalling pathways communicate and how different types of death-associated damage contribute to cellular dysfunction in yeast is likely to prove informative about mammalian stress responses and PCD induced by acetic acid. Although acetic acid has been shown to target mitochondria and induce several damages in CRC cells, so far no stress signalling pathways have been identified.

The non-dissociated form of the acid enters glucose repressed cells by simple diffusion. Once inside the cell acetic acid dissociates and if the extracellular pH is lower than the intracellular pH this will lead to the accumulation of acetate and to the acidification of the intracellular environment. The effects promoted by these events are diverse but include inhibition of amino acid uptake and carbohydrate metabolism. Acetic acid also affects mitochondria function and dynamics triggering the release to the cytosol of mitochondria resident apoptogenic molecules such as cyt *c* and Aifp, the latter leading to DNA fragmentation. Partial permeabilization of the vacuolar membrane and release of Pep4p also occur during acetic acid-induced PCD. The link between vacuolar release of Pep4p and mitochondria degradation is still elusive. Once released, Pep4p could be directly or indirectly involved in the proteolytic degradation of mitochondria. However, mitochondria could be the first organelle affected by acetic acid, promoting a subsequent vacuole permeabilization followed by Pep4p release. Finally, acetic acid could act directly and simultaneously at the vacuole and mitochondria, which may further amplify the signal by a positive feedback mechanism. Broken arrows represent hypothetical pathways.

S. cerevisiae, the common baker's yeast, is a unicellular fungus of the ascomycete family. This unicellular eukaryote is one of the best established experimental model organisms and has been extensively used to unveil the mechanisms of eukaryotic cellular processes and their modes of regulation, including cell cycle control (McInerny, 2011), ageing, stress responses (Gasch & Werner-Washburne, 2002) and programmed cell death (Carmona-Gutierrez et al., 2010). More recently, it has been used as a valuable tool to understand the development of some human diseases such as degenerative disorders, including Huntington's and Parkinson's disease (Petranovic et al., 2008). The features that make this simple eukaryote a model system of choice are numerous: i) its cultivation is simple, fast and inexpensive; ii) it is a haplodiplont enabling sexual crossing and clonal division; iii) genetic manipulations, such as gene disruption, insertion or mutation are easy to perform, owing to the presence of a very efficient homologous recombination pathway; iv) heterologous expression of genes either from an episomal plasmid or from chromosomal integration can be achieved; v) the genome sequence and a collection of single deletion mutants for diploid cells and for non-essential genes of haploid cells is available. Moreover, *S. cerevisiae* is the organism with the most comprehensive datasets gathered from high-throughput data collected by functional genomic tools including transcriptome, proteome, metablome, interactome, locasome and flux analysis (see Petranovic et al., 2008). Moreover, there is a high degree of conservation between yeast and higher eukaryotes, with respect to numerous basic biological processes, and approximately half of the genes involved in human hereditary diseases are predicted to have yeast homologues (Hartwell, 2004). Nonetheless, it has to be mentioned that yeast also harbours some relevant differences comparatively with other organisms. For instance, yeast is highly specialized to grow on glucose medium and, unlike higher eukaryote cells, it easily forms respiratory deficient mutants which have little or none mitochondrial respiration. This ability to survive under conditions in which mitochondrial respiration is impaired or absent represents an ideal model, specifically with regard to the study of the involvement of mitochondrial proteins in PCD. Despite the vast differences in complexity between yeast and humans, the study of stress responses and PCD in yeast has provided insights into pathways that modulate response to stress and PCD in mammals. Thus, advances in the elucidation of cell death in yeast not only assume relevance in terms of biotechnological processes, but also in biomedicine. The studies that many yeast biologists have performed contributed to establish *S. cerevisiae* as an accepted model and will allow it to continue at the forefront of model organisms used to answer to many relevant biological questions.

6. Acknowledgments

This work was supported by Fundação para a Ciência e Tecnologia (FCT), Portugal grants FCOMP-01-0124-FEDER-007047 and PTDC/AGR-ALI/102608/2008.

7. References

Abbott, D.A., Suir, E., van Maris, A.J. & Pronk, J.T. (2008). Physiological and transcriptional responses to high concentrations of lactic acid in anaerobic chemostat cultures of *Saccharomyces cerevisiae*. *Appl Environ Microbiol*, Vol. 74, No. 18, pp. 5759-5768

Adrian, C. & Martin, S.J. (2001). The mitochondrial apoptosome: a killer unleashed by the cytochrome seas. *Trends Biochem Sci*, Vol. 26, pp. 390-397

Alexandre, H. & Charpentier, C. (1998). Biochemical aspects of stuck and sluggish fermentation in grape must. *J Ind Microbiol Biotechnol*, Vol. 20, pp. 20-27

Almeida, B., Ohlmeier, S., Almeida, A.J., Madeo, F., Leão C., Rodrigues, F. & Ludovico, P. (2009). Yeast protein expression profile during acetic acid-induced apoptosis indicates causal involvement of the TOR pathway. *Proteomics*, Vol. 9, pp. 720-732

Arneborg, N., Jesperson, L. & Jakobsen, M. (2000). Individual cells of *Saccharomyces cerevisiae* and *Zygosaccharomyces bailii* exhibit different short-term intracellular pH responses to acetic acid. *Arch Microbiol*, Vol. 174, pp. 125-128

Arneborg, N., Moos, M.K. & Jakobsen, M. (1995). The effect of acetic acid and specific growth rate on acetic acid tolerance and trehalose content of *Saccharomyces cerevisiae*. *Biotechnol Lett*, Vol. 17, pp. 1299-1304

Arneborg, N., Moos, M.K. & Jakobsen, M. (1997). Induction of acetic acid tolerance and trehalose accumulation by added and produced ethanol in *Saccharomyces cerevisiae*. *Biotechnol Lett*, Vol. 19, No 9, pp. 931-933

Bauer, B.E., Rossington, D., Mollapour, M., Mamnun, Y., Kuchler, K. & Piper, P.W. (2003). Weak organic acid stress inhibits aromatic amino acid uptake by yeast, causing a strong influence of amino acid auxotrophies on the phenotypes of membrane transporter mutants. *Eur J Biochem*, Vol. 270, pp. 3189–3195

Bely, M., Rinaldi, A. & Dubourdieu, D. (2003). Influence of assimilable nitrogen on volatile acidity production by *Saccharomyces cerevisiae* during high sugar fermentation. *J Biosci Bioeng*, Vol. 96, pp. 507-512

Benes, P., Vetvicka, V. & Fusek, M. (2008). Cathepsin D - Many functions of one aspartic protease. *Crit Rev Oncol Hematol*, Vol. 68, pp. 12-28

Boya, P. & Kroemer G. (2008). Lysosomal membrane permeabilization in cell death. *Oncogene*, Vol. 27, pp. 6434-6451

Brett, C.L., Tukaye, D.N., Mukherjee, S. & Rao, R. (2005). The yeast endosomal Na+(K+)/H+ exchanger Nhx1 regulates cellular pH to control vesicle trafficking. *Mol Biol Cell*, Vol. 16, pp. 1396–1405

Bussey, H. (1988). Proteases and the processing of precursors to secreted proteins in yeast. *Yeast*, Vol. 4, pp. 17–26

Burtner, C.R., Murakami, C.J., Kennedy, B.K. & Kaeberlein, M. (2009). A molecular mechanism of chronological aging in yeast. *Cell Cycle*, Vol. 8, pp. 256-270

Buttner, S., Eisenberg, T., Carmona-Gutierrez, D., Ruli, D., Knauer, H., Ruckenstuhl, C., Sigrist, C., Wissing, S., Kollroser, M., Fröhlich, K.U., Sigrist, S. & Madeo, F. (2007). Endonuclease G regulates budding yeast life and death. *Mol Cell*, Vol. 25, pp 233–246

Cai, J. & Jones, D.P. (1998). Superoxide in apoptosis. Mitochondrial generation triggered by cytochrome c loss. *J Biol Chem*, Vol. 273, pp. 11401–11404

Cardoso, H. & Leão, C. (1992). Mechanisms underlying low and high enthalpy death induced by short-chain monocarboxylic acids in *Saccharomyces cerevisiae*. *Appl Microbiol Biotechnol*, Vol. 38, pp. 388-392

Carmelo, V., Santos, H. & Sá-Correia, I. (1997). Effect of extracellular acidification on the activity of plasma membrane ATPase and on the cytosolic and vacuolar pH of *Saccharomyces cerevisiae*. *Biochim Biophys Acta*, Vol. 1325, pp. 63-70

Carmona-Gutierrez, D., Eisenberg, T., Büttner, S., Meisinger, C., Kroemer, G. & Madeo, F. (2010). Apoptosis in yeast: triggers, pathways, subroutines. *Cell Death Differ*, Vol. 17, pp. 763-773

Carmona-Gutierrez, D., Bauer, M.A., Ring, J., Knauer, H., Eisenberg, T., Büttner, S., Ruckenstuhl, C., Reisenbichler, A., Magnes, C., Rechberger, G.N., Birner-Gruenberger, R., Jungwirth, H., Fröhlich, K.U., Sinner, F., Kroemer, G. & Madeo, F. (2011). The propeptide of yeast cathepsin D inhibits programmed necrosis. *Cell Death Dis*, 2, e161; doi:10.1038/cddis.2011.43

Casal, M. & Leão, C. (1995). Utilization of short-chain monocarboxylic acids by the yeast *Torulaspora delbrueckii*: specificity of the transport systems and their regulation. *Biochim Biophys Acta*, Vol. 1267, pp. 122-130

Casal, M., Cardoso, H. & Leão, C. (1996). Mechanisms regulating the transport of acetic acid in *Saccharomyces cerevisiae*. *Microbiology*, Vol. 142, pp. 1385-1390

Cássio, F., Côrte-Real, M. & Leão, C. (1993). Quantitative analysis of proton movements associated with the uptake of weak-carboxylic acids. The yeast *Candida utilis* as a model. *Biochim Biophys Acta*, Vol. 1153, pp. 59-66

Cássio, F., Leão, C. & van Uden, N. (1987). Transport of lactate and other short-chain monocarboxylates in the yeast *Saccharomyces cerevisiae*. *Appl Environ Microbiol*, Vol. 53, pp. 509–513

Cheng, W.C., Teng, X., Park, H.K., Tucker, C.M., Dunham, M.J. & Hardwick JM. (2008). Fis1 deficiency selects for compensatory mutations responsible for cell death and growth control defects. *Cell Death Differ*, Vol. 15, pp. 1838–1846

Chipuk, J.E., Moldoveanu, T., Llambi, F., Parsons, M.J. & Green, D.R. (2010). The Bcl-2 family reunion. *Mol Cell*, Vol. 37, pp. 299-310

Delavallée, L., Cabon, L., Galán-Malo, P., Lorenzo, H.K. & Susin, S.A. (2011). AIF-mediated caspase-independent necroptosis: a new chance for targeted therapeutics. *IUBMB Life*, Vol. 63, pp. 221-232

Du Toit, W.J. & Lambrechts, M.G. (2002). The enumeration and identification of acetic acid bacteria from south african red wine fermentations. *Int J Food Microbiol*, Vol. 74, pp. 57-64

Fabrizio, P., Pletcher, S.D. Minois, N., Vaupel, J.W. & Longo, V.D. (2004). Chronological aging independent replicative life span regulation by Msn2/Msn4 and Sod2 in *Saccharomyces cerevisiae*. *FEBS Lett*, Vol. 557, pp. 136–142

Fannjiang, Y., Cheng, W.C., Lee, S.J., Qi, B., Pevsner, J., McCaffery, J.M., Hill, R.B., Basanez, G. & Hardwick, J.M. (2004). Mitochondrial fission proteins regulate programmed cell death in yeast. *Genes Dev*, Vol. 18, pp. 2785-2797

Fernandes, A.R., Mira, N.P., Vargas, R.C., Canelhas, I. & Sá-Correia, I. (2005). *Saccharomyces cerevisiae* adaptation to weak acids involves the transcription factor Haa1p and Haa1p-regulated genes. *Biochem Biophys Res Comm*, Vol. 337, pp. 95–103

Fernandes, L., Côrte-Real, M., Loureiro, V., Loureiro-Dias, M.C. & Leão, C. (1997). Glucose respiration and fermentation in *Zygosaccharomyces bailii* and *Saccharomyces cerevisiae* express different sensitivity patterns to ethanol and acetic acid. *Lett Appl Microbiol*, Vol. 25, No. 4, pp. 249-253

Fernandes, L., Côrte-Real, M. and Leão, C. (1999). A peculiar behaviour for cell death induced by weak carboxylic acids in the wine spoilage yeast *Zygosaccharomyces bailii*. *Lett Appl Microbiol*, Vol. 28, pp. 345-349

Frank, S., Gaume, B., Bergmann-Leitner, E.S., Leitner, W.W., Robert, E.G., Catez, F., Smith, C.L. & Youle, R.J. (2001). The role of dynamin-related protein 1, a mediator of mitochondrial fission, in apoptosis. *Dev Cell*, Vol 1, pp. 515-525

Fuller, R.S., Sterne, R.E. & Thorner, J. (1988). Enzymes required for yeast pheromone processing. *Annu Rev Physiol*, Vol. 50, pp. 345–362

Gasch, A.P. & Werner-Washburne, M. (2002). The genomics of yeast responses to environmental stress and starvation. *Funct Integr Genomics*, Vol. 2, pp. 181-192

Gasch, A.P., Spellman, P.T., Kao, C.M., Carmel-Harel, O., Eisen, M.B., Storz, G., Botstein, D. & Brown, P.O. (2000). Genomic expression programs in the response of yeast cells to environmental changes. *Mol Biol Cell*, Vol. 11, pp. 4241-4257

Gewies, A. (2003). *ApoReview - Introduction to Apoptosis*, retrieved from www.celldeath.de/encyclo/aporev/aporev.htm

Gomes, P., Sampaio-Marques, B., Ludovico, P., Rodrigues, F. & Leão, C. (2007). Low auxotrophy-complementing amino acid concentrations reduce yeast chronological life span. *Mech Ageing Dev*, Vol. 128, pp. 383-391

Gourlay, C.W. & Ayscough, K.R. (2006). Actin-induced hyperactivation of the Ras signaling pathway leads to apoptosis in *Saccharomyces cerevisiae*. *Mol Cell Biol*, Vol. 26, pp. 6487-501

Green, D.R. & Evan, G.I. (2002). A matter of life and death. *Cancer Cell*, Vol. 1, pp. 19-30

Guaragnella, N., Antonacci, L., Giannattasio, S., Marra, E. & Passarella, S. (2008). Catalase T and Cu, Zn-superoxide in the acetic acid-induced programmed cell death in *Saccharomyces cerevisiae*. *FEBS Lett*, Vol. 582, pp. 210–214

Guaragnella, N., Antonacci, L., Passarella, S., Marra, E. & Giannattasio, S. (2007). Hydrogen peroxide and superoxide anion production during acetic acid-induced yeast programmed cell death. *Folia Micrbiol*, Vol. 52, pp. 237–240

Guaragnella, N., Antonacci, L., Passarella, S., Marra, E. & Giannattasio, S. (2010a). Knock-out of metacaspase and/or cytochrome *c* results in the activation of a ROS-independent acetic acid-induced programmed cell death pathway in yeast. *FEBS Lett*, Vol. 584, pp. 3655–3660

Guaragnella, N., Bobba, A., Passarella, S., Marra, E. & Giannattasio, S. (2010b). Yeast acetic acid-induced programmed cell death can occur without cytochrome *c* release which requires metacaspase *YCA1*. *FEBS Lett*, Vol. 584, pp. 224–228

Guaragnella, N., Pereira, C., Sousa, M.J., Antonacci, L., Passarella, S., Côrte-Real, M., Marra, E. & Giannattasio, S. (2006). YCA1 participates in the acetic acid induced yeast programmed cell death also in a manner unrelated to its caspase-like activity. *FEBS Lett*, Vol. 580, pp. 6880–6884

Guldfeldt, L.U. & Arneborg, N. (1998). Measurement of the Effects of acetic acid and extracellular pH on intracellular pH of nonfermenting individual *Saccharomyces cerevisiae* cells by fluorescence microscopy. *Appl Environ Microbiol*, Vol. 64, pp. 530–534

Hartwell, L.H. (2004). Yeast and cancer, *Biosci. Rep*. Vol. 24 523–544

Hauptmann, P., Riel, C., Kunz-Schughart, L.A., Frohlich, K.U, Madeo, F. & Lehle, L. (2006). Defects in N-glycosylation induce apoptosis in yeast. *Mol Microbiol*, Vol. 59, pp. 765-778

Hauptmann, P. & Lehle, L. (2008). Kex1 protease is involved in yeast cell death induced by defective N-glycosylation, acetic acid, and chronological aging. *J Biol Chem*, Vol. 283, pp. 19151-19163

Hengartner, M.O. (2000). The biochemistry of apoptosis. *Nature*, Vol. 407, pp. 770–776

Herker, E., Jungwirth, H., Lehmann, K.A., Maldener, C., Frohlich, K.U., Wissing, S. Büttner, S., Fehr, M., Sigrist, S. & Madeo, F. (2004). Chronological aging leads to apoptosis in yeast. *J Cell Biol*, Vol. 164, pp. 501–507

James, D.I., Parone, P.A., Mattenberger, Y. & Martinou, J.C. (2003). hFis1, a novel component of the mammalian mitochondrial fission machinery. *J Biol Chem*, Vol. 278, pp. 36373-36379

Jan, G., Belzacq, A.S., Haouzi, D., Rouault, A., Métivier, D., Kroemer, G. & Brenner, C. (2002). Propionibacteria induce apoptosis of colorectal carcinoma cells via short-chain fatty acids acting on mitochondria. *Cell Death Differ*, Vol. 9, pp. 179-88

Johansson, A.C., Appelqvist, H., Nilsson, C., Kågedal, K., Roberg, K. & Öllinger, K. (2010). Regulation of apoptosis-associated lysosomal membrane permeabilization. *Apoptosis*, Vol. 15, pp. 527-540

Johansson, A.C., Steen, H., Öllinger, K. & Roberg, K. (2003). Cathepsin D mediates cytochrome *c* release and caspase activation in human fibroblast apoptosis induced by staurosporine. *Cell Death Differ*, Vol. 10, pp. 1253-1259

Jungwirth, H., Ring, J., Mayer, T., Schauer, A., Buttner, S., Eisenberg, T., Carmona-Gutíerrez, D., Kuchler, K. & Madeo, F. (2008) Loss of peroxisome function triggers necrosis. *FEBS Lett*, Vol. 582, pp. 2882–2886

Kawahata, M., Masaki, K., Fujii, T. & Iefuji, H. (2006). Yeast genes involved in response to lactic acid and acetic acid: acidic conditions caused by the organic acids in *Saccharomyces cerevisiae* cultures induce expression of intracellular metal metabolism genes regulated byAft1p. *FEMS Yeast Res*, Vol. 6, pp. 924–936

Kerr, J.F., Wyllie, A.H. & Currie, A.R. (1972). Apoptosis: a basic biological phenomenon with wide-ranging implications in tissue kinetics. *Br J Cancer*, Vol. 26, pp. 239-257

Kim, I., Rodriguez-Enriquez, S. & Lemasters, J.J. (2007). Selective degradation of mitochondria by mitophagy. *Arch Biochem Biophys*, Vol. 462, pp. 245-253

Kim, M., Ahn J.W., Jin, U.H., Choi, D., Paek, K.H. & Pai, H.S. (2003). Activation of the programmed cell death pathway by inhibition of proteasome function in plants. *J Biol Chem*, Vol. 278, pp. 19406–19415

Kinnally, K.W., Peixoto, P.M., Ryu, S.Y. & Dejean, L.M. (2011). Is mPTP the gatekeeper for necrosis, apoptosis, or both? *Biochim Biophys Acta*, Vol. 1813, No. 4, pp. 616-622

Kirkegaard, T. & Jäättelä, M. (2008). Lysosomal involvement in cell death and cancer. *Biochim Biophys Acta*, Vol. 1793, pp. 746-754

Kissova, I., Plamondon, L.T. Brisson, L., Priault, M., Renouf, V., Schaeffer, J., Camougrand, N. & Manon, S. (2006). Evaluation of the roles of apoptosis, autophagy, and mitophagy in the loss of plating efficiency induced by Bax expression in yeast. *J Biol Chem*, Vol. 281, pp. 36187-36197

Kissova, I., Salin, B., Schaeffer, J., Bhati, S., Manon, S. & Camougrand, N. (2007). Selective and non-selective autophagic degradation of mitochondria in yeast. *Autophagy*, Vol. 3, pp. 329-336

Kitagaki, H., Araki, Y., Funato, K. & Shimoi, H. (2007). Ethanol-induced death in yeast exhibits features of apoptosis mediated by mitochondrial fission pathway. *FEBS Lett*, Vol. 581, pp. 2935-2942

Kroemer, G., Galluzzi, L., Vandenabeele, P., Abrams, J., Alnemri, E.S., Baehrecke, E.H., Blagosklonny, M.V., El-Deiry, W.S., Golstein, P., Green, D.R., Hengartner, D.R., Knight, R.A., Kumar, S., Lipton, S.A., Malorni, W., Nunez, G., Peter, M.E., Tschopp, J., Yuan, J., Piacentini, M., Zhivotovsky, B. & Melino, G. (2009). Classification of cell death: recommendations of the Nomenclature Committee on Cell Death 2009. *Cell Death Differ*, Vol. 16, pp. 3-11

Lan, A., Lagadic-Gossmann, D., Lemaire, C., Brenner, C. & Jan, G. (2007). Acidic extracellular pH shifts colorectal cancer cell death from apoptosis to necrosis upon exposure to propionate and acetate, major end-products of the human probiotic propionibacteria. *Apoptosis*, Vol. 12, pp. 573-591

Lawrence, C.L., Botting, C.H., Antrobus, R. & Coote, P.J. (2004). Evidence of a new role for the high-osmolarity glycerol mitogen-activated protein kinase pathway in yeast: regulating adaptation to citric acid stress. *Mol Cell Biol*, Vol. 24, pp. 3307–3323

Leão, C. & van Uden, N. (1986). Transport of lactate and other monocarboxylates in the yeast *Candida utilis*. *Appl Microbiol Biotechnol*, Vol. 23, pp. 389-393

Lee, Y.Y., Iyer, P. & Torget, R.W. (1999). Dilute-acid hydrolysis of lignocellulosic biomass. *Adv Biochem Eng Biotechnol*, Vol. 65, pp. 93–115

Lemasters, J.J. (2005). Selective mitochondrial autophagy, or mitophagy, as a targeted defense against oxidative stress, mitochondrial dysfunction, and aging. *Rejuvenation Res*, Vol. 8, pp. 3-5

Li, B.-Z. & Yuan, Y.-J. (2010). Transcriptome shifts in response to furfural and acetic acid in *Saccharomyces cerevisiae*. *Appl Microbiol Biotechnol*, Vol. 86, pp. 1915–1924

Ligr, M., Madeo, F., Frohlich, E., Hilt, W., Frohlich, K.U., Wolf, D.H. (1998). Mammalian Bax triggers apoptotic changes in yeast. *FEBS Lett*, Vol. 438, pp 61-65

Ludovico, P. (1999). Efeitos do ácido acético no potencial de membrana mitocondrial e sua relação com a perda de integridade e viabilidade celular em *Zygosaccharomyces bailii* e *Saccharomyces cerevisiae*. Estudos por citometria de fluxo e espectrofluorimetria. Tese de Mestrado, Universidade do Minho

Ludovico, P., Sansonetty, F., Silva, M.T. & Côrte-Real, M.. (2003). Acetic acid induces a programmed cell death process in the food spoilage yeast *Zygosaccharomyces bailii*, *FEMS Yeast Res*, Vol. 3, pp. 91–96

Ludovico, P., Rodrigues, F., Almeida, A., Silva, M.T., Barrientos, A. & Côrte-Real, M. (2002). Cytochrome *c* release and mitochondria involvement in programmed cell death induced by acetic acid in *Saccharomyces cerevisiae*. *Mol Biol Cell*,Vol. 13, pp. 2598-2606

Ludovico, P., Sousa, M.J., Silva M.T., Leão, C. & Côrte-Real, M. (2001). *Saccharomyces cerevisiae* commits to a programmed cell death process in response to acetic acid. *Microbiology*, Vol. 147, pp. 2409-2415

Madeo, F., Frohlich, E. & Frohlich, K.U. (1997). A yeast mutant showing diagnostic markers of early and late apoptosis. *J Cell Biol*, Vol. 139, pp. 729-734

Madeo, F., Frohlich, E., Ligr, M., Grey, M., Sigrist, S.J., Wolf, D.H. & Frohlich, K.U. (1999). Oxygen stress: a regulator of apoptosis in yeast. *J Cell Biol*, Vol. 145, pp. 757-767

Madeo, F., Herker, E., Maldener, C., Wissing, S., Lächelt, S., Herlan, M., Fehr, M., Lauber, K., Sigrist, S.J., Wesselborg, S. & Fröhlich, K.U. (2002). A caspase related protease regulates apoptosis in yeast. *Mol Cell*, Vol. 9, pp. 911–917

Maiorella, B., Blanch, H.W. & Wilke, C.R. (1983). By-product inhibition effects on ethanolic fermentation by *Saccharomyces cerevisiae*. *Biotechnol Bioeng*, Vol. 25, pp. 103–121

McInerny, C.J. (2011), Cell cycle regulated gene expression in yeasts. Adv Genet, Vol. 73, pp. 51-85

Mason, D.A., Shulga, N., Undavai, S., Ferrando-May, E., Rexach, M.F. & Goldfarb, D.S. (2005). Increased nuclear envelope permeability and Pep4p-dependent degradation of nucleoporins during hydrogen peroxide-induced cell death. *FEMS Yeast Res*, Vol. 5, pp. 1237-1251

Masson, O., Bach A.S., Derocq, D., Prébois, C., Laurent-Matha, V., Pattingre, S. & Liaudet-Coopman, E. (2010). Pathophysiological functions of cathepsin D: targeting its catalytic activity versus its protein binding activity? *Biochemie*, Vol. 92, pp. 1635-1643

Matsui, M., Yamamoto, A., Kuma, A., Ohsumi, Y. & Mizushima, N. (2006). Organelle degradation during the lens and erythroid differentiation is independent of autophagy. *Biochem Biophys Res Commun*, Vol. 339, pp. 485–489

Matsuyama, S., Llopis, J., Deveraux, Q.L., Tsien, R. & Reed, J.C. (2000). Changes in mitochondrial and cytosolic pH: early events that modulate caspase activation during apoptosis. *Nat Cell Biol,* Vol. 2, pp. 318–325

Mira, N.,P., Lourenço, A.B., Fernandes, A.R., Becker, J.D. & Sá-Correia, I. (2009). The RIM101 pathway has a role in *Saccharomyces cerevisiae* adaptive response and resistance to propionic acid and other weak acids. *FEMS Yeast Res,* Vol. 9, No. 2, pp. 202-216

Mira, N.P., Palmam, M., Guerreiro, J.F. & Sá-Correia, I. (2010). Genome-wide identification of *Saccharomyces cerevisiae* genes required for tolerance to acetic acid. *Microb Cell Fact,* Vol. 9, pp. 79

Mollapour, M. & Piper, P.W. (2006). Hog1p mitogen-activated protein kinase determines acetic acid resistance in *Saccharomyces cerevisiae. FEMS Yeast Res,* Vol. 6, No. 8, pp. 1274-1280

Mollapour, M. & Piper, P.W. (2007). Hog1 mitogen-activated protein kinase phosphorylation targets the yeast Fps1 aquaglyceroporin for endocytosis, thereby rendering cells resistant to acetic acid. *Mol Cell Biol,* Vol. 27, pp. 6446–6456

Mollapour, M., Fong, D., Balakrishnan, K., Harris, N., Thompson, S., Schuller, C., Kuchler, K. & Piper, P.W. (2004). Screening the yeast deletant mutant collection for hypersensitivity and hyperresistance to sorbate, a weak organic acid food preservative. *Yeast,* Vol. 21, pp. 927–946

Mollapour, M., Shepherd, A. & Piper, P.W. (2009). Presence of the Fps1p aquaglyceroporin channel is essential for Hog1p activation, but suppresses Slt2(Mpk1)p activation, with acetic acid stress of yeast. *Microbiology,* Vol. 155, pp. 3304–3311

Mozdy, A.D., McCaffery, J.M. & Shaw, J.M. (2000). Dnm1p GTPase-mediated mitochondrial fission is a multi-step process requiring the novel integral membrane component Fis1p. *J Cell Biol,* Vol. 151, pp. 367-380

Nowikovsky, K., Reipert, S., Devenish, R.J. & Schweyen, R.J. (2007). Mdm38 protein depletion causes loss of mitochondrial K+/H+ exchange activity, osmotic swelling and mitophagy. *Cell Death Differ,* Vol. 14, pp. 1647-1656

Palmqvist, E. & Hahn-Hägerdal, B. (2000). Fermentation of lignocellulosic hydrolysates. I: inhibition and detoxification. *Bioresour Technol,* Vol. 74, pp. 17-24

Pampulha, M.A. & Loureiro-Dias, M.C. (1989). Combined effect of acetic acid, pH and ethanol on intracellular pH of fermenting yeast. *Appl Microbiol Biotechnol,* Vol. 31, pp. 547–550

Pampulha, M.A. & Loureiro-Dias, M.C. (1990). Activity of glycolytic enzymes of *Saccharomyces cerevisiae* in the presence of acetic acid. *Appl Microbiol Biotechnol,* Vol. 34, pp. 375–380

Pampulha, M.A. & Loureiro-Dias, M.C. (2000). Energetics of the effect of acetic acid on growth of *Saccharomyces cerevisiae. FEMS Microbiol Lett,* Vol. 184, pp. 69–72

Parone, P.A. & Martinou, J.C. (2006). Mitochondrial fission and apoptosis: an ongoing trial, *Biochim Biophys Acta,* Vol. 1763, pp. 522-530

Pereira, C., Camougrand, N., Manon, S., Sousa, M.J. & Côrte-Real, M. (2007). ADP/ATP carrier is required for mitochondrial outer membrane permeabilization and cytochrome *c* release in yeast apoptosis. *Mol Microbiol,* Vol. 66, pp. 571-582

Pereira, C., Silva, R.D., Saraiva, L., Johansson, B., Sousa, M.J. & Côrte-Real, M. (2008). Mitochondria dependent apoptosis in yeast. *Biochim Biophys Acta,* Vol. 1783, 1286-1302

Pereira, C., Chaves, S., Alves, S., Salin, B., Camougrand, N., Manon, S., Sousa, M.J. & Côrte-Real, M. (2010). Mitochondrial degradation in acetic acid-induced yeast apoptosis: The role of Pep4 and the ADP/ATP carrier. *Mol Microbiol*, Vol. 76, pp. 1398-1410

Petranovic, D. & Nielsen, J. (2008). Can yeast systems biology contribute to the understanding of human disease? *Trends Biotechnol*, Vol. 26, pp. 584-590

Phillips, A.J., Crowe, J.D. & Ramsdale, M. (2006). Ras pathway signaling accelerates programmed cell death in the pathogenic fungus *Candida albicans. Proc Natl Acad Sci USA*, Vol. 103, pp. 726–731

Phillips, A.J., Sudbery, I. & Ramsdale, M. (2003). Apoptosis induced by environmental stresses and amphotericin B in *Candida albicans. Proc Natl Acad Sci* U S A, Vol. 100, 25 pp. 14327-14332

Phowchinda, O., Délia-Dupuy, M.L. & Strehaiano, P. (1995). Effects of acetic acid on growth and fermentative activity of *Saccharomyces cerevisiae. Biotechnol Lett*, Vol. 17, pp. 237–242

Pinto, I., Cardoso, H. & Leão, C. (1989). High enthalpy and low enthalpy death in *Saccharomyces cerevisiae* induced by acetic acid. *Biotechnol Bioeng*, Vol. 33, pp. 1350-1352

Piper, P., Mahé, Y., Thompson, S., Pandjaitan, R., Holyoak, C., Egner, R., Mühlbauer, M., Coote, P. & Kuchler, K. (1998). The Pdr12 ABC transporter is required for the development of weak organic acid resistance in yeast. *EMBO J*, Vol. 17, pp. 4257-4265

Pozniakovsky, A.I., Knorre, D.A., Markova, O.V., Hyman, A.A, Skulachev, V.P. & Severin, F.F. (2005). Role of mitochondria in the pheromone- and amiodarone-induced programmed death of yeast. *J Cell Biol*, Vol. 168, pp. 257-269

Priault, M., Salin, B., Schaeffer, J., Vallette, F.M., di Rago, J.P. & Martinou, J.C. (2005). Impairing the bioenergetic status and the biogenesis of mitochondria triggers mitophagy in yeast. *Cell Death Differ*, Vol. 12, pp. 1613–1621

Prudêncio, C., Sansonetty, F. & Côrte-Real, M. (1998). Flow cytometric assessment of cell structural and functional changes induced by acetic acid in the yeasts *Zygosaccharomyces bailii* and *Saccharomyces cerevisiae. Cytometry*, Vol. 31, pp. 307-313

Ramsdale, M. (2006). Programmed Cell Death and Apoptosis in Fungi, In: *The Mycota XIII, Fungal Genomics,* Alistair J.P. Brown, pp. 113-146, Springer-Verlag, Berlin Heidelberg

Repnik, U., & Turk, B. (2010). Lysosomal-mitochondrial cross-talk during cell death. *Mitochondrion*, Vol. 10, pp. 662-669

Ribeiro, G.F., Côrte-Real, M. & Johansson, B. (2006). Characterization of DNA damage in yeast apoptosis induced by hydrogen peroxide, acetic acid, and hyperosmotic shock. *Mol Biol Cell*, Vol. 17, pp. 4584-4591

Rodrigues, F., Côrte-Real, M., Leão, C., van Dijken, J.P. & Pronk, J.T. (2001). Oxygen requirements of the food spoilage yeast *Zygosaccharomyces bailii* in synthetic and complex media. *Appl Environ Microbiol*, Vol. 67, pp. 2123-2128

Rodrigues, F., Ludovico, P. & Leão, C. (2005). Sugar Metabolism in Yeasts: an Overview of Aerobic and Anaerobic Glucose Catabolism. In: *Biodiversity and Ecophysiology of Yeasts,* Carlos A. Rosa & Gábor Péter, pp. 101-121, Springer Springer Lab Manuals, Germany

Rodriguez-Enriquez, S., Kim, I., Currin, R.T., & Lemasters, J.J. (2006) Tracker dyes to probe mitochondrial autophagy (mitophagy) in rat hepatocytes. *Autophagy* Vol. 2, pp. 39-46

Roset, R., Ortet, L. & Gil-Gomez, G. (2007). Role of Bcl-2 family members on apoptosis: what we have learned from knock-out mice. *Front Biosci,* Vol. 12, pp. 4722-4730

Sagulenko, V., Muth, D., Sagulenko, E., Paffhausen, T., Schwab, M. & Westermann, F. (2008). Cathepsin D protects human neuroblastoma cells from doxorubicin-induced cell death. *Carcinogenesis,* Vol. 29, pp. 1869-1877

Santos, J., Sousa, M.J., Cardoso, H., Inácio, J., Silva, S., Spencer-Martins, I. & Leão, C. (2008). Ethanol tolerance of sugar transport and the rectification of stuck wine fermentations. *Microbiology,* Vol. 154, pp. 422-430

Saraiva L., Silva R.D., Pereira G., Gonçalves J. & Côrte-Real M. (2006). Specific modulation of apoptosis and Bcl-xL phosphorylation in yeast by distinct mammalian protein kinase C isoforms. *J Cell Sci,* Vol. 119, pp. 3171-3181

Schauer, A., Knauer, H., Ruckenstuhl, C., Fussi, H., Durchschlag, M., Potocnik, U. & Frohlich, K.U. (2009). Vacuolar functions determine the mode of cell death. *Biochim Biophys Acta,* Vol. 1793, pp. 540-545

Schuller, C., Mamnun, Y.M., Mollapour, M., Krapf, G., Schuster, M., Bauer, B.E., Piper, P.W. & Kuchler, K. (2004). Global phenotypic analysis and transcriptional profiling defines the weak acid stress response regulon in *Saccharomyces cerevisiae. Mol Biol Cell,* Vol. 15, pp. 706-720

Scorrano, L. (2005). Proteins that fuse and fragment mitochondria in apoptosis: con-fissing a deadly con-fusion? *J Bioenerg Biomembr,* Vol. 37, pp. 165-170

Severin, F.F. & Hyman, A.A. (2002). Pheromone induces programmed cell death in *S. cerevisiae. Curr Biol,* Vol. 12, pp. 233-235

Shinohara, K., Tomioka, M., Nakano, H., Tone, S., Ito, H. & Kawashima, S. (1996). Apoptosis induction resulting from proteasome inhibition. *Biochem J,* Vol. 317, pp. 385-388

Silva, R.D., Sotoca, R., Johansson, B., Ludovico, P., Sansonetty, F., Silva, M.T., Peinado, J.M. & Côrte-Real, M. (2005). Hyperosmotic stress induces metacaspase- and mitochondria-dependent apoptosis in *Saccharomyces cerevisiae. Mol Microbiol,* Vol. 58, pp. 824-834

Sokolov, S., Knorre, D., Smirnova, E., Markova, O., Pozniakovsky, A., Skulachev, V. & Severin, F. (2006). Ysp2 mediates death of yeast induced by amiodarone or intracellular acidification, *Biochim Biophys Acta,* Vol. 1757, pp. 1366-1370

Sousa, M.J., Miranda, L., Côrte-Real, M. & Leão, C. (1996). Transport of acetic acid in *Zygosaccharomyces bailii*: effects of ethanol and their implications on the resistance of the yeast to acid environments. *Appl Environ Microbiol,* Vol. 62, pp. 3152-3157

Sousa, M.J., Rodrigues, F., Côrte-Real, M. & Leão, C. (1998). Mechanisms underlying the transport and intracellular metabolism of acetic acid in the presence of glucose by the yeast *Zygosaccharomyces bailii. Microbiology,* Vol. 144, pp. 665-670

Susin, S.A., Lorenzo, H.K., Zamzami, N., Marzo, I., Snow, B.E., Brothers, G.M., Mangion, J., Jacotot, E., Costantini, P., Loeffler, M., Larochette, N., Goodlett, D.R., Aebersold, R., Siderovski, D.P., Penninger, J.M. & Kroemer, G. (1999). Molecular characterization of mitochondrial apoptosis-inducing factor. *Nature,* Vol. 397, pp. 441-446

Tal, R., Winter, G., Ecker, N., Klionsky, D.J. & Abeliovich, H. (2007). Aup1p, a yeast mitochondrial protein phosphatase homolog, is required for efficient stationary phase mitophagy and cell survival. *J Biol Chem*, Vol. 282, pp. 5617-5624

Thomas, S. & Davenport, R.R. (1985). *Zygosaccharomyces bailii*, a profile of characteristics and spoilage activities. *Food Microbiol*, Vol. 2, pp. 157–169

Tolkovsky, A.M., Xue, L., Fletcher, G.C. & Borutaite, V. (2002). Mitochondrial disappearance from cells: A clue to the role of autophagy in programmed cell death and disease? *Biochimie,* Vol. 84, pp. 233-240

Vahsen, N., Cande, C., Briere, J.J., Benit, P., Joza, N., Larochette, N., Mastroberardino, P.G., Pequignot, M.O., Casares, N., Lazar, V., Feraud, O., Debili, N., Wissing, S., Engelhardt, S., Madeo, F., Piacentini, M., Penninger, J.M., Schagger, H., Rustin, P. & Kroemer, G. (2004). AIF deficiency compromises oxidative phosphorylation. *EMBO J*, Vol. 23, pp. 4679–4689

Valenti, D., Vacca, R.A., Guaragnella, N., Passarella, S., Marra, E. & Giannattasio, S. (2008). A transient proteasome activation is needed for acetic acid-induced programmed cell death to occur in *Saccharomyces cerevisiae*. *FEMS Yeast Res*, Vol. 8, pp. 400-404

Valle, E., Bergillos, L., Gascon, S., Parra, F. & Ramos, S. (1986). Trehalase activation in yeasts is mediated by an internal acidification. *Eur J Biochem*, Vol. 154, pp. 247–251

van Uden, N. (1984). Temperature profiles of yeasts. *Adv Microb Physiol*, Vol. 25, 195-251

Vilela-Moura, A., Schuller, D., Mendes-Faia, A., Silva, R.D., Chaves, S.R., Sousa, M.J. & Côrte-Real, M. (2011). The impact of acetate metabolism on yeast fermentative performance and wine quality: reduction of volatile acidity of grape musts and wines. *Appl Microbiol Biotechnol*, Vol. 89, pp. 271–280

Wei, M.C., Zong, W.X., Cheng, EH., Lindsten, T., Panoutsakopoulou, V., Ross, A.J., Roth, K.A., MacGregor, G.R., Thompson, C.B. & Korsmeyer, S.J. (2001). Proapoptotic BAX and BAK: a requisite gateway to mitochondrial dysfunction and death. *Science*, Vol. 292, pp. 727-730

Wissing, S., Ludovico, P., Herker, E., Buttner, S., Engelhardt, S.M., Decker, T., Link, A., Proksch, A., Rodrigues, F., Côrte-Real, M., Frohlich, K.U., Manns, J., Cande, C., Sigrist, S.J., Kroemer, G. & Madeo, F. (2004). An AIF orthologue regulates apoptosis in yeast. *J Cell Biol*, Vol. 166, pp. 969-974

Youle, R.J. & Karbowski, M. (2005). Mitochondrial fission in apoptosis, *Nat Rev Mol Cell Biol,* Vol. 6, pp. 657-663

Zhang, J.G., Liu, X.Y., He, X.P., Guo, X.N., Lu, Y. & Zhang, B.R. (2011). Improvement of acetic acid tolerance and fermentation performance of *Saccharomyces cerevisiae* by disruption of the *FPS1* aquaglyceroporin gene. *Biotechnol Lett*, Vol. 33, pp. 277-284

5

Metabolic Optimization by Enzyme-Enzyme and Enzyme-Cytoskeleton Associations

Daniela Araiza-Olivera et al.*
Instituto de Fisiología Celular, Universidad Nacional Autónoma de México, Mexico

1. Introduction

Probably enzymes are not dispersed in the cytoplasm, but are bound to each other and to specific cytoskeleton proteins. Associations result in substrate channeling from one enzyme to another. Multienzymatic complexes, or metabolons have been detected in glycolysis, the Krebs cycle and oxidative phosphorylation. Also, some glycolytic enzymes interact with mitochondria. Metabolons may associate with actin or tubulin, gaining stability. Metabolons resist inhibition mediated by the accumulation of compatible solutes observed during the stress response. Compatible solutes protect membranes and proteins against stress. However, when stress is over, compatible solutes inhibit growth, probably due to the high viscosity they promote. Viscosity inhibits protein movements. Enzymes that undergo large conformational changes during catalysis are more sensitive to viscosity. Enzyme association seems to protect the more sensitive enzymes from viscosity-mediated inhibition. The association-mediated protection of the enzymes in a given metabolic pathway would constitute a new property of metabolons: that is, to enhance survival during stress. It is proposed that resistance to inhibition is due to elimination of non-productive conformations in highly motile enzymes.

2. Metabolons: Enzyme complexes that channel substrates

The cytoplasm should not be regarded as a liquid phase containing a large number of soluble enzymes and particles. Instead, it has become evident that there is a high degree of organization where different lipid and protein structures associate among themselves and with other molecules. The high molecule concentration found in the cytoplasm promotes macromolecule associations such as protein-protein, protein-membrane, protein-nucleic acid, protein-polysaccharide and thus is a control factor for all biological processes (Srere & Ovadi, 1990). Indeed, the classical studies by Green (Green *et al.*, 1965), Clegg (Clegg, 1964)

* Salvador Uribe-Carvajal[1,**], Natalia Chiquete-Félix[1], Mónica Rosas-Lemus[1], Gisela Ruíz- Granados[1], José G. Sampedro[2], Adela Mújica[3] and Antonio Peña[1]
[1]*Instituto de Fisiología Celular, Universidad Nacional Autónoma de México,*
[2]*Instituto de Física, Universidad Autónoma de San Luís Potosí and*
[3]*CINVESTAV, Instituto Politécnico Nacional*
Mexico
** Corresponding Author

and Fulton (Fulton, 1982) have suggested that enzymes are not dispersed in the cytoplasm. Instead, enzymes are localized at specific sites where they are associated between them and with the cytoskeleton. The cytoskeleton is a trabecular network of fibrous proteins that micro-compartmentalizes the cytoplasm (Porter *et al.*, 1983). Associated enzymes channel substrates from one to another preventing their diffusion to the aqueous phase (Gaertner *et al.*, 1978; Minton & Wilf, 1981; Ovadi *et al.*, 1996).

In a multienzyme complex, intermediaries can be channeled more than once from the active site of an enzyme to the next to obtain the final product (Al-Habori, 2000; Robinson *et al.*, 1987). Channeling requires stable interactions of the multienzymatic metabolons (Al-Habori, 2000; Cascante *et al.*, 1994; Ovadi & Srere, 1996; Ovadi & Saks, 2004; Srere & Ovadi, 1990; Srere, 1987). The metabolon stability is facilitated by the compartmentalization of the cell in different organelles and structures (Jorgensen *et al.*, 2005).

There are many advantages inherent to metabolons (Jorgensen *et al.*, 2005) (Fig. 1): I) Improved catalytic efficiency of the enzymes. This is obtained by channeling an intermediary from the active site of an enzyme directly to the active site of the next. II) Channeling optimizes kinetic constants. III) Labile or toxic intermediates are retained within the metabolon. IV) Inhibitors are excluded from the active site of enzymes. V) Control and coordination of the enzymes in a given pathway is enhanced. VI) Finally, alternative metabolons may favor different pathways (Fig. 1). Most metabolons seem to be transient, opening the possibility for a quick change in some elements that allows them to redirect metabolism (Jorgensen *et al.*, 2005).

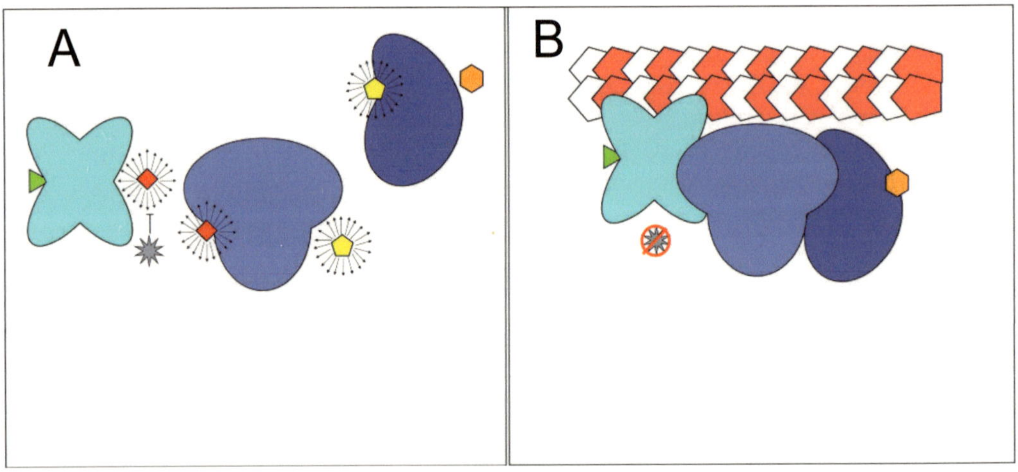

Fig. 1. Advantages of Metabolons. (A) In isolated enzymes the substrate (green), intermediaries (red and yellow) and product (orange) diffuse into the aqueous phase (little arrows). Toxic intermediaries and inhibitors (grey) are free to exit/enter the active site in each enzyme. (B) In metabolons (we show filamentous actin in red and white) channeling allows transfer of the substrate (green) from the active site of an enzyme direct to the next to obtain a final product (orange) without diffusion to the cytoplasm of intermediaries (not-depicted) are prevented, while inhibitors (grey) are excluded from the active sites.

The enzymes from the Krebs cycle are attached to the mitochondrial membrane in an enzymatic complex; this was the first "metabolon" described (Srere, 1987). In oxidative phosphorylation, multiprotein complexes seem to associate in supercomplexes and eventually in respiratory chains resulting in controlled electron channeling and proton-pumping stoichiometry (Guerrero-Castillo *et al.*, 2011). It has been proposed that these supercomplexes constitute an exquisite mechanism to regulate the yield of ATP (Guerrero-Castillo *et al.*, 2009; 2011; Schägger *et al.*, 2001). In addition, in some organisms such as trypanosomes, glycolytic enzymes are contained in small organelles called glycosomes, where channeling is highly efficient (Aman *et al.*, 1985). Tumor cells also produce aggregates containing glycolytic enzymes (Coe & Greenhouse, 1973). Interactions between organelles such as the endoplasmic reticulum and mitochondria have been described (Dorn & Scorrano, 2010; Kornmann *et al.*, 2009; Lebiedzinska *et al.*, 2009). Mitochondria are both, the main source of ATP and inducers of cellular death (Anesti & Scorrano, 2006). Mitochondrial functions are regulated by interactions with other organelles and cytoplasmic proteins (Kostal & Arriaga, 2011). Cytoskeletal proteins such as actin and tubulin, direct mitochondria to specific sites in the cell (Senning & Marcus, 2010) and control coupling of phosphorylation by interacting with mitochondrial porin (Xu *et al.*, 2001; Lemasters & Holmuhamedov, 2006; Rostovtseva *et al.*, 2008; Rostovtseva *et al.*, 2004; Xu *et al.*, 2001). In addition to cytoeskeletal proteins, hexokinase, a glycolytic enzyme binds mitochondria in mammalians (Pastorino & Hoek, 2008), yeast and plants (Balasubramanian *et al.*, 2008) regulatin the energy yield of mitochondria as well as the induction of programmed cell death (Kroemer *et al.*, 2005; Pastorino & Hoek, 2008; Xie & Wilson, 1988). All the above data suggest that enzymes are highly organized (Clegg & Jackson, 1989) and the cytoskeleton plays an important role (Minaschek *et al.*, 1992; Keleti *et al.*, 1989; Porter *et al.*, 1983).

3. The cytoskeleton: A scaffold where metabolons are bound

The eukaryotic cytoplasm is supported by the cytoskeleton, a network of structural proteins that shapes the cell and has binding sites for different enzymes. Such sites have been identified in filamentous actin (F-actin), in microtubules and in the cytoplasmic domain of the erythrocyte band 3, which is also an anion exchanger. Glycolytic enzyme binding to actin usually results in stimulation, whereas binding to microtubules or to band 3 inhibits activity (Real-Hohn *et al.*, 2010). Actin is involved in a variety of cell functions that include contractility, cytokinesis, maintenance of cell shape, cell locomotion and organelle localization. In addition, glycolytic enzymes and F-actin co-localize in muscle cells, probably reflecting compartmentation of the glycolytic pathway (Waingeh *et al.*, 2006).

Actin is highly conserved in eukaryotic cells. It may be found as a monomer (G-actin) or as a polymeric filament (F-actin) that is interconverted in an extremely dynamic, highly controlled process. The polar actin monomers polymerize head-to-tail to yield a polar filament. Actin filaments are constituted by 8 nm diameter, double-helical structures formed by assemblies of monomeric actin with a barbed end (or plus end) and a pointed end (or minus end). The spontaneous polymerization of actin monomers occurs in three phases: nucleation, elongation and maintenance. Nucleation consists in the formation of a dimer, followed by the addition of a third monomer to yield a trimer; this process is very slow. Further monomer addition becomes thermodynamically favorable and the filament elongates rapidly: much faster at the plus end than at the minus end. In the maintenance phase, there is no net filament growth and the concentration of ATP-G-actin is kept stationary (Fig. 2)..

Upon incorporation to a filament, G-actin-bound ATP is hydrolyzed. ADP and Pi remain non-covalently bound. Then Pi is released slowly. Thus, the elongating filaments contain: the barbed end, rich in ATP-actin, the center, rich in ADP-Pi-actin and the pointed end containing ADP-actin. Many actin-binding proteins regulate actin polymerization. Profilin is an actin monomer-binding protein; Arp 2/3 complex are nucleation proteins; CapZ and gelsolin regulate the length of the actin filament and the cofilin/ADP family cuts F-actin and accelerates depolymerization (Kustermans *et al.*, 2008). However, protein functions may vary; in *Dictyostelium*, CapZ prevents filament elongation and increases the concentration of unpolymerized actin; in contrast, in yeast this same protein prevents depolymerization increasing F-actin concentration (Welch *et al.*, 1997). The cytoskeleton can be rapidly remodeled by the small RhoGTPases (Rho, Rac and Cdc42), which act in response to extracellular stimuli (Kustermans *et al.*, 2008). There are exogenous natural compounds that can disturb actin dynamics (Kustermans *et al.*, 2008).

4. The glycolytic metabolon

The association of enzymes with the cytoskeleton probably stabilizes metabolons. In this regard, glycolytic enzymes such as fructose 1,6-bisphosphate aldolase (aldolase), glyceraldehyde 3-phosphate dehydrogenase (GAPDH), piruvate kinase (PK), glucose phosphate isomerase (GPI), and lactate dehydrogenase (LDH) associate with actin. Other glycolytic enzymes such as triose phosphate isomerase and phosphoglycerate mutase bind indirectly through interactions with other enzymes. Enzyme-enzyme-actin complexes are called piggy-back interactions. Also, aldolase and GAPDH compete for binding sites (Knull & Walsh, 1992; Waingeh *et al.*, 2006).

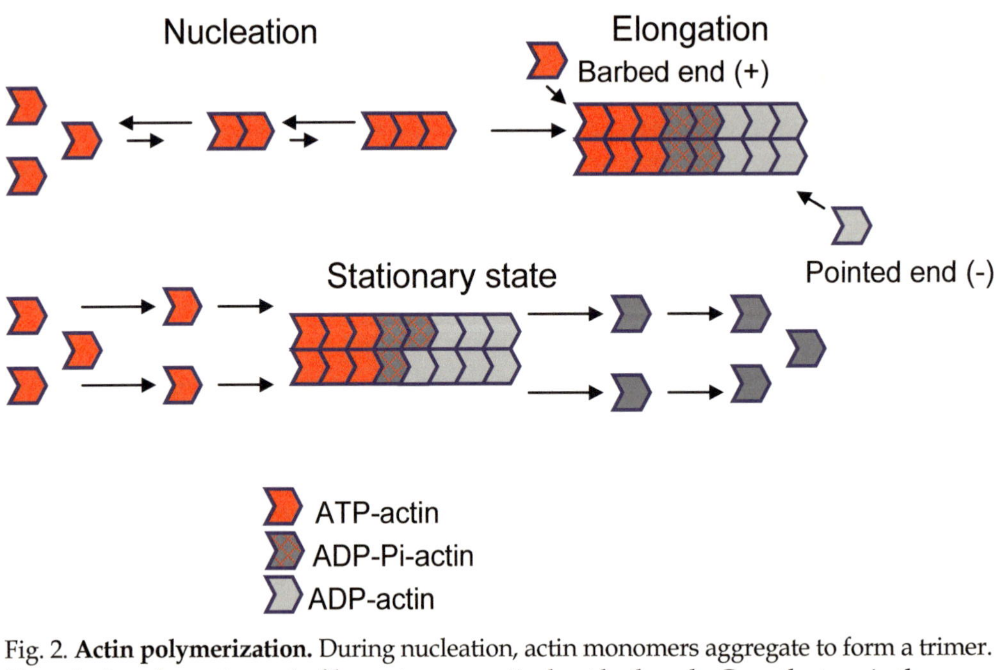

Fig. 2. **Actin polymerization.** During nucleation, actin monomers aggregate to form a trimer. Then during elongation actin filaments grow actively at both ends. Growth stops in the maintenance phase, also known as stationary phase. (Modified from Kustermans *et al.*, 2008)

Enzyme/actin interaction is regulated by ionic strength (Waingeh *et al.*, 2006). In homogenates of muscle tissue suspended in isosmotic sucrose, proteins such as F-actin, myosin, troponin and tropomyosin associate with glycolytic enzymes (Brooks & Storey, 1991). Glycolytic enzyme association to actin is not accepted universally, for instance, the F-actin/glycolytic enzyme interaction has been modeled mathematically at physiological ionic strength and protein concentrations. The results suggest that under cellular conditions only a small percentage of TPI, GAPDH, PGK and LDH would be associated with F-actin (Brooks & Storey, 1991).

Protein dynamics seem important for their interactions. Brownian dynamics (BD) simulations detect that rabbit F-actin has different binding modes/affinities for aldolase and GAPDH (Forlemu *et al.*, 2006). Some metabolites such as ATP and ADP modulate enzyme interactions and the resulting substrate channeling (Forlemu *et al.*, 2006).

A barely explored effect of the association of enzymes with the cytoskeleton is the modulation of the dynamics of actin polymerization. Such an effect has been reported for aldolase (Chiquete-Felix *et al.*, 2009; Schindler *et al.*, 2001). An interesting finding is that some growth factors, such as PGF and EGF enhance the GAPDH/cytoskeleton interaction, possibly increasing keratinocyte migration (Tochio *et al.*, 2010). Indeed, GAPDH seems to participate in cytoskeleton dynamics processes such as endocytosis, membrane fusion, vesicular transport and nuclear tRNA transport (Cueille *et al.*, 2007).

In red blood cell membranes, GAPDH, aldolase and PFK interact with an acidic sequence at the amino-terminal extreme of band 3 with high affinity (Campanella *et al.*, 2005). Under physiological conditions, the binding of glycolytic enzymes to band 3 results in inhibition of the glycolytic flux (Real-Hohn *et al.*, 2010).

Association to microtubules regulates the energetic metabolism (Keleti *et al.*, 1989; Keller *et al.*, 2007; Walsh *et al.*, 1989) at the level of some glycolytic enzymes such as pyruvate kinase, phosphofructokinase (Kovács *et al.*, 2003) and enolase (Keller *et al.*, 2007). When the glycolytic enzymes are associated and anchored to the sarcomere, ATP is produced more efficiently (Keller *et al.*, 2007). The interaction of enzymes with themselves and with the cytoskeleton confers more stability to the enzyme activity and to the whole network (Keleti *et al.*, 1989; Volker *et al.*, 1995; Walsh *et al.*, 1989). F-actin stabilizes some glycolytic enzymes of muscle and sperm (Walsh & Knull, 1988; Ovadi & Saks, 2004). That is the case of the phosphofructokinase (PFK) and aldolase where the dilution-mediated inactivation of PFK is stopped upon aldolase addition. If PFK is associated with microtubules, it still loses activity when diluted, however, in these conditions it recovers the lost activity upon aldolase addition (Raïs *et al.*, 2000; Vértessy *et al.*, 1997). All this evidence supports the existence of a cytoskeleton-bound glycolytic metabolon.

5. Compatible solutes protect cellular structures during stress

Compatible solutes are defined as molecules that reach high concentrations in the cell without interfering with metabolic functions (Brown & Simpson, 1972). These are mostly amino acids and amino acid derivatives, polyols, sugars and methylamines. Compatible solutes are typically small and harbor chemical groups that interact with protein surfaces. Indeed, some authors have proposed to call them "chemical or pharmacological chaperones" as they stabilize native structures (Loo & Clarke, 2007; Romisch, 2004). Some compatible solutes are:

glycine betaine, a thermoprotectant in *B. subtilis* (Chen & Murata, 2011; Holtmann & Bremer, 2004). Ectoine, that in halophile microorganisms confers resistance to salt and temperature stress (Pastor *et al.*, 2010). Glycerol is accumulated in yeast under high osmotic pressure (Blomberg, 2000). Glycerol stabilizes thermolabile enzymes preventing their inactivation (Zancan & Sola-Penna, 2005). The disaccharide trehalose protects against environmental injuries (heat, cold, desiccation, and anoxia) and nutritional limitations (Argüelles, 2000; Crowe *et al.*, 1984) in bacteria, yeast, fungi, plants and invertebrates. In biotechnology, trehalose is one of the best protein stabilizing known (Jain & Roy, 2008; Sampedro *et al.*, 2001).

6. Effect of compatible solutes on the activity of enzymes

Compatible solute synthesis and accumulation is triggered by harsh conditions and results in protein stabilization and enhanced survival. Proteins may be unfolded, partially unfolded or native (Chilson & Chilson, 2003). In the absence of stress, high compatible solute concentrations inhibit cellular growth, metabolism and division (Wera *et al.*, 1999), e.g. a trehalase-deficient mutant of *S. cerevisiae* subjected to heat or saline stress accumulated high amounts of trehalose and survived. However, when these mutants were returned to normal conditions they are unable to grow or sustain metabolic activity (Garre & Matallana, 2009; Wera *et al.*, 1999).

6.1 Inhibition of isolated enzymes; possible role of viscosity

Under stress, high compatible solutes change the physicochemical properties of the cytoplasm. However, the effect of the high viscosity generated by molar concentrations of compatible solutes on enzyme activity has drawn little attention. Trehalose and other polyols protect proteins from thermal unfolding via indirect interactions (Liu *et al.*, 2010). Therefore the stabilizing mechanism must rely in the modified physicochemical properties of aqueous media.

Large-scale conformational changes in proteins involve the physical displacement of associated solvent molecules and solutes. The resistance to the movement or displacement of solvent molecules is a frictional process. Kramers theory provides the mathematical basis to understand and analyze reactions at high viscosity (Kramers, 1940). The application of Kramer´s theory to proteins indicates that the movements involved in folding or in enzyme-substrate association and processing must be highly sensitive to viscosity (Jacob and Schmid, 1999; Jacob *et al.*, 1999; Sampedro and Uribe, 2004).

Studies on cellular viscosity in yeast cytoplasm showed a value of 2 cP at 30°C (Williams *et al.*, 1997). Also, in vitro determinations for 0.6 M trehalose solutions showed a viscosity of 1.5 cP at 30°C (Table 1). Therefore, one may infer that yeast cytoplasm viscosity with 0.6 M trehalose should be in the vicinity of 2.5-3 cP.

The plasma membrane H^+-ATPase from yeast depends on large domain motion for catalysis (Kulbrandt, 2004), was inhibited at all trehalose concentrations tested (Sampedro *et al.*, 2002). The rate constant for the ATPase reaction ($V_{max} = k_{cat}$ [E_t]) was inversely dependent on solution viscosity; as higher the viscosity lower the reaction rate of catalysis (Sampedro *et al.*, 2002). Notably, when temperature was raised inhibition disappeared, in agreement with the fact that viscosity decreases when temperature increases (Table 1). Similar results have been obtained with Na^+/K^+-ATPase and Na^+-ATPase in the presence of polyethylene glycol and glycerol (Esmann *et al.*, 2008). In glucose oxidase, activity inhibition by varying concentrations of trehalose was due to the promotion of a highly compact state, which correlated with the increased viscosity of the medium (Paz-Alfaro *et al.*, 2009).

TREHALOSE (M)	0.2	0.4	0.5	0.6	0.8
TEMP (°C)	VISCOSITY (cP)				
20	1.35	1.59	1.81	2.04	2.58
25	1.20	1.37	1.51	1.74	2.20
30	1.08	1.18	1.33	1.50	1.91
35	0.94	1.03	1.18	1.31	1.67
40	0.86	0.94	1.04	1.13	1.49
45	0.75	0.81	0.90	1.04	1.31

Data modified from Sampedro *et al.*, 2002.

Table 1. Viscosity of trehalose solutions at different concentrations and temperatures.

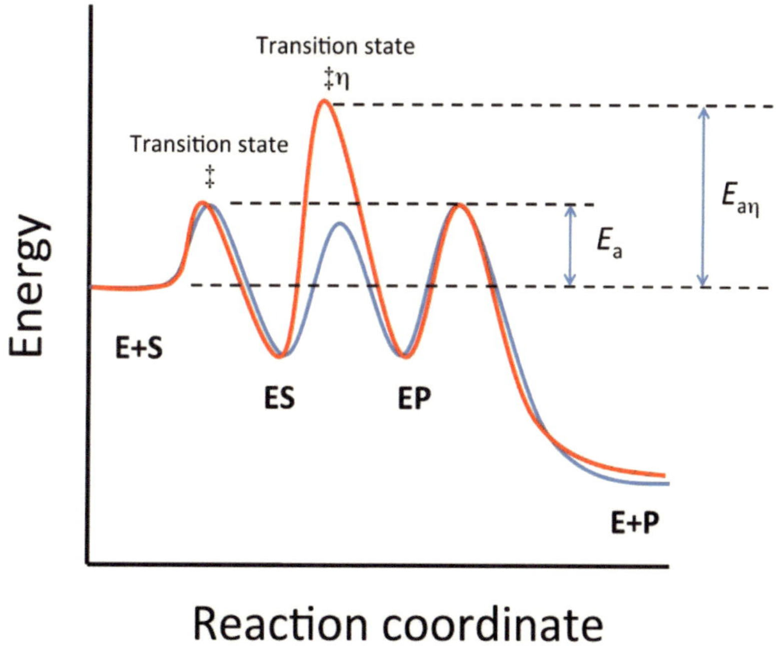

Fig. 3. Reaction coordinate diagram, comparing an enzyme reaction at normal viscosity (blue) and at high viscosity (h; red). When a diffusive protein domain process is present in the catalytic cycle, it becomes rate limiting when viscosity is high. Therefore the overall activation energy (E_a) increases.

Many enzymes are inhibited by viscosity. Glutathione reductase is inhibited at 25°C, by trehalose (70% inhibition at 1.5 M trehalose) although inhibition disappears at 40°C (Sebollela *et al.*, 2004). Also pyrophosphatase and glucose 6-phosphate dehydrogenase show temperature dependence of trehalose-mediated inhibition (Sebollela *et al.*, 2004).

Aminoglycoside nucleotidyltransferase 2''-I is inhibited by glycerol in a temperature-dependent way (Gates & Northrop, 1988). The hyaluronan-synthase from *Streptococcus equisimilis*is is inhibited by of PEG, ethylene glycol, glycerol or sucrose (Tlapak-Simmons *et al.*, 2004). At high viscosities (greater than 4 mPa s-1) different carbohydrates inhibit egg-white lysozyme (Lamy *et al.*, 1990; Monkos, 1997).

Detailed studies on diffusive protein-structural components demonstrated that for β-lactam synthase a conformational change is rate-limiting on k_{cat}. Therefore, the rate for catalysis shows a high inhibition by medium viscosity (Raber *et al.*, 2009). Crystallographic analysis of adhesion kinase-1 shows a large conformational motion of the activation loop upon ATP binding. This is an essential step during catalysis and explains the viscosity inhibitory effect (Schneck *et al.*, 2010). In the plasma membrane H^+-ATPase, the enzyme fluctuates between two structural conformations (E1↔E2) during catalysis. The N-domain (nucleotide binding) rotates 73° towards the phosphorylation site to deliver ATP to the phosphorylation site (Kuhlbrandt, 2004). In all cases, the rate-limiting step is a conformational change that seems to be the one inhibited by viscosity (Fig. 3).

6.2 Enzyme association results in protection against inhibition

Compatible solute-mediated inhibition does not seem to uniformly affect all enzymes. Furthermore, in the face of both the stress condition and the compatible solute, catabolic pathways seem to resist inhibition, thus providing the energy needed for survival (Hoffmann & Holzhütter, 2009; Hounsa *et al.*, 1998). In our hands, in a yeast cytoplasmic extract, compatible solutes inhibit the whole glycolytic pathway much less than many of its individual, isolated enzymes (Araiza-Olivera *et al.*, 2010). In contrast, anabolism seems to be shot both during the stress situation and later (Attfield, 1987). Inhibition of anabolism would explain the inability of cells to reproduce (Wera *et al.*, 1999). The mechanism for resistance to inhibition, exhibited by the catabolic enzymes is a matter of study (Marcondes *et al.*, 2011; Raïs *et al.*, 2000).

The effect of a compatible solute (trehalose) on the activity of some yeast glycolytic enzymes such as GAPDH, HXK, ALD and PGK has been analyzed. These enzymes were tested individually or in mixtures (Araiza-Olivera *et al.*, 2010). When isolated, GAPDH and HXK were inhibited by trehalose while others, such as ALD and PGK were resistant. Probably GAPDH and HXK are more motile than ALD and PGK. Remarkably, when the sensitive enzymes were mixed with the resistant enzymes a protection effect was observed. This led to analyze the whole glycolytic pathway and again, inhibition was minimal in comparison with the individual, isolated enzymes (Araiza-Olivera *et al.*, 2010). Thus, it was decided to explore the possible mechanisms underlying this effect, i.e, why some metabolic pathways, such as glycolysis resist the viscosity-mediated inhibition promoted by compatible solutes, even if they contain several viscosity-sensitive enzymes.

The protection effect was specific for each protein couple, as GAPDH was not protected by neither HXK, albumin or lactate-dehydrogenase. Also, the pentose pathway enzyme glucose 6-phosphate dehydrogenase (G6PDH) was not protected by ALD against inhibition by trehalose. Once in the complexes, probably the more flexible enzymes that are more sensitive to viscosity (Sampedro & Uribe 2004) are stabilized by the more resistant, more rigid enzymes forming a less motile, more resistant complex.

The proposal that enzyme association favors a more stable folded state would require the motile enzymes to eliminate some non-productive conformations (Villali & Kern, 2010). These associations are probably further stabilized by some elements of the cytoskeleton, such as tubulin (Raïs *et al.*, 2000; Walsh *et al.*, 1989) or F-actin (Minaschek *et al.*, 1992; Waingeh *et al.*, 2006). Thus, it is proposed that another function of enzyme association into metabolons, in addition to substrate channeling and metabolic control might be to resist compatible solute-mediated inhibition.

7. Concluding remarks

Under stress, compatible solutes accumulate to very high levels in the cytoplasm. This results in enhanced viscosity. As revised in section 6.1, viscosity is known to inhibit diverse enzymes. Indeed, high viscosity may be the mechanism by which diverse cell functions are inhibited in the presence of high compatible solute concentrations, e.g. cells are unable to. In contrast, catabolism remains active even in the presence of compatible solutes. One possible mechanism for this resistance to inhibition is probably the specific association of glyolytic enzymes among themselves and probably with the cytoskeleton. Resistance to viscosity-mediated inhibition is proposed as a novel, important property of enzyme association into metabolons. The mechanism of protection that association confers against viscosity still has to be defined. Protection of activity is needed for survival during stress.

8. References

Al-Habori M. Microcompartmentation, metabolic channelling and carbohydrate metabolism. *Int. J. Biochem. Cell. Biol.* 1995; 27(2):123-32.

Aman RA & Wang CC. An improved purification of glycosomes from the procyclic trypomastigotes of Trypanosoma brucei. *Mol. Biochem. Parasitol.* 1986; 21(3):211-20.

Anesti V & Scorrano L. The relationship between mitochondrial shape and function and the cytoskeleton. *Biochim Biophys Acta.* 2006; 1757(5-6): 692-9.

Araiza-Olivera D, Sampedro JG, Mújica A, Peña A & Uribe-Carvajal S. The association of glycolytic enzymes from yeast confers resistance against inhibition by trehalose. *FEMS Yeast Res.* 2010; 10 (3):282-9.

Argüelles JC. Physiological roles of trehalose in bacteria and yeasts: a comparative analysis. *Arch Microbiol.* 2000; 174(4):217-24.

Attfield PV. Trehalose accumulates in Saccharomyces cerevisiae during exposure to agents that induce heat shock response. *FEBS Lett.* 1987; 225(1-2):259-63.

Balasubramanian R, Karve A & Moore BD. Actin-based cellular framework for glucose signaling by Arabidopsis hexokinase1. *Plant Signal Behav.* 2008; 3(5):322-4.

Blomberg A. Metabolic surprises in Saccharomyces cerevisiae during adaptation to saline conditions: questions, some answers and a model. *FEMS Microbiol Lett.* 2000; 82(1):1-8.

Brooks SP & Storey KB. Where is the glycolytic complex? A critical evaluation of present data from muscle tissue. *FEBS Lett.*1991; 278(2):135-8.

Brown AD & Simpson JR. Water relations on sugar-tolerant yeast: the role of intracellular polyols. *J. Gen Microbiol.* 1972; 72:589-591.

Campanella ME, Chu H & Low PS. Assembly and regulation of a glycolytic enzyme complex on the human erythrocyte membrane. *Proc Natl Acad Sci.* 2005; 102(7):2402-7.

Cascante M, Sorribas A & Canela EI. Enzyme-enzyme interactions and metabolite channelling: alternative mechanisms and their evolutionary significance. *Biochem. J.* 1994; 298 (Pt 2):313-20

Chen TH & Murata N. Glycinebetaine protects plants against abiotic stress: mechanisms and biotechnological applications. *Plant Cell Environ.* 2011; 34(1):1-20.

Chilson OP & Chilson AE. Perturbation of folding and reassociation of lactate dehydrogenase by proline and trimethyl amine oxide. *Eur.J.Biochem.* 2003; 270,4823–4834.

Chiquete-Felix N, Hernández JM, Méndez JA, Zepeda-Bastida A, Chagolla-López A & Mújica A. In guinea pig sperm, aldolase A forms a complex with actin, WAS, and Arp2/3 that plays a role in actin polymerization. *Reproduction.* 2009; 137(4):669-78.

Clegg JS. The control of emergence and metabolism by external osmotic pressure and the role of free glycerol in developing cysts of *Artemia salina*. *J. Exp. Biol.* 1964; 41:879-92.

Clegg JS & Jackson SA. Evidence for intermediate channelling in the glycolytic pathway of permeabilized L-929 cells. *Biochem. Biophys. Res. Commun.* 1989; 160(3):1409-14.

Coe EL & Greenhouse WV. Possible regulatory interactions between compartmentalized glycolytic systems during initiation of glycolysis in ascites tumor cells. *Biochim. Biophys. Acta.* 1973; 329(2):171-82.

Crowe JH & Crowe LM, Chapman D. Preservation of Membranes in Anhydrobiotic Organisms: The Role of Trehalose. *Science.* 1984; 223(4637):701-703.

Cueille N, Blanc CT, Riederer IM & Riederer BM. Microtubule-associated protein 1B binds glyceraldehyde-3-phosphate dehydrogenase. *J Proteome Res.* 2007; 6(7):2640-7.

Dorn GW 2nd &, Scorrano L. Two close, too close: sarcoplasmic reticulum-mitochondrial crosstalk and cardiomyocyte fate. *Circ Res.* 2010; 107(6):689-99. Review.

Esmann M, Fedosova NU & Marsh D. Osmotic Stress and Viscous Retardation of the Na,K-ATPase Ion Pump. *Biophysical J.* 2008; 94:2767-2776.

Forlemu NY, Waingeh VF, Ouporov IV, Lowe SL & Thomasson KA. Theoretical study of interactions between muscle aldolase and F-actin: insight into different species. *Biopolymers.* 2007; 85(1):60-71.

Fulton AB. How crowded is the cytoplasm? *Cell.* 1982; 30(2):345-7.

Gaertner FH. Unique catalytic properties of enzyme clusters. *Trends Biochem. Sci.* 1978; 3, 63.

Garre E & Matallana E. The three trehalases Nth1p, Nth2p and Ath1p participate in the mobilization of intracellular trehalose required for recovery from saline stress in Saccharomyces cerevisiae. *Microbiology.* 2009; 155:3092–3099.

Gates CA & Northrop DB. Determination of the rate-limiting segment of aminoglycoside nucleotidyltransferase 2''-I by pH and viscosity-dependent kinetics. *Biochemistry.* 1988; 27(10):3834-3842.

Green DE, Murer E, Hultin HO, Richardson SH, Salmon B, Brierley GP & Baum H. Association of integrated metabolic pathways with membranes. I. Glycolytic enzymes of the red blood corpuscle and yeast. *Arch. Biochem. Biophys.* 1965; 112(3):635-47.

Guerrero-Castillo S, Vázquez-Acevedo M, González-Halphen D & Uribe-Carvajal S. In Yarrowia lipolytica mitochondria, the alternative NADH dehydrogenase interacts specifically with the cytochrome complexes of the classic respiratory pathway. *Biochim Biophys Acta.* 2009; 1787 (2):75-85.

Guerrero-Castillo S, Araiza-Olivera D, Cabrera-Orefice A, Espinasa-Jaramillo J, Gutiérrez-Aguilar M, Luévano-Martínez LA, Zepeda-Bastida A & Uribe-Carvajal S. Physiological uncoupling of mitochondrial oxidative phosphorylation. Studies in different yeast species. *J Bioenerg Biomembr.* 2011; 43(3):323-31.

Hoffmann S & Holzhütter HG. Uncovering metabolic objectives pursued by changes of enzyme levels. *Ann. N. Y. Acad. Sci.* 2009; 1158:57-70

Holtmann G & Bremer E. Thermoprotection of Bacillus subtilis by exogenously provided glycine betaine and structurally related compatible solutes: involvement of Opu transporters. *J Bacteriol.* 2004; 186(6):1683-93.

Hounsa CG, Brandt EV, Thevelein J, Hohmann S & Prior BA. Role of trehalose in survival of Saccharomyces cerevisiae under osmotic stress. *Microbiology* 1998; 144:671-80.

Jacob M, Geeves M, Holterman G, & Schmid FX. Diffusional crossing in a two-state protein folding reaction. 1999 *Nat. Struct. Biol.* 1999; 6:923-926.

Jacob M & Schmid FX. Protein folding as a diffusional process. *Biochemistry* 1999; 38:13773-13779.

Jain NK & Roy I. Role of trehalose in moisture-induced aggregation of bovine serum albumin. *Eur J Pharm Biopharm.* 2008; 69(3):824-34.

Jørgensen K, Rasmussen AV, Morant M, Nielsen AH, Bjarnholt N, Zagrobelny M, Bak S & Møller BL. Metabolon formation and metabolic channeling in the biosynthesis of plant natural products. *Curr. Opin. Plant Biol.* 2005; 8(3):280-91.

Keleti T, Ovádi J & Batke J. Kinetic and physico-chemical analysis of enzyme complexes and their possible role in the control of metabolism. *Prog. Biophys. Mol. Biol.* 1989; 53(2):105-52.

Keller A, Peltzer J, Carpentier G, Horváth I, Oláh J, Duchesnay A, Orosz F & Ovádi J. Interactions of enolase isoforms with tubulin and microtubules during myogenesis. *Biochim. Biophys. Acta.* 2007;1770 (6):919-26.

Knull HR & Walsh JL. Association of glycolytic enzymes with the cytoskeleton. *Curr Top Cell Regul.* 1992; 33:15-30. Review.

Kornmann B, Currie E, Collins SR, Schuldiner M, Nunnari J, Weissman JS & Walter P. An ER-mitochondria tethering complex revealed by a synthetic biology screen. *Science.* 2009; 325(5939):477-81.

Kostal V & Arriaga EA. Capillary Electrophoretic Analysis Reveals Subcellular Binding between Individual Mitochondria and Cytoskeleton. *Anal Chem.* 2011.

Kovacs J, Low P, Pacz A, Horvath I, Olah J & Ovadi J. Phosphoenolpyruvate-dependent tubulin-pyruvate kinase interaction at different organizational levels. *J. Biol. Chem.* 2003; 278(9):7126-30.

Kramers HA. Brownian motion in a field of force and the diffusion model of chemical reactions.*Physica.*1940; 7:284–304.

Kroemer G, El-Deiry WS, Golstein P, Peter ME, Vaux D, Vandenabeele P, Zhivotovsky B, Blagosklonny MV, Malorni W, Knight RA, Piacentini M, Nagata S & Melino G. Classification of cell death: recommendations of the Nomenclature Committee on Cell Death. *Cell Death Differ.* 2005, Suppl 2: 1463-7.

Kühlbrandt W. Biology, structure and mechanism of p-type ATPases. *Nature*. 2004; 5:282-295.

Kustermans G, Piette J & Legrand-Poels S. Actin-targeting natural compounds as tools to study the role of actin cytoskeleton in signal transduction. *Biochem Pharmacol*. 2008; 76(11):1310-22.

Lamy L, Portmann MO, Mathlouthi M & Larreta-Garde V. Modulation of egg-white lysozyme activity by viscosity intensifier additives. *Biophys Chem*. 1990; 36(1):71-76.

Lebiedzinska M, Szabadkai G, Jones AW, Duszynski J & Wieckowski MR. Interactions between the endoplasmic reticulum, mitochondria, plasma membrane and other subcellular organelles. *Int J Biochem Cell Biol*. 2009; 41(10):1805-16. Review.

Lemasters JJ & Holmuhamedov E. Voltage-dependent anion channel (VDAC) as mitochondrial governator--thinking outside the box. *Biochim Biophys Acta*. 2006; 1762(2): 181-90.

Liu FF, Ji L, Zhang L, Dong XY & Sun Y . Molecular basis for polyol-induced protein stability revealed by molecular dynamics simulations. *J Chem Phys*. 2010; 132(22):225103.

Loo TW & Clarke DM. Chemical and pharmacological chaperones as new therapeutic agents. *Expert Rev Mol Med*. 2007; 9(16):1-18.

Marcondes MC, Sola-Penna M, Torres Rda S & Zancan P. Muscle-type 6-phosphofructo-1-kinase and aldolase associate conferring catalytic advantages for both enzymes. *IUBMB Life*. 2011; 63(6):435-45.

Minaschek G, Gröschel-Stewart U, Blum S & Bereiter-Hahn J. Microcompartmentation of glycolytic enzymes in cultured cells. *Eur J Cell Biol*. 1992; 58(2):418-28.

Minton AP & Wilf J. Effect of macromolecular crowding upon the structure and function of an enzyme: glyceraldehyde-3-phosphate dehydrogenase. *Biochemistry* 1981; 20(17):4821-6.

Monkos K. Concentration and temperature dependence of viscosity in lysozyme aqueous solutions. *Biochim Biophys Acta*. 1997; 1339(2):304-10.

Ovádi J & Srere PA. Metabolic consequences of enzyme interactions. *Cell. Biochem. Funct*. 1996; 14(4):249-58.

Ovádi J & Saks V. On the origin of intracellular compartmentation and organized metabolic systems. *Mol. Cell Biochem*. 2004; 256-257(1-2):5-12.

Pastor JM, Salvador M, Argandoña M, Bernal V, Reina-Bueno M, Csonka LN, Iborra JL, Vargas C, Nieto JJ & Cánovas M. Ectoines in cell stress protection: uses and biotechnological production. *Biotechnol Adv*. 2010; 28(6):782-801. Review.

Pastorino JG & Hoek JB. Regulation of hexokinase binding to VDAC. *J Bioenerg Biomembr*. 2008; 40(3): 171-82.

Paz-Alfaro K, Ruiz-Granados YG, Uribe-Carvajal S & Sampedro JG. Trehalose-mediated stabilization of glucose oxidase from *Aspergillus niger*. *J Biotechnol*. 2009; 141:130-136.

Porter ME & Johnson KA. Transient state kinetic analysis of the ATP-induced dissociation of the dynein-microtubule complex. *J. Biol. Chem*. 1983; 258(10):6582-7.

Raber ML, Freeman MF & Townsend CA. Dissection of the stepwise mechanism to beta-lactam formation and elucidation of a rate-determining conformational change in beta-lactam synthetase. *J Biol Chem*. 2009; 284(1):207-217.

Raïs B, Ortega F, Puigjaner J, Comin B, Orosz F, Ovádi J & Cascante M. Quantitative characterization of homo- and heteroassociations of muscle phosphofructokinase with aldolase. *Biochim Biophys Acta.* 2000; 1479(1-2):303-14.

Real-Hohn A, Zancan P, Da Silva D, Martins ER, Salgado LT, Mermelstein CS, Gomes AM & Sola-Penna M. Filamentous actin and its associated binding proteins are the stimulatory site for 6-phosphofructo-1-kinase association within the membrane of human erythrocytes. *Biochimie.* 2010; 92(5):538-44.

Robinson JB Jr, Inman L, Sumegi B & Srere PA. Further characterization of the Krebs tricarboxylic acid cycle metabolon. *J. Biol. Chem.* 1987; 262 (4):1786-90.

Römisch K. A cure for traffic jams: small molecule chaperones in the endoplasmic reticulum. *Traffic.* 2004; 5(11):815-820.

Rostovtseva TK, Antonsson B, Suzuki M, Youle RJ, Colombini M & Bezrukov S. M. Bid, but not Bax, regulates VDAC channels. *J Biol Chem.* 2004; 279(14): 13575-83.

Rostovtseva TK, Sheldon KL, Hassanzadeh E, Monge C, Saks V, Bezrukov SM & Sackett DL. Tubulin binding blocks mitochondrial voltage-dependent anion channel and regulates respiration. *Proc Natl Acad Sci U S A.* 2008; 105(48): 18746-51.

Sampedro JG, Muñoz-Clares RA & Uribe S. Trehalose-mediated inhibition of the plasma membrane H+-ATPase from Kluyveromyces lactis: dependence on viscosity and temperature. *J Bacteriol.* 2002; 184(16):4384-91.

Sampedro JG, Cortés P, Muñoz-Clares RA, Fernández A & Uribe S. Thermal inactivation of the plasma membrane H+-ATPase from Kluyveromyces lactis. Protection by trehalose. *Biochim. Biophys. Acta* 2001; 1544(1-2):64-73.

Sampedro JG & Uribe S. Trehalose-enzyme interactions result in structure stabilization and activity inhibition. The role of viscosity. *Mol. Cell. Biochem.* 2004; 256-257(1-2):319-27.

Schägger H & Pfeiffer K. The ratio of oxidative phosphorylation complexes I-V in bovine heart mitochondria and the composition of respiratory chain supercomplexes. *J Biol Chem.* 2001; 276 (41):37861-7.

Schindler R, Weichselsdorfer E, Wagner O & Bereiter-Hahn J. Aldolase-localization in cultured cells: cell-type and substrate-specific regulation of cytoskeletal associations. *Biochem Cell Biol.* 2001; 79(6):719-28.

Schneck JL, Briand J, Chen S, Lehr R, McDevitt P, Zhao B, Smallwood A, Concha N, Oza K, Kirkpatrick R, Yan K, Villa JP, Meek TD & Thrall SH. Kinetic mechanism and rate-limiting steps of focal adhesion kinase-1. *Biochemistry.* 2010; 49(33):7151-7163.

Sebollela A, Louzada PR, Sola-Penna M, Sarone-Williams V, Coelho-Sampaio T & Ferreira ST. Inhibition of yeast glutathione reductase by trehalose: possible implications in yeast survival and recovery from stress. *Int J Biochem Cell Biol.* 2004; 36(5):900-908.

Senning EN & Marcus AH. Actin polymerization driven mitochondrial transport in mating S. cerevisiae. *Proc Natl Acad Sci.* 2010; 107(2):721-5.

Srere PA. Complexes of sequential metabolic enzymes. *Annu. Rev. Biochem.* 1987; 56:89-124.

Srere PA & Ovadi J. Enzyme-enzyme interactions and their metabolic role. *FEBS Lett.* 1990; 268(2):360-4.

Tlapak-Simmons VL, Baron CA & Weigel PH. Characterization of the purified hyaluronan synthase from Streptococcus equisimilis. *Biochemistry.* 2004; 43(28):9234-9242.

Tochio T, Tanaka H, Nakata S & Hosoya H. Fructose 1,6-bisphosphate aldolase A is involved in HaCaT cell migration by inducing lamellipodia formation. *J Dermatol Sci.* 2010; 58(2):123-9.

Villali J & Kern D. Choreographing an enzyme's dance. *Curr Opin Chem Biol.* 2010; 14(5):636-43.

Vértessy BG, Orosz F, Kovács J & Ovádi J. Alternative binding of two sequential glycolytic enzymes to microtubules. Molecular studies in the phosphofructokinase/ aldolase/ microtubule system. *J. Biol. Chem.* 1997; 272(41): 25542-6.

Volker KW, Reinitz CA & Knull HR. Glycolytic enzymes and assembly of microtubule networks. *Comp Biochem Physiol B Biochem Mol Biol.* 1995;112(3):503-14.

Waingeh VF, Gustafson CD, Kozliak EI, Lowe SL, Knull HR & Thomasson KA. Glycolytic enzyme interactions with yeast and skeletal muscle F-actin. *Biophys J.* 2006; 90(4):1371-84.

Walsh JL & Knull HR. Heteromerous interactions among glycolytic enzymes and of glycolytic enzymes with F-actin: effects of poly(ethylene glycol). *Biochim Biophys Acta.* 1988; 952(1):83-91.

Walsh JL, Keith TJ & Knull HR. Glycolytic enzyme interactions with tubulin and microtubules. *Biochim. Biophys. Acta* 1989; 999(1):64-70.

Welch MD, Mallavarapu A, Rosenblatt J & Mitchison TJ. Actin dynamics in vivo. *Curr Opin Cell Biol.* 1997; 9(1):54-61.

Wera S, De Schrijver E, Geyskens I, Nwaka S & Thevelein JM. Opposite roles of trehalase activity in heat-shock recovery and heat-shock survival in Saccharomyces cerevisiae. *Biochem J.* 1999; 343Pt 3:621-626.

Williams SP, Haggie PM & Brindle KM. 19F-NMR Measurements of the rotational mobility of proteins in vivo. *Biophys J.* 1997; 72:490-498.

Xie GC & Wilson JE. Rat brain hexokinase: the hydrophobic N-terminus of the mitochondrially bound enzyme is inserted in the lipid bilayer. *Arch Biochem Biophys* 1988; 267(2): 803-10.

Xu X, Forbes JG & Colombini M. Actin modulates the gating of Neurospora crassa VDAC. *J Membr Biol.* 2001; 180(1): 73-81.

Zancan P & Sola-Penna M. Trehalose and glycerol stabilize and renature yeast inorganic pyrophosphatase inactivated by very high temperatures. *Arch. Biochem. Biophys.* 2005; 444(1):52-60.

6

Intracellular Metabolism of Uranium and the Effects of Bisphosphonates on Its Toxicity

Debora R. Tasat[1,2], Nadia S. Orona[1], Carola Bozal[2], Angela M. Ubios[2] and Rómulo L. Cabrini[2,3]
[1]*Universidad Nacional de Gral San Martín, Escuela de Ciencia y Tecnología,*
[2]*Universidad de Buenos Aires, Facultad de Odontología,*
[3]*National Commission of Atomic Energy,*
Argentina

1. Introduction

Uranium is the heaviest naturally occurring element found in the Earth's crust. It is an alpha-emitter radioactive element that present both radiotoxicant and chemotoxicant properties. Uranium is present in environment as a result of natural deposits and releases by human applications (mill tailings, nuclear industry and military army). The release of uranium or its by-products into the environment (air, soil and water) presents a threat to human health and to the environment as a whole. Uranium can enter the body by ingestion, inhalation or dermal contact yet, the primary route of entry into the body is inhalation. Research on inhaled, ingested, percutaneous and subcutaneous industrial uranium compounds has shown that solubility influences the target organ, the toxic response, and the mode of uranium excretion. The overall clearance rate of uranium compounds from the lung reflects both mechanical and dissolution processes depending on the morphochemical characteristics of uranium particles. In this review we emphasize on one of the principal physical characteristics of uranium particles, its size. As is known, based on uranium chemical composition, three different kinds are defined: natural, enriched (EU) and depleted (DU) uranium. The radiological and chemical properties of natural uranium and DU are similar. In fact, natural uranium has the same chemotoxicity, but its radiotoxicity is 60% higher. DU, being a waste product of uranium enrichment, has several civilian and military applications. Lately, it was used in international military conflicts (Gulf and recently as the Balkan Wars) and was claimed to contribute to health problems. Herein, we reviewed the toxicological data in vivo and in vitro on both natural and depleted uranium and renewed efforts to understand the intracellular metabolism of this heavy toxic metal. The reader will find this chapter divided in three sections. The first section, describes the presence of the uranium in the environment, the routes of entrance to the body and its impact on health. The second section which is committed to uranium cytotoxicity and its mechanism of action stressed on the oxidative metabolism and a third section dedicated to the effect of different compounds, mainly bisphosphonates, as substances with the ability to restrain uranium toxicity.

2. Uranium in the environment, routes of entrance to the body and impact on health

Uranium is a natural and commonly occurring radioactive element to be found ubiquitous in rock, soil, and water. Uranium concentrations in natural ground water can be more than several hundreds µg/l without impact from mining, nuclear industry, and fertilizers. It is a reactive metal, so it is not found as free uranium in the environment. Besides natural uranium, antopogenic activities such as uranium mining and further uranium processing to nuclear fuel, emissions form burning coal and oil, and the application of uranium containing phosphate fertilizers may enrich the natural uranium concentrations in the soil, water and air by far. The wide dispersal of pollutants in the environment (heavy metals, pesticides, fuel particles, and radionuclides) created by various human activities are of increasing concern. In particular, the release of harmful constituents from uranium or its by-products into the environment presents a threat to human health and the environment in many parts of the world. For instance, the civilian and military use of uranium, as well as fuel in nuclear power reactors, counterweights in aircraft and penetrators in shrapnel, may lead to the release of this radionuclide into the environment. This was the case in Amsterdam after the aircraft crash in 1992 (Uijt de Haag et al., 2000), around uranium processing areas (Pinney et al., 2003) or following the drop of some 300 tons of depleted uranium (DU) during the Gulf War (Bem & Bou-Rabee, 2004). This uranium dispersion may cause pollution of the air and water wells and/or into the food chain (Di Lella et al., 2005), which may lead to a chronic contamination by inhaltion or ingestion of local populations.

Radioactive elements are those that undergo spontaneous decay, in which energy is emitted either in the form of particles or electromagnetic radiation with energies sufficient to cause ionization. This decay results in the formation of different elements, some of which may themselves be radioactive, in which case they will also decay. Uranium exists in several isotopic forms, all of which are radioactive. The most toxicologically important of the 22 currently recognized uranium isotopes are uranium-234 (^{234}U), uranium-235 (^{235}U), and uranium-238 (^{238}U). When an atom of any of these three isotopes decays, it emits an alpha particle and transforms into a radioactive isotope of another element. The process continues through a series of radionuclides until reaching a stable, non-radioactive isotope of lead. There are three kinds of mixtures (based on the percentage of the composition of the three isotopes): natural uranium, enriched uranium (EU), and depleted uranium(DU). Enriched uranium is quantified by its ^{235}U percentage. Uranium enrichment for a number of purposes, including nuclear weapons, can produce uranium that contains as much as 97.3% ^{235}U and has a specific activity of ~50 µCi/g .The residual uranium after the enrichment process is "depleted" uranium and possesses a specific activity of 0.36 µCi/g, even less radioactivity than natural uranium (Research Triangle Institute 1997).

There are three things that determine the toxicity of radioactive materials: its radiological effect, its chemical effect and its particle size. Regarding its radiological effect uranium releases alpha particles (1gr DU releases 13.000 alpha particles per second), chemically is a very toxic heavy metal, and regarding its size, uranium particles within the air fit in the nanometer range (aerodinamic diameter of 0.1 microns or less), being this third characteristic far more biologically toxic than the first two. It is because uranium is both a heavy metal and a radioactive element that it is considered among the elements an unusual one. The hazards associated with this element are dependent upon uranium's chemical form (solubility, level of enrichment), physical form (morphology and size) and route of intake.

2.1 Chemical form

Uranium is a heavy metal that forms compounds and complexes of different varieties and solubilities. The chemical action of all isotopes and isotopic mixtures of uranium is identical, regardless of the specific activity (i.e., enrichment), because chemical action depends only on chemical properties. Thus, the chemical toxicity of a given amount or weight of natural, depleted, and enriched uranium will be identical. However, the chemical form of uranium determines its solubility and thus, transportability in body fluids as well as retention and deposit in various organs. On the basis of the toxicity of different uranium compounds in animals, it was concluded that the relatively more water-soluble compounds (uranyl nitrate, uranium hexafluoride, uranyl fluoride, uranium tetrachloride) were the most potent systemic toxicants. The poorly water-soluble compounds (uranium tetrafluoride, sodium diuranate, ammonium diuranate) were of moderate-to-low systemic toxicity, and the insoluble compounds (uranium trioxide, uranium dioxide, uranium peroxide, triuranium octaoxide) had a much lower potential to cause systemic toxicity. Harrison et al. (1981) studied the gastrointestinal absorption in animals of two uranium compounds with different solubilities. They showed that uranyl nitrate (soluble) was absorbed seven times more than uranium dioxide (insoluble). Generally, hexavalent uranium, which tends to form relatively soluble compounds, is more likely to be considered a systemic toxicant. However, particles with very low solubility could accumulate within biological systems and persist there for long durations.

Uranium is a reactive element that is able to combine with, and affect the metabolisms of: lactate, citrate, pyruvate, carbonate and phosphate. Uranyl cations bind tenaciously to protein, nucleotides, and as it can be absorbed by phosphate or carbonate compounds. In so, all different forms have singular biological activities and thus, different toxicities. As was already mentioned depleted uranium (DU) is a byproduct of the enrichment process of uranium, highly toxic to humans both radiologically as an alpha particle emitter and chemically as a heavy metal. Still, the major toxicological concern of U^{238} excess is biochemical rather than radiochemical. In fact uranium, in the form of uranyl nitrate hexahydrate, is considered the most potent toxicant (Stokinger et al., 1953; Tannenbaum et al., 1951). The variety of the molecular forms in which uranium can be presented extends by the ability of the uranium atom to form complex connections.

2.2 Physical form

It is very well known that for any kind of particles whatever their composition is (ordinary carbon, metallic-nonradioactive, etc), the smaller the particle the more harmful they are. This is exactly the case of micro or fine particles (aerodynamic diameter between 100 - 0.1 microns) and nano or ultrafine particles (aerodynamic diameter less than 0.1 micron). Reduction in size to the nanoscale level results in an enormous increase of surface to volume ratio, so relatively more molecules of the chemical are present on the surface, thus enhancing the intrinsic toxicity (Donaldson et al., 2004). Mankind has lived with low-level background radiation for as long as we have existed but, the uranium in a DU weapon explodes on impact as it penetrates a target. It burns at extremely high temperatures (above 5,000 degrees centigrade) and in the process vaporizes into very small (micro and nano) particles. These particles become airborne like a gas, polluting the atmosphere and getting transported around the world being able to enter by inhalation to the population at large. Therefore, there are concerns regarding its potential health effects on the general population and due to internalization of DU during military operations, particularly on this subpopulation. The micro and nanometer size uranium particles released after impact are biologically dangerous and undoubtly a growing part of our world since 1991. It has been reported that inhaled nanoparticles reach the blood and may then be distribuited to target sites such as the liver, kidney, brain, lung, heart or blood cells (Oberdörster et al., 1994; MacNee et al., 2000; Kreyling et al., 2004). Still, the hazard from inhaled uranium aerosols or any noxious agent is determined by the likelihood that the agent will reach the site of its toxic action. The two main factors that influence the degree of hazard from toxic airborne particles are: the site of deposition in the respiratory tract and, the fate of the particles within the lungs. The deposition site within the lungs depends mainly on the particle size of the inhaled aerosol, while the subsequent fate of the particle depends on the physico-chemical properties of the inhaled particle as well as of the physiological status of the lung and target organs of the individual. For humans, inhalation is the most frequent route of access and therefore, the process of aggregation of the nanoparticles in the inhaled air has to be taken into account. Nanoparticles may translocate through membranes and there is little evidence for an intact cellular or sub-cellular protection mechanism. The typical path within the organ and/or cell which may be the result of either diffusion or active intracellular transportation is also of relevance. Very little information on these aspects is presently available and this implies that there is an urgent need for toxicokinetic data for nanoparticles.

2.3 Health effects by route of exposure

Uranium health effects studies derive largely from epidemiology and toxicological animal models. This contaminant can enter the body through inhalation, ingestion or by dermal contact and its toxicity has been demonstrated for different organs. Health effects associated with oral or dermal exposure to natural and depleted uranium (DU) appear to be solely chemical in nature and not radiological, while those from inhalation exposure may also include a slight radiological component, especially if the exposure is chronic. In general, ingested uranium is less toxic than inhaled uranium, which may be also partly attributable to the relatively low gastrointestinal absorption of uranium compounds. Because natural uranium and DU produce very little radioactivity per mass, the renal and respiratory effects from exposure of humans and animals to uranium is usually attributed to its chemical properties. Thus, the toxicity of uranium varies according to its chemical form as well as to the route of exposure.

2.3.1 Inhalation route

Inhalation represents one of the most important occupational risk of uranium exposure especially for workers at the uranium mines. Workers are exposed to both, natural uranium (moderately radioactive) as enriched uranium (highly radioactive). However, to a lesser extent, uranium dust can also enter percutaneously (direct contact or through contaminated clothes), subcutaneously (through wounds in the skin and mucous) and orally (ingestion). Epidemiological studies indicate that routine exposure of humans to airborne uranium (in the workplace and the environment at large) is not associated with increased mortality. In fact, data of several mortality assessments of populations living near uranium mining and milling operations have not demonstrated significant associations between mortality and exposure to uranium (Boice et al., 2003, 2007, 2010). However, it has been reported in humans, that brief accidental exposures to very high concentrations of uranium hexafluoride have caused fatalities. In addition, laboratory studies in animals indicate that inhalation exposure to certain uranium compounds can be fatal (ATSDR). It has to be pointed out that these deaths are believed to result from renal failure caused by absorbed uranium.

The toxicity of uranium compounds to the lungs and distal organs varies when exposed by the inhalation route. The respiratory tract acts as a serial filter system and in each of its compartments (nose, larynx, airways, and alveoli). The mechanisms of particle deposition may change for each compartment as well as for the particle size that entered. Nanoparticles are primarily displaced by Brownian motion and therefore underlie diffusive transport and deposition mechanisms. It means that the smaller the particle, higher the probability of a particle to reach the epithelium of the lung. In general, by the inhalation route, the more soluble compounds (uranyl fluoride, uranium tetrachloride, uranyl nitrate hexahydrate) are less toxic to the lungs but more toxic systemically. Early studies with UF_6 demonstrated that this uranium type may present both chemical and radiological hazards. UF_6 is one of the most highly soluble industrial uranium compounds and when airborne, hydrolyzes rapidly on contact with water to form hydrofluoric acid (HF) and uranyl fluoride (UO_2F_2) as follows: $UF_6 + 2H_2O \longrightarrow UO_2F_2 + 4HF$. Thus, an inhalation exposure to UF_6 is actually an inhalation exposure to a mixture of fluorides. Chemical toxicity may involve pulmonary irritation, corrosion or edema from the HF component and/or renal injury from the uranium component (Fisher et al., 1991). The acute-duration LC50 (lethal concentration, 50% death) for uranium hexafluoride has been calculated for rats and guinea pigs (Leach et al., 1948). In these experiments, animals were exposed to uranium hexafluoride in a nose-only exposure for periods of up to 10 minutes and observed during 14 days. Lethality data suggested that rats are more resistant to UF_6 -induced lethality than are guinea pigs (total mortality of 34% and 46% respectively), proving that the biological response depends also on the host being species specific. It is worth to note that although animals were exposed to uranium via inhalation, histopathological examination indicated that renal injury, but not lung injury, was the primary cause of death (Leach et al., 1948, 1970). However, animals that died during or shortly after exposure had congestion, acute inflammation, and focal degeneration of the upper respiratory tract. The tracheas, bronchi, and lungs exhibited acute inflammation with epithelial degeneration, acute bronchial inflammation, and acute pulmonary edema and inflammation, respectively.

On the contrary, though inhalation exposure insoluble salts and oxides (uranium tetrafluoride, uranium dioxide, uranium trioxide, triuranium octaoxide) are more toxic to the lungs due to the longer retention time in the lung tissue, they are less toxic to distal organs. Harris et al. (1961) found prolonged half lives (120 days or more) for both dioxide and trioxide uranium insoluble compounds. Although insoluble uranium compounds are also lethal to animals by the inhalation route, it occurs at higher concentrations than soluble compounds.

Three different mechanisms are involved in the removal of particles from the respiratory tract. The first is mucociliary action in the upper respiratory tract (trachea, bronchi, bronchioles, and terminal bronchioles), which sweeps particles deposited there into the throat, where they are either swallowed into the gastrointestinal tract or spat out. The second mechanisms is the dissolution (which leads to absorption into the bloodstream) and the third one, the phagocytosis of the particles deposited in the deep respiratory tract (respiratory bronchioles, alveolar ducts, and alveolar sacs). After deposition of insoluble particles in the respiratory tract, translocation may potentially occur to the lung interstitium, the brain, liver, spleen and possibly to the foetus in pregnant females (MacNee et al., 2000; Oberdörster et al., 2002). It as to be emphasized that up to date there is extremely limited data available on these pathways. Several studies demonstrated that particles, whatever the element, triggered pro-inflammatory response characterized by upregulation of cytokine levels and/or immune cell density in lungs after inhalation of particulate matter. This inflammation was induced by particles of various sizes such as nanoparticles or ultra fine particles (Inoue et al., 2005; Stoeger et al., 2006), or by soluble transition metals (McNeilly et al., 2005). Induction of diverse inflammatory reactions was also reported following uranium contamination in different tissues. For instance, activation of cytokine expression and/or production was noted either in pulmonary tissues following uranium exposure by inhalation (Monleau et al., 2006) or in macrophages after in vitro contamination (Gazin et al., 2004; Wan et al., 2006)

2.3.2 Oral route (ingestion)

Experimental studies in humans consistently show that absorption of uranium by the oral route is <5%. Still, this is for the population at large, the main route of uranium entry to the body. UNSCEAR (United Nations Scientific Committee on the Effects of Atomic Radiation, 1993) has considered that limits for natural and depleted uranium in drinking water (the most important source of human exposure) should be based on the chemical toxicity rather than on a radiological toxicity, which has not been observed in either humans or animals. Evidence from several animal studies showed that the amount of uranium absorbed from the gastrointestinal tract was about 1% (La Touche et al., 1987), although other studies have reported even lower absorption efficencies. The most sensitive target of uranium toxicity to mammals, and perhaps humans, is the kidney. While acute, high-level exposure to uranium compounds can clearly cause nephrotoxicity in humans (Lu & Zhao, 1990; Pavlakis et al., 1996), the evidence for similar toxicity as the result of long-term, lower-level occupational exposures is equivocal. In 1987 ATSDR (Agency for Toxic Substances and Disease Registry U.S.) established a minimum level of risk (MRL) for uranium ingestion. Several epidemiology studies (Kurttio et al., 2002; Zamora et al., 1998, 2009) examined the possible association between chronic exposure to elevated levels of uranium in drinking water and alterations in kidney function. These effects may represent a subclinical manifestation of uranium toxicity not necessarily leading to renal dysfunction. By contrast, chronic ingestion of this toxicant could be the starting point of an irreversible renal injury (Wise Uranium Project 1999). Mao et al. (1995) found a significant association between cumulative uranium exposure (product of uranium concentration in drinking water) and urine albumin levels (expressed as mg/mmol creatinine) in adults living in households with elevated uranium levels in drinking water. In accordance, Zamora et al. (1998, 2009) found a significant association between β2-microglobulin, and alkaline phosphatase levels observed in residents living in an area of high uranium levels in the drinking water.

Besides drinking water, uranium can entered the body through the ingestion of contaminated meat and/or fish. Smith & Black (1975) measured the uranium content in the tissue of cattle that graze near the uranium mines being slightly higher than the amount found in control non-contaminated animals. In humans, a study comparing uranium absorption between subjects primarily exposed to uranium in the diet and subjects exposed to elevated levels of uranium in the drinking water (Zamora et al., 2002) did not find significant differences in fractional absorption between these two subroutes.

Studies in rats suggest that the primary pathway for gastrointestinal absorption of soluble uranium is through the small intestinal epithelium (Dublineau et al., 2005, 2006) via the transcellular pathway (Dublineau et al., 2005). In the event of ingestion, the digestive tract is the first biological system exposed to uranium intake via the intestinal lumen. However, little research has addressed the biological consequences of a contamination with uranium on intestinal properties such as the barrier function and/or the immune status of this tissue. Dublineau et al. (2006, 2007) studied both acute contamination with DU at high doses and chronic contamination at low doses on inflammatory reactions in the intestine when orally delivered. The authors found that acute and chronic ingestion of DU modulated expression and/or production of cytokines in the intestine and had similar effects than those observed with lead on the nitric oxide pathway.

2.3.3 Dermal contact

2.3.3.1 Percutaneous entry route

For uranium workers, either in the mines or involved in the mining processes, the percutaneous route is after inhalation is the second main route of uranium contamination. The dust of uranium compounds can permeate clothes and, depending on its solubility, penetrate through the skin. Orcutt et al. (1949) reported that the percutaneous route is an effective mean of entry for soluble uranium compounds.

Our group demonstrated (de Rey et al., 1983), the existence of a differential percutaneous absorption for soluble and insoluble uranium compounds after topical application in rats. By transmission electron microscopy (TEM) we were able to localize these heavy compounds within the tissues. Almost immediately, dense deposits of soluble uranyl nitrate were observed at the epidermal barrier level, 24 h later these deposits were seen close to the endothelium and 72 h no traces of uranium was found neither in the epidermis nor in the dermis indicating that, the uranium had been absorbed into the blood. Mortality (due to renal failure) and body weight measurements indicated the high toxicity of uranyl nitrate and other soluble uranium compounds tested. On the contrary, no variations on these parameters were found when uranium dioxide was employed. Later, changes in skin thickness and permeability after percutaneously chronic exposure of uranium industrial products Peccorini et al. (1990), and of uranium trioxide (Ubios et al., 1997) were reported. We found that, in addition to the systemic effects, such as loss of body weight and nephropathy, the transepithelial penetration and the subcutaneous implantation of uranium induced structural alterations in the stratified squamous epithelium which lead to epidermic atrophy and increased permeability of the skin. In 2000, we demonstrated that there is an inverse relation between the area of the surface exposed to uranium and the time of exposure with, the subsequent percutaneous toxicity (López et al., 2000). We concluded that the larger the area exposed to uranium or the longer the exposure time, the lower was the rate of survival.

2.3.3.2 Subcutaneous entry route

Subcutaneous or intradermal uranium contamination takes place in the presence of a wound. This possibility becomes a real risk to workers daily handling uranium dust and nowadays it also includes soldiers who fought in the modern wars (Balkan, Gulf, etc). Penetration of DU shrapnel bullets into the skin became an issue of increasing attention. In fact, the only cases in which there were documented exposures to uranium are those of the Gulf War veterans who retained depleted uranium shrapnel fragments (McDiarmid et al., 2000, 2004, 2007).

In an experimental model of subcutaneous implantation of uranium dioxide (insoluble) in rats de Rey et al. (1984) showed that animals receiving doses greater than 0.01 g / kg died within the first six days due to acute renal failure. Uranium contamination by this route of exposure showed no differences regarding the type of particle. Histological analysis revealed the presence of deposits of uranium taken up by macrophages at 24 and 48 h post exposure. Deposits were found between the endothelial cells and the renal parenchyma, suggesting that the transport and deposition of uranium insoluble compound implanted subcutaneously occurs.

2.4 Uranium biokinetics

As was mentioned before, uranium can enter the human body through inhalation, ingestion or through the skin. Measurement of the quantities of uranium and its biokinetics can be performed in vivo, ex vivo and in vitro. In vivo techniques measure the quantities of internally deposited uranium directly using a whole-body counter, ex vivo techniques permit estimation of internally deposited uranium by analysis of body fluids (urine, blood, feces), or (in rare instances) tissues obtained through biopsy or postmortem tissue sectioning and in vitro allows to study the mechanism by which uranium interferes with cellular organelles and molecules (USTUR 2011).

The large majority of uranium (>95%) that enters the body is not absorbed and is eliminated from the body via the feces. Excretion of absorbed uranium is mainly via the kidney. Absorption of inhaled uranium compounds takes place in the respiratory tract via transfer across cell membranes. The deposition of inhalable uranium dust particles in the lungs depends on the particle size, and its absorption depends on its solubility in biological fluids (ICRP 1996). Estimates of systemic absorption from inhaled uranium-containing dusts in occupational settings based on urinary excretion of uranium range from 0.76 to 5%.

Gastrointestinal absorption of uranium can vary from <0.1 to 6%, depending on the solubility of the uranium compound. Studies in volunteers indicate that approximately 2% of the uranium from drinking water and dietary sources is absorbed in humans (Leggett & Harrison, 1995), while a comprehensive review indicates that the absorption is 0.2% for insoluble compounds and 2% for soluble hexavalent compounds (ICRP 1996).

Data on dermal absorption of uranium is limited. In hairless rats, dermal exposure to uranyl nitrate resulted in 0.4% of the dose being absorbed (Petitot et al., 2007a, 2007b); damage to the skin resulted in higher absorption efficiencies.

Although no data are currently available regarding the metabolism or biotransformation of uranium in vivo, for either humans or animals it is known that following absorption, uranium forms soluble complexes with bicarbonates, citrates and proteins, all of which are present in high concentrations in the body (Cooper et al., 1982). Regardless of the route of entry, the absorbed uranium is distributed widely but preferably is deposited in bone, kidney and liver. Uranium, once in the bloodstream, has a very short plasma half-life. Approximately 60% is eliminated in the first 24h in the urine (Walinder et al., 1967). Laboratory animal data indicate that a fraction of the uranium in the plasma is associated with low molecular weight proteins ultrafilterable while the rest is bound to transferrin and other plasma proteins (ICRP 1995). In body fluids, tetravalent uranium tends to oxidize to the hexavalent form, followed by the formation of the uranyl ion. Wrenn et al. (1985) showed that 90% of the uranium is excreted in feces and the remaining 10% in urine while the uranium deposited in external soft tissues is removed very slowly (Hursh et al., 1969).

As the general population and workers involved in uranium mining and manufacture of uranium devices are exposed to this heavy metal toxicant, not only the study on health should be encourage but, , effective management of waste uranium compounds is necessary to prevent uranium exposure. In the next section, a more detailed description on the risks associated with uranium exposure is presented.

3. Toxicity of uranium: Cellular mechanisms

The primary purpose of this second section of the chapter is to provide an overall perspective on the toxicology of uranium. It contains descriptions and evaluations of toxicological studies in vivo and in vitro and provides conclusions, where possible, on the relevance of uranium toxicity and toxicokinetic data to public health.

As described in section I uranium, depending on the route of entry and the dose, produces structural and functional alterations in target organs mainly in bone, kidney, and lung, and may even compromise the individual's life.

3.1 Uranium in vivo toxicological effects: Uranium as a heavy metal particle

In general, when uranium enters the organism, it accumulates in a non soluble form in hepatocytes, kidney proximal tubule cells and macrophages or macrophages-like cells present in tissues throughout the body (lung, liver, spleen, skin and bone). In each of the cells mentioned above, uranium is specifically concentrated by lysosomes where the actinides are precipitated as insoluble phosphates. The mechanism of intralysosomal concentration may be explained by the high phosphatase activity of these organelles. Moreover, uranium and phosphate have a strong chemical affinity for each other thus, as the DNA and mitochondria are loaded with phosphate, uranium may be considered a DNA and mitochondria deep penetration bomb attacking on fundamental cellular levels. Limited data exists regarding in vivo genotoxicity in humans following exposure to uranium. The only cases in which there were documented exposures to uranium are those of the Gulf War veterans who retained depleted uranium shrapnel fragments (McDiarmid et al., 2000, 2001, 2004, 2007, 2009)

It is well known that the biochemical reaction to heavy metals can alter cellular mechanisms, principally oxidative metabolism, leading to genetic mutations which in turn, can restrain cell growth and cause cancer. Heavy metals, that are also radioactive, amplify these effects. Several reports have shown that uranium, both toxic and radioactive, induces oxidative stress causing adverse biological effects which include as was seen for heavy metals DNA damage, cancer and other neurological defects (Miller et al., 2002; Abou-Donia, 2002; Barber et al., 2007). Among heavy metals lead, aluminum and mercury have been shown to dramatically increase cytogenotoxicity. Interestingly, lead is the final end product of the step by step radioactive decay of uranium. Therefore, it would not be farfetched to imagine that uranium and lead may have very similar chemical characteristics though, uranium is twice as dense. In fact, some results regarding the effect of depleted uranium (DU) on the nitric oxide (NO) pathway almost mimic the observed with lead (see below). The interaction of lead with sulfhydryl (SH) sites causes most of its toxic effects, which include impaired synthesis of RNA, DNA and protein, diminished antioxidants (glutathione), and interferes with the metabolism of vitamin D. Lead may also affect the body's ability to utilize the essential elements calcium, magnesium, and zinc.

3.1.1 Cancer

Generally reports examine lung cancer mortality among two subpopulations: smokers and non-smokers uranium miners and, soldiers who participated in the unfortunate modern armed conflicts during and after 1991. During the mining process, uranium particles and its decay products such as ^{222}Ra are released into the environment. Workers in uranium mines and the people living nearby are likely to inhale and ingest suspended air particles containing uranium and radon. Inhalation of uranium particle increases the frequency of chromosomal aberrations (WISE Uranium Project, 1999) and the risk of lung cancer (WISE Uranium Project, 1999). Therefore, uranium aerosolized nanoparticles, both as a heavy metal particle and due to its radioactivity when enter the respiratory system and deposited in the respiratory mucosa, are responsible for the induction of this pathology (lung cancer). Saccomano et al. (1996) in order to evaluate the incidence of tumors; their cell types; and the relationship of particulate size to their position in the bronchial tree conducted a retrospective and comparative study from 1947-1991. They studied a cohort of 467 uranium miners and 311 non-miners with lung cancer and concluded that inhaled uranium containing dust, radon, and cigarette smoke combine to form large particulates that deposit in the central bronchial tree. Furthermore, they show that the proportion of lung cancers in the central zone was significantly greater in miners than in non-miners presumably due to the deposition of radon decay products attached to the silica dust particles. More recently, two new reports show association between uranium miners and smokers. The first study took place in France revealing significant association between the relative risk of lung cancer and silicosis. Amabile et al. (2009) demonstrated that the relation between radon and lung cancer persisted even after adjusting the data for smoking and silicotic status but, these authors remind us of the complexity involved in assessing occupational risks in the case of multiple sources of exposure. The second investigation was done among German uranium miners (Schnelzer et al., 2010). Adversely in this study, the authors concluded that stability of the uranium-related lung cancer risks with and without adjustment for smoking could suggest that smoking does not act as a major confounder at least in the cohort study. In brief, a number of studies reported death from lung cancers from occupational inhalation exposure of mine workers however; the available studies document no lung cancers solely from inhaled uranium- bearing dust. It is generally accepted that lung cancers developed subsequent to inhalation of uranium- containing dusts were principally due to radon daughters and long-term cigarette smoking, and not to uranium metallotoxicity or uranium radioactive emissions.

In the months and years following the Balkan (1999) and Gulf (1991) wars a large number of soldiers, UN peacekeepers, and civilians have exhibited unexpected and unexplained health problems, including excess leukemias and other cancers, neurological disorders, birth defects, and a constellation of symptoms loosely gathered under the rubric "Gulf War Illnesses", suggesting that the use of DU during these conflicts could be considered a possible cause. Thus, this is another subpopulation where the action of uranium and its possible link to cancer is important to be studied. In 2004, Tirmarch et al. (2004) reviewed the epidemiological knowledge of uranium, the means of exposure and the associated risk of cancer. These authors concluded that only studies with a precise reconstruction of doses and sufficient numbers of individuals could allow a better assessment of the risks associated with uranium exposure at levels encountered during conflicts using depleted uranium weapons. Nevertheless, it is well known that when uranium binds to DNA it can damage DNA directly, or indirectly by altering uranium related DNA signaling mechanism. Cell mutations can either result in cell death or may trigger a whole slew of protein replication errors, some of which can lead to different cancer types. In fact, the incidence of cancer has increased markedly in Iraq following the Gulf War. As was reported by Aitken et al. (1999) there are some areas in southern Iraq that have experienced a two- to five fold increase in reported cancers. In most of these cases the lung, bronchial tubes, bladder, and skin are damaged. In addition, increased incidence of stomach cancer in males and breast cancer in females has also been reported, as well as an overall increase in leukemia cases.

3.1.2 The respiratory system

Exposure by inhalation to uranium dust particles can lead, as function of its solubility, to uranium accumulation predominantly in the lungs and tracheobronchial lymph nodes, as well as the development of neoplasia and fibrosis at the pulmonary level (ATSDR, 1999). Particle genotoxicity can be caused by direct actions or by indirect mechanisms often mediated by reactive oxygen species (ROS) produced mainly by the inflammatory cells (Kirsch-Volders et al., 2003; Martin et al., 1997). In agreement with these observations, Monleau et al., (2006), demonstrated DNA strand breaks in lung rats after DU acute and chronic exposure by inhalation, was a consequence of oxidative stress and induction of pro inflammatory IL8 and TNFα gene expression. These effects seemed to be linked to the DU doses and independent of the solubility of uranium compound.

3.1.3 Excretory systems

As in the case of other heavy metals, a considerable body of evidence suggests that overexposure to uranium may cause pathological alterations to the kidneys in both humans and animals. Studies have shown that the solubility of uranium compounds plays a significant role in the amount of damage occurred in the kidneys. Inhaled uranium compounds with slow- to medium-dissolution rates are relatively insoluble, and are therefore retained longer in the lungs, resulting in lower toxicity to the kidneys and other distal organs. In 1949, under the Manhattan Project, Voegtlin & Hodge observed, lesions in the renal tubules regardless the route of entry for uranium. In 1982 Haley showed that all uranium compounds, uranyl nitrate, proved to be the most nephrotoxic being the most obvious effect the damage of the proximal convoluted tubules (ATSDR, 1999). As detailed previously, uranium can link to carbonates. Human renal effects following acute exposure to DU includes proteinuria and abnormal phenol sulfonphthalein excretion. In addition, increased urinary catalase activity and diuresis have also been found (ATSDR, 1999). McDiarmid et al. (2000) reported an increase in a variety of renal function parameters such as serum creatinine, β2-microglobulin, retinol binding protein, serum uric acid, urine creatinine, and urine protein in patient veterans. Although β2-microglobulin concentrations were higher and urine protein concentrations were lower in patients exposed to DU, no significant relationships were found between these parameters in neither control and uranium exposed groups.

In agreement with McDiarmid et al., we found that the incorporation of uranium by oral exposure in mice (Martinez et al., 2000) provoke a markedly increase of urea and creatinine levels when compared to controls. These biochemical parameters corresponded to uranium-severe alterations such as widening of the urinary space, cell vacuolization in the convoluted tubules and a large amount of hyaline casts as was seen by the histopathological study of the kidneys (Martínez et al., 2003).

3.1.4 The bone skeletal system

The 25% of the systemically administered uranium deposits in the skeleton linked to the newly formed bone. It is possible to find uranium in bone formation fronts building a critical deposit organ in chronic intoxications. An initial deposit of uranium was demonstrated by autoradiography at the endosteal and periosteal surfaces and haversian bone, areas where particularly calcification took place (Neuman et al., 1948). Our group was the first to demonstrate that acute poisoning with uranyl nitrate, inhibits endochondral ossification with reduced bone surfaces covered by active osteoblasts and a consequent increase in inactive osteoblasts (Guglielmotti et al., 1984). In this case, we proposed that the toxic effect of uranium would be causing an alteration of the differentiation process of osteoblasts and/or their precursors, resulting in the formation of sealing trabeculae on metaphyseal bone. Guglielmotti et al. (1985) proposed alveolar wound healing as another useful model for the study of bone formation as it is considered a sensitive indicator of bone damage under various experimental conditions. In this model we observed after uranium acute intoxication, not only inhibition of bone formation but, inhibition of alveolar bone healing after extraction (Guglielmotti et al., 1987). Later, our group studied the toxic effect of uranyl nitrate on bone modeling and remodeling by performing histomorphometric measurements in the periodontal cortical bone in dental alveolus of mandibles of rats (Ubios et al., 1991). Our results revealed a decrease in bone formation in rats treated with uranium. On the remodeling side the decrease in bone formation was coupled to an increase in bone resorption where on the modeling side no bone resorption was observed and the decrease in bone formation was linked to an increase in resting bone zones. Because osteoblasts play a significant role in bone formation, it is possible that uranyl nitrate can directly affect these cells and their precursors by binding to cell membranes. Based on these data, uranium toxicity may be viewed as a potential contributor to osteoporosis or other osteopenic diseases in exposed individuals (Ubios et al., 1991). Bone growth was found to be impaired in tibiae (Ubios et al 1995) and mandibles (Ubios et al 1998) after exposure to uranyl nitrate and to uranium dioxide (Diaz Sylvester et al 2002). In addition, we found a lower degree of eruption and tooth development in lactant rodents exposed orally to acute doses of uranyl nitrate (Pujadas Bigi et al., 2003). More recently, we demonstrated that uranyl nitrate induced severe ultrastructural alterations both in active and inactive osteoblast revealing fragmented and swollen RER, Golgi and puffy nuclei with fine granular content (Tasat et al., 2007).

It is well known that vitamin D (1,25(OH)(2)D(3)) is a hormone essential in mineral and bone homeostasis. The effect of acute and chronic DU exposure on the active vitamin D metabolism was investigated in an experimental animal model by Tissandie et al. (2006, 2007). In acute DU intoxicated-rats, cytochrome P450 (CYP27A1, CYP2R1, CYP27B1, CYP24A1), enzymes involved in vitamin D metabolism and, two vitamin D(3)-target genes (ECaC1, CaBP-D9K) were assessed by real time RT-PCR in liver and kidneys. It was seen that DU modulated both activity and expression of CYP enzymes involved in vitamin D metabolism and consequently affected vitamin D target genes levels. In DU chronic-intoxicated rats, through drinking water, active vitamin D (3) significantly decreased in plasma level. In kidney, a decreased gene expression was observed for cyp24a1, the principal regulators of CYP24A1. Similarly, mRNA levels of vitamin D target genes ecac1, cabp-d28k and ncx-1, involved in renal calcium transport were decreased in this organ. Then, it is clear that DU affected both the vitamin D active form, its receptor expression, and consequently modulated the expression of cyp24a1 and vitamin D target genes involved in calcium homeostasis.

3.1.5 The reproductive system

In vivo heavy metal studies have demonstrated that carcinogenic effects can occur in unexposed offspring. Similarly, preconception paternal irradiation has been implicated as a causal factor in childhood cancer and it has been suggested that this paternal exposure to radiation may play a role in the occurrence of leukemia and other cancers to offspring.

There are few studies that have examined the effects of DU on human reproductive tract and development.

Back in 1967, Muller et al. reported that male uranium miners were found to have more firstborn female children than expected. In early 2000, decreased fertility, embryo/fetal toxicity including teratogenicity and reduced growth of the offspring was observed following uranium exposure at different gestation periods by Domingo et al. (2001). In 2002, when monitoring veterans of Gulf War, McDiarmid et al. reported adverse health effects on the reproductive and central nervous systems of DU-exposed veterans. More recently, Miller et al. (2010) investigated the possibility that chronic preconception paternal DU exposure could lead to transgenerational transmission of genomic instability. They showed that implantation of DU pellets in male mice for seven months followed by mating with untreated females resulted in transmission of genetic damage to somatic cells of offspring. Several issues remain unknown: the exact mechanism by which this occurred and if it is DU-radiation or DU-chemical effects theresponsible for transgenerational transmission of factor(s) leading to genomic instability in F1 progeny from DU-exposed fathers.

3.1.6 The nervous system

Despite evidence suggesting a link between neurological toxicity and DU exposure, as was reported by McDiarmid et al. (2002), there is no data clearly demonstrating that excess neurologic disease/mortality risk is associated with uranium exposure. Preliminary animal studies failed to demonstrate any performance deficiencies in locomotor activity, discrimination learning, and general functional observations (Pellmar et al., 1999). At that moment and because the methods used to examine those endpoints could have been insensitive to reveal subtle cognitive effects, the Gulf War Executive Summary recommended further studies of cognitive function, neurophysiological responses, and brain DU concentrations in Gulf War veterans (Fulco et al., 2000). Several follow-up studies on Gulf War veterans with DU fragments embedded in their soft tissues have led to suggest the possible role of DU as an inductor of neurotoxicity. At present, some experimental studies suggest a positive link between neurological toxicity and uranium. In 2005, studies in rodents indicated that DU readily traverses the blood-brain barrier, accumulates in specific brain regions, and results in increased oxidative stress, altered electrophysiological profiles, and sensorimotor deficits (Briner & Murray, 2005). In this case, after uranium crosses the blood-brain barrier behavioral changes and lipid oxidation were observed in rats, in as little as 2 weeks. In agreement Monleau et al. (2005) proved that after repeated exposure to uranium dioxide particles by inhalation, uranium bioaccumulates in the brain producing behavioral changes (Monleau et al., 2005). A possible mechanism of this neurotoxicity could be the oxidative stress induced by reactive oxygen species imbalance. After chronic ingeston of DU Lestaevel et al. (2009) analyzed the expression and /or the activity of the main antioxidant enzymes, superoxide dismutase (SOD), catalase (CAT) and glutathione peroxidase (GPx) in cerebral cortex and found that all of them increased significantly. These results illustrate that oxidative stress plays a key role in the mechanism of uranium neurotoxicity. On the contrary, the nitic oxide (NO) pathway was almost unperturbed.

3.1.7 Are there common metabolic pathways for in vivo uranium exposure?

From the above studies at least two common metabolic pathways for uranium exposure could be proposed.

3.1.7.1 The oxidative and antioxidative pathway

As was mentioned above, exposure to metallic environmental toxicants has been demonstrated to induce a variety of oxidative stress responses in mammalian cells. Environmental stressors such as chemical toxicants can create oxidizing imbalances in the cellular redox state resulting in a loss of reducing potential, a condition termed "oxidative stress". Heavy metals are environmental persistent toxicants that have been shown to exert oxidative stress on living systems through the production of reactive oxygen species (ROS), which overwhelm the cell's capacity to maintain a reduced state thus, damaging various cellular components (Ercal et al., 2001; Stohs & Bagchi, 1995). The main threats to human health from heavy metals are associated with exposure to lead, cadmium, mercury and arsenic (arsenic is a metalloid, but is usually classified as a heavy metal). These metals have been extensively studied and their effects on human health regularly reviewed by the international research community and organizations. Once again there are few studies exploring the effect of uranium on the oxidative metabolism. Nevertheless, as uranium and lead are both heavy metals, the mechanism/s by which they induce cell damage could be probably alike.

Studies on the pathogenetic effect of lead showed its action is multifactorial since it directly interrupts the activity of enzymes, competitively inhibits absorption of important trace minerals and deactivates antioxidant sulphydryl pools (Patrick et al., 2006). In 2001, Ercal et al. proposed two independent but related mechanisms for free radical-induced damage by lead (Ercal et al., 2001). The first involved the direct formation of reactive oxigen species (ROS) and the second, the depletion of the cellular antioxidant pool. Still, interrelations between these two mechanisms exist so that the increase in ROS on one side simultaneously leads to depletion of antioxidant pools on the other (Gurer & Ercal, 2000). The major effect of lead is on glutathione metabolism (Hunaiti & Soud, 2000) with glutathione representing more than 90% of the non-tissue sulphur pool of human body. The sulphydryl groups of glutathione bind effectively to toxic metals such as arsenic and mercury. Therefore, an organism exposed to lead has significantly lowered levels of glutathione, with respect to control groups, which may in turn induce an imbalance in the oxidative metabolism concomitantly increasing the production of ROS. Metal-induced ROS cause damage to cellular proteins, nucleic acids and lipids, resulting in a variety of cellular dysfunctions including cell death. Intracellularly, ROS can modulate gene expression by interfering with signal transduction cell proliferation pathway activating various transcription factors, thus controlling cell cycle progression and apoptosis (Evan & Vousden, 2001). The most important pathway involves the nuclear factors NF-kB, AP-1, etc., and the tumour suppressor protein p53. Furthermore, deregulation of cell growth and differentiation is a typical characteristic of the cancer phenotype, and cancer is a multifactorial disease, it could be hypothesized that uranium, as a heavy metal, could be one of the factors implicated. Mammalian cells have developed multiple homeostatic systems to counteract the effects of metal induced- oxidative stress by scavenging free radicals and repairing oxidant damage to biomolecules. There are two specific enzymes, glutathione reductase (GR) and deltaaminolevulinic acid dehydrogenase (ALAD) that are both inhibited by lead (Hoffman et al., 2000). In large part, in response to oxidative stress, the transcription factor, nuclear factor erythroid 2-related factor 2 (Nrf2) is activated and coordinates the expression of antioxidant gene products (Osburn & Kensler, 2008). The Nrf2 activation has been demonstrated in response to a variety of metals including lead (Korashy & El-Kadi, 2006). In this context, given that several metals are known to generate ROS, and that mammalian cells activate Nrf2-mediated transcription in response to ROS, it could be expected that uranium exposure could activate the Nrf2 pathway. Metal exposure has been shown to exert a number of effects on the Nrf2 pathway including reduction of sulfhydryl groups in Keap-1, MAPK activation, and inhibition of proteasomal pathways which stabilizes Nrf2. The cumulative impact of these events is stabilization and activation of Nrf2 and transcriptional upregulation of anti-oxidant genes. Effective chelators of lead like meso-2,3-dimercapto succinic acid and calcium disodium ethylenediaminetetraacetic acid are used in treatment of lead toxicity (Gurer & Ercal , 2000). Similarly, several chelators have been assayed to inhibit uranium toxic effects in vivo on target organs. This aspect will be described in part 3 of this chapter.

3.1.7.2 The nitric oxide (NO) pathway

Although there are very few studies exploring into uranium on the nitric oxide (NO) pathway there is an extensive series of publications on the effects of lead on this signal pathway (Barbosa et al., 2006; Kong et al., 2000; Zhu et al., 2005). All these studies showed evidence that the NO pathway is a target for lead exposure. Lead-induced inhibition of NO production in central nervous system (Zhu et al., 2005), blood (Barbosa et al., 2006), kidneys (Dursun et al., 2005), inmmune cells (Bishayi & Sengupta, 2006) and in intestine (Kong et al., 2000), suggesting a common mechanism present for these tissues. Therefore, the NO pathway seemed to be a preferential target for uranium effects. DU-induced decreases gene expression, stimulation of enzyme activity of eNOS and slight decrease in iNOS activity, as well as diminution of NO metabolite content (Dublineau et al., 2007). Dublineau et al. (2007) postulated several hypotheses to explain these inhibitory effects of uranium on the NO pathway. The first one involved direct interference of uranium with calcium. An exchange of uranyl ion with calcium has been assumed at bone surface and inhibition of different calcium transporters by uranium was reported several years ago (Desmedt et al., 1993; Thompson & Nechay, 1981). Interactions between metals and calcium were already suggested for lead with transporters or calcium-binding proteins (Kern & Audesirk, 2000; Simons, 1993). Several authors demonstrated that Pb^{+2} could be substituted for Ca^{+2} in the activation of calmodulin, leading to negative effects on iNOS activity (Gribovskaja et al., 2005; Simons, 1993). It can be thus suggested that uranium inhibits the NO production via interaction with calmodulin, but further experiments have to be performed to validate this hypothesis. The second hypothesis for uranium-NO inhibition is the activation of inhibitors of iNOS, such as IL-4 and IL-10 (Simmons & Murphy, 1993). The third hypothesis is an inhibitory effect of uranium induced oxidative stress. In fact, oxidative stress has been observed following uranium contamination in different systems (Linares et al., 2006; Periyakaruppan et al., 2007; Tasat et al., 2007). Here again, similarity between uranium and lead may be noticed because lead-induced oxidative stress has been shown as causing NO inactivation (Vaziri & Ding, 2001).

3.2 In vitro uranium cytotoxicity

Although in vivo studies demonstrate that uranium toxicity removes the body's supply of antioxidant enzymes as glutathione, superoxide dismutase (SOD), catalase (CAT) allowing free radicals to run uncontrolled through tissues and organs, in vitro studies must be performed in order to understand the intracellular mechanisms of cell response to this toxicant (Periyakaruppan et al., 2007).

Unlike the abundant research on the effect of uranium in various organs in vivo, the cellular effects have only been studied in a limited number of cell culture models in vitro. After inhalation and deposition of uranium, particles principally reach two main target cells: macrophages and epithelial cells (Schins & Borm, 1999). Uranium uptake by alveolar macrophages has been shown to occur after inhalation of soluble and insoluble compounds

(Tasat & de Rey, 1987; Berry et al., 1997) and subsequent uranium accumulation was proved to activate macrophages which in turn, secrete different mediators like pro- and anti-inflammatory cytokines (Driscoll, 2000; Driscoll et al., 1997). Thus, herein we first describe uranium possible intracellular mechanism on these two cell types.

In rat lung epithelial cells Periyakaruppan et al. (2007) found that uranium induced a significant oxidative stress and a decrease in cell proliferation. These findings were attributed to a reduction in the antioxidant potential of the cells due specifically to loss of total glutathione and superoxide dismutase. This investigation pointed out the ineffectiveness of antioxidant system's response to induced -uranium oxidative stress in the cells. More recently, these same researchers (Periyakaruppan et al., 2009) showed that oxidative stress may lead to apoptotic signaling pathways. Epithelial lung cells treated with DU resulted in a dose and time-dependent increase in the activity of caspases-3 and -8, both enzymes involved in the apoptotic process. Xie et al. (2010) observed that DU human bronchial lung cells exposure can transform and induce significant chromosome instability. Most investigators agreed in that chronic DU contamination increases chemokine CCL-2, pro-inflammatory cytokine IL-1b and anti-inflammatory cytokine IL-10 (Wan et al., 2006).

Macrophages or macrophage-like cells are present in large numbers in tissues throughout the body including the lung, liver, bone, lymph nodes, brain, kidney, skin, and spleen. More often than not, alveolar macrophages are the cells in charge of cleaning organic and inorganic inhaled particles from the lungs (Fels & Cohn, 1986; Lohmann-Matthes et al., 1994). As we (Tasat & de Rey, 1987) previously showed by scanning electron microscopy (SEM) and transmission electron microscopy (TEM), after in vitro phagocytosis of uranium dioxide, macrophages are implicated in the retention of these particles resulting in what is known as "activated macrophages". Even more, we observed that uranium rat alveolar macrophage exposure resulted in time and concentration dependent uptake of uranium particles, citotoxicity, and induction of cell death. In vitro studies showed that these activated macrophages respond by secreting diverse intracellular mediators: reactive oxygen species (Tasat & de Rey, 1987), proinflammatory cytokines (TNFa, IL6) (Driscoll, 2000) and MAPK activation (Gazin et al., 2004) among others. In 2002, Kalinich et al. reported on cultured mouse macrophages that DU in the form of uranyl acetate induced cell death by apoptosis. In agreement with Kallinich we found that uranyl nitrate (12.5-200mM) on cultured rat alveolar macrophages impaired cell viability, induced secretion of superoxide anion (O_2^-), TNFα and apoptosis via caspase -3 and its clivated PARP substrate (Orona, 2009). Furthermore, we explored into the oxidative metabolism and the pro-inmflamatory cytokine profile and suggested that apoptosis could be reached by different intracellular signaling pathways depending on the uranium concentration. In our latest work (Orona et al., unpulished) we showed that when exposed to low doses of uranyl nitrate, high increases of superoxide anion generation may act as the principal mediator and directly damage alveolar macrophage DNA. On the other hand, when macrophages were treated with high DU doses, cell death seems to be encountered as the result of the TNFα receptor activation, also known as the cell death receptor. We suggested that the signaling pathway mediated by O_2^- may be blocked, prevailing damage to DNA by the TNFα route (figure 1).

In accordance Wan et al. (2006) showed in murine macrophages and CD4+ T lymphocytes, that DU cytotoxicity was concentration dependent. By microarray and real-time reverse-transcriptase polymerase chain reaction (RT-PCR) analyses these authors revealed that DU

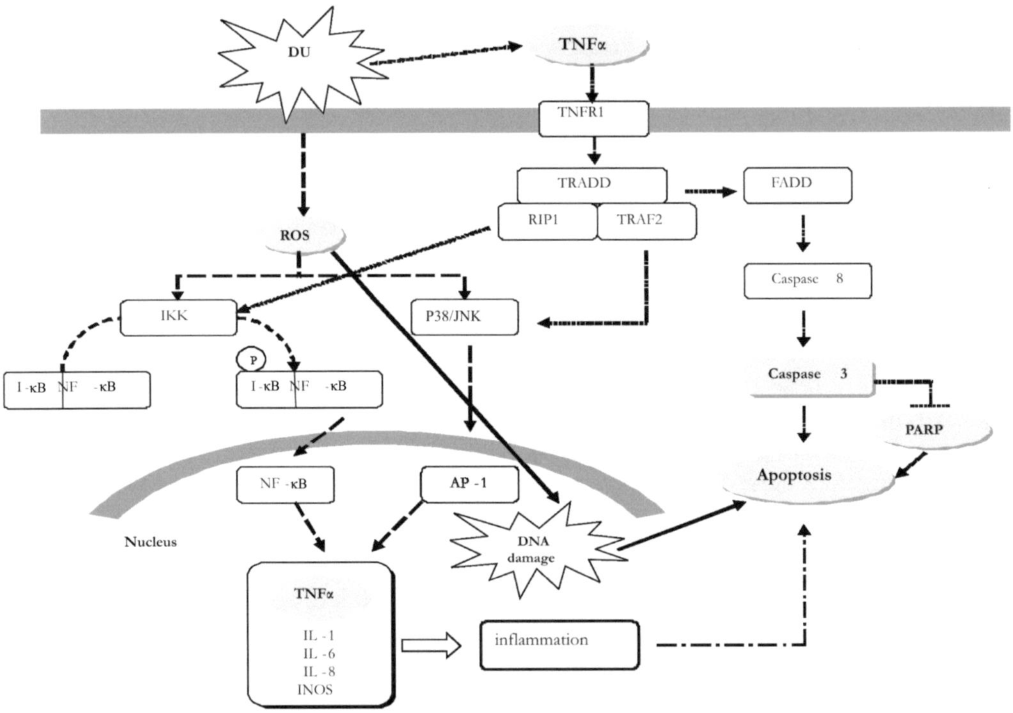

Fig. 1. Proposed model for the effect of low and high depleted uranium (DU) concentrations in cultured rat alveolar macrophages. Orona et al. (unpublished, 2011)

altered gene expression patterns in both cell types being the most differentially expressed genes the ones related to signal transduction, such as c-jun, NF- kappa Bp65, neurotrophic factors (e.g., Mdk), chemokine and chemokine receptors (e.g., TECK/CCL25), and interleukins such as IL-10 and IL-5, indicating a possible involvement of DU in cancer development and autoimmune diseases.

As previously mentioned, bone and kidney are main target organs for uranium deposition, so next we describe the effect of uranium on normal fetal and tumoral osteoblasts and renal cells in cultured. To evaluate the effects of DU in vitro, Miller et al. (2002) employed immortalized human osteoblast cells (HOS) demonstrating its genotoxic and neoplastic activity. Furthermore Miller et al. (2003) showed that DU exposure impacted genomic instability manifested as delayed lethality and micronuclei formation. Still, up to 2003 the cellular and molecular pathways of uranium toxicity in osteoblast cells were unknown. In 2007, we studied (Tasat et al., 2007) on human fetal osteoblasts cells (HFob) in culture, cell proliferation, generation of reactive oxygen species (ROS), apoptosis, and alkaline phosphatase (APh) activity. We found that 1-100 µM of urnayl nitrate in vitro failed to modify cell proliferation ratio, induced apoptosis, increased ROS generation in a dose-dependent manner and decreased APh activity. We then suggested that ROS could play a more complex role in osteoblast physiology than simply causing oxidative damage.

Concentration-dependent cytotoxicity was also observed in NRK-52E cells, an immortalized cell culture model representative of rat kidney proximal epithelium cells (Thiébault et al.,

2007). Researchers have also evaluated the transcriptomic and proteomic responses of HEK293 kidney cells, and renal tissue from rats exposed to DU and found that there were several oxidative-response-related transcripts that were up-regulated, and significantly increased peroxide levels that support the implication of oxidative stress (Prat et al., 2005). Recently, Prat et al. (2010) in a study based on DNA microarrays, analyzed gene expression after acute uranium exposure in several human cell lines taken from kidneys or lungs. They found no common gene between cells originating from lungs and kidneys but, higlighted a gene (SPP1) coding for osteopontin, a secreted protein linked to bone mineralization. They concluded that uranyl ions affected the excretion of osteopontin in a time- and dose-dependent manner and therefore proponed this protein as a potential biomarker of uranyl mineralization effects in vivo.

Other in vitro studies showed toxicity of uranium. On reproductive cells like Chinese Hamster Ovary (CHO), Lin et al. (1993) showed that DU decreased cell viability in a dose dependent manner, reduced cell cycle kinetics (CCK) and increased micronuclei (MN), chromosome aberrations (CA) and sister-chromatid exchanges (SCE) frequencies. This array of DU cytogenetic altered parametes could probably establish a biological basis for the potential teratogenic effect of uranium on developing fetal mice. In 2005, Stearns DM et al., showed in this same mammalian cell line that uranyl acetate could be chemically genotoxic and mutagenic through the formation of strand breaks and covalent U-DNA adducts. Once more health risks for uranium exposure could go beyond those for radiation exposure.

4. Inhibition of uranium toxicity

The general purpose of this third section is to review the studies on chelating agents employed as treatment for acute uranium exposure, and the ability of these compounds to prevent or counteract the effects of uranium on target organs. As was described in the second section, ultrafine uranium particles enter the body mainly by inhalation, and depending on its solubility they remain in the lungs or slowly dissolved and are transported into the blood, getting distributed to bone, kideny, brain, liver, lymph, spleen, testes and other organs. Uranium remains in plasma for a very short period of time (Walinder et al., 1967) while instead in kidney and bone, the highest concentration of this heavy toxic metal can be found (Hursh & Sopor, 1973). In the kidneys, it causes destruction of the proximal convoluted tubules resulting in potentially life threatening renal failure (Adams & Spoor, 1974; Domingo et al., 1987). As regards its effect on bone, acute exposure to uranium was found to inhibit endochondral ossification (Guglielmotti et al., 1984), alveolar wound healing (Guglielmotti et al, 1985), bone formation in the tooth alveolus (Ubios et al., 1991), and maxillary growth (Ubios et al., 1998).

The first detailed study of the action of the salts of uranium was made by Chittenden & Hutchinson in two publications which appeared in 1887 and 1888. The first of these publications was concerned with the effect of various soluble salts of uranium. The authors showed that in very high dilutions especially the uranium nitrate, exerted an inhibitory effect upon the action of saliva on the amylolytic action and the proteolytic action of pepsin and trypsin. The second paper, although it was chiefly concerned with the effect of uranium salts upon protein metabolism, gave a brief account of the toxic effect of uranium in liver and kidney. The ability of uranium to induce a glycosuria was recorded, and the observation was made that the glycosuria usually did not occur until after the appearance of albumin in the urine. The toxic effect of uranium for the liver and for the kidney is usually attributed to the action of the metal as such. Nevertheless, back in 1916, MacNider showed that the toxicity of uranium runs parallel with its ability to lead to the formation of various acid bodies, and that if the appearance of these substances in the urine is delayed and their amount diminished there is less evidence of the toxic action of the metal. Even more, MacNider (1916) showed that uranium nitrate toxic action may be inhibited by the use of an alkali. Since then several has been the chemicals studied in order to prevent the effects of uranium poisoning in its target organs, mainly kidneys and bones.

During the Manhattan Project in the early 1940s, the toxic properties of uranium were thorouly investigated finding out that uranium oxides stick very well to cotton cloth, but did not wash out with soap or laundry detergent (Orcutt, 1949) but, with a 2% solution of sodium bicarbonate. Donnelly & Holman (1942) found that giving sodium citrate to animals protected them from an otherwise lethal toxic dose of uranium. Sodium citrate is a neutralized form of citric acid. The sodium citrate caused the animals to excrete uranium faster, resulting in less uranium deposited in the body. Citrate salts, such as potassium citrate and magnesium citrate, are available for treatment of kidney stones and would probably be useful in the treatment of suspected uranium poisoning. There have been no clinical trials done with this type of treatment, but evidence for the efficacy of citrate treatment from animal studies when intoxicated with uranium are very convincing. As citrate salts and citric acid are readily available simple treatments would offer the best protection for those exposed to uranium with v ery few side effects (possibly nausea, diarrhea, gas). Several compounds ranging from natural and simple such as sodium citrate, to more sophisticated synthetic chelators agents have been tested in order to treat uranium contamination. To date, several studies have been performed to determine the effectiveness of different chelating agents in preventing the toxic effets of uranium on the body (Domingo et al., 1990; Cabrini et al., 1984; Stradling et al., 1991; Martinez et al., 2000; Bozal et al., 2005). However, not all chelating compound tested showed to be effective to counterbalance the action of uranium. In animal models tetracyclines, proved to be effective in neutralizing the uranium inhibition of bone formation but were unable to prevent nephrotoxic side effects (Guglielmotti et al., 1989). For acute uranium contamination Ortega et al. (1989) assayed 16 different chelating agents administered intraperitoneally (ip). They showed that Tiron (sodium 4,5-dihydroxybenzene-1,3-disulfonate) was the most effective from all the compounds tested increasing urinary and faecal excretion of uranium 24 hours post-administration and lowering the toxic concentration in bone and kidney, thus allowing survival of 100 and 70% depending on the interval between the administration of the toxic and the antidote (Domingo et al., 1992). The increase in the interval between exposure to uranium and the administration of chelating agent drastically reduced the mobilization of uranium from the target organs, proving to be a critical point that must be taken into account. Thus, the faster the chelator is administered after lethal uranium intoxication the greater the likelihood of increased survival. Therefore, if time frame for the administration of any chelating compounds is beyond 24 hours post-intoxication, none of them will be able to reverse the deposition of uranium in the bones. Durbin et al. (1997) found that daily injections of 5-Li or TREN-(Me-3.2-HOPO) were able to reduce by 50% retention of uranium in the tissues, accelerating the renal excretion of the poison. This compound is able to greatly reduce the damage at the renal tubular epithelium but not proved to be effective in removing uranium on skeletal tissues. In 2000, Martinez et al.

showed that ethane-1-hydroxy-1,1-biphosphonate (EHBP) administered both orally or subcutaneously, was highly effective as an antidote in uranium contamination increasing survival rate from 45% up to 100% of success, depending on the age of the animal. The longest survival time was observed for suckling animals (Ubios et al., 1994) when compared to adult animals (Martinez et al., 2000). Again, the efficiency of the treatment depended on the time lapse between exposure to the toxicant and the administration of the bisphosphonate proving that efficency is inversely proportional to the time interval (Ubios et al., 1998).

The effectiveness of EHBP focuses on the prevention of the adverse effects of uranium on renal function and the uranium-inhibition reversal on bone formation. Therefore, after uranium contamination, EHBP can not only act as a therapeutic agent to improve survival but, as an agent capable to prevent uranium toxicity on target organs.

Once ingested, EHBP is absorbed mainly through the small intestine. From the absorbed EHBP, 60% is deposited in bone and the remaining 40% excreted in urine (Fleisch, 1995). Also in 1995, Henge-Napoli et al. showed that the chelating effect of EHBP prevents uranium from reaching the kidney. Therefore it was assumed that EHBP could "intercept" uranium while in the circulatory system, preventing tissue damage of the target organs (Martinez et al., 2003). Bioavailable uranium could link to EHBP in blood conforming chelates that allow uranium to be excreted supressing its effect as a life-threatening compound. Our group has extensive evidence on the preventive effect of EHBP for acute

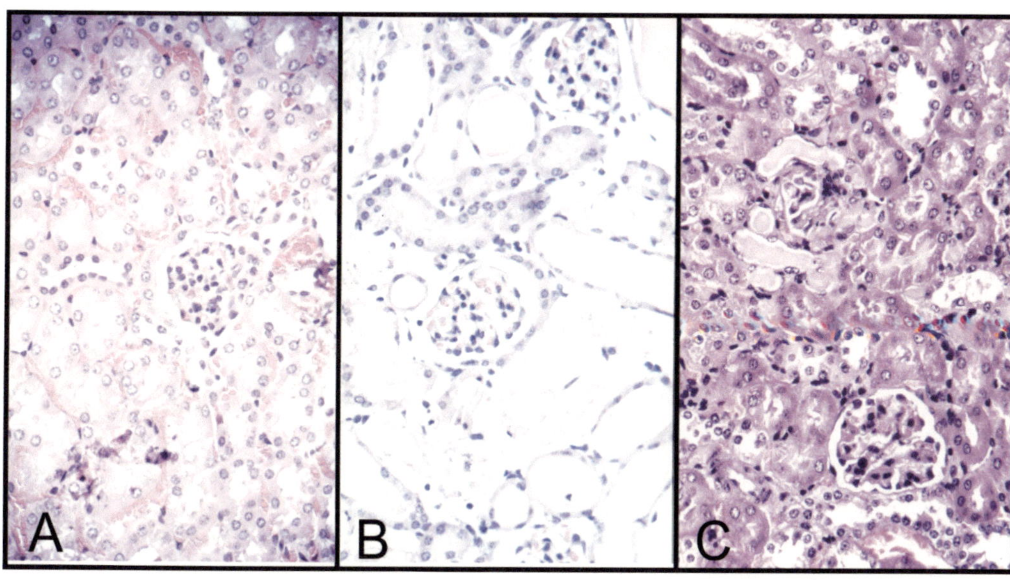

Fig. 2. **Histologic sections of renal cortex**. A): control, B): exposed to a lethal oral dose of uranyl nitrate 48 h post-intoxication, and C) intoxicated with uranyl nitrate and treated-EHBP after 14 days. A) Note integrity of tubule and glomerular structure. B) Note the marked vacuolization of the tubules, abundant hyaline cylinders, and extensive areas of necrosis. Widening of the uriniferous tubule and Bowman´s capsule is also evident. C) Note that although areas presenting necrosis and hyaline cylinders are evident, the proportions of these areas are much lesser than in B). (HE 400X)

uranium intoxications administered intraperitoneally (Ubios et al., 1994), subcutaneously (Ubios et al., 1998; Martinez et al., 2000) or orally (Martinez et al., 2000). In all cases we demonstrated that EHBP is able to ameliorate the structural and functional damages induced by uranium. On kidneys EHBP causes a decrease in the renal concentration of uranium (Henge-Napoli et al., 1999) and prevents from functional alterations resulting in urea and creatinine values similar to those of control non-intoxicated animals (Martinez et al., 2003). At 14 days post- uranium intoxication, kidney lesions from EHBP treated animals (orally or subcutaneously), showed clear signs of tissue recovery featuring areas of apparent normal parenchyma (figure 2) and urea and creatinine levels remained within normal range. A marked reduction in hyaline casts in kidney sections might suggest that uranium-induced damage could be reversible (Martínez et al., 2003). In the same work we demonstrated that the efficacy of EHBP in non-fasted animals.

The effect of EHBP on bone was similar to the observed in kidney. EHBP reverses induced-uranium alterations on the metaphyseal cartilage and endochondral ossification. Although EHBP causes a decrease in the height of the metaphyseal cartilage, it does not alter the activity or the formation of bone subchondral trabeculae indicating that though at a slower rate, the endochondral ossification process continues (figure 3) (Bozal et al., 2005). During growth, catch-up growth has been defined as growth with a velocity above the statistical limits of normality for age during a defined period of time, which follows a period of impaired growth (Williams et al., 1974). We reported that tooth eruption, dental development and mandibular growth of suckling rats, impaired by acute uranium exposure

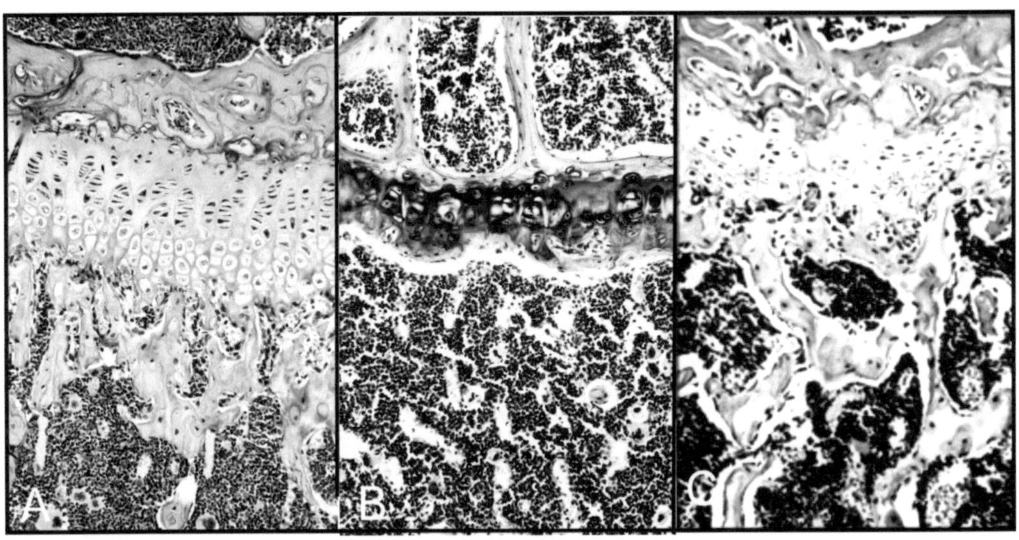

Fig. 3. **Histologic sections of the metaphyseal bone.** A): control, B) exposed to an oral dose of 350 mg/kg of uranyl nitrate 48 h-post intoxicaton. Note the marked diminution in cartilage width and the absence of proliferation cells in the animal exposed to uranyl nitrate, as compared to A). The absence of trabeculae and formation of sealed bone in uranyl nitrate exposed animals can be observed, and the diminution in subchondral bone volume is evident. C): intoxicated with uranyl nitrate and treated-EHBP after 14 days (C). Note that there is some reduction in growth cartilage width but, no differences in subchondral bone volume are observed when comared to A). (HE 200X)

catch up to controls 27 days post - EHBP treatment (Pujadas Bigi & Ubios, 2007). As for metaphyseal cartilage, a decrease in mandibular growth in uranium-exposed animals was observed. This finding is in agreement with the inhibition on bone formation and endochondral ossification (Díaz Sylvester et al., 2002; Bozal et al., 2005). The reduction in dental development is associated to cementum and dentin of the dental root. Therefore, the delay in dental growth in uranium-exposed animals and the fact that we observed catch-up growth after 27 days, may indicate that osteoblast and odontoblast precursor cells, which are genetically determined to form bone and dentin respectively, did not suffer irreversible damage during uranium exposure or that uranium affected only the cells that were determined at the time of the exposure (Pujadas Bigi & Ubios, 2007).

It must be kept in mind that longterm oral EHBP therapy was widely used to treat osteopenia and is still used to treat Paget´ s disease and hypercalacemia. It is worth to note that oral intake is a simple and fast route of entry to the body, which is essential in case of accidental uranium ingestion.

5. Conclusion

As is known uranium, natural and depleted (DU) present similar chemical and radiological properties. Although DU is not classified as a dangerous substance radiologically, for its emissions are very low, it is, in large quantities, a potential toxicological hazard. Despite much study, epidemiologycal, in vivo on experimental animal models and in vitro on cultured cells, uranium impact on health remains controversial. As uranium decay proceeds its final product, lead, increases in relative abundance in nature. Both elements have similar chemical toxicity, so inhaled fume or ingested particles of any of them is considered a health hazard. In this context, in this chapter we made an attempt to consider uranium chemotoxicity and its intracellular mechanism of action similar to that of lead, a heavy persistent metallic toxic particle. The current knowledge in the field of metallo-biochemistry of oxidative stress indicates that metal-induced and metal-enhanced formation of free radicals and other reactive species can be regarded as a common factor in determining metal-induced toxicity and carcinogenicity. Metals interfere with cell signalling pathways and affect growth receptors, tyrosine and serine/threonine kinases, and nuclear transcription factors by ROS-dependent and ROS-independent mechanisms. Many of the DNA base modifications caused by free radicals are pro-mutagenic, pointing to a strong link between oxidative damage, declined antioxidant mechanisms and carcinogenesis. As occupational exposures to heavy metals (lead, arsenic) primarily by inhalation are causally associated with lung cancer, natural uranuim and DU could follow the same path. Neverhteless, further studies have to be done in order to conclude uranium implication in this matter. Different compounds are used to treat heavy metal poisoning and chelate redox active metals. For this last purpose, antioxidants are the selected substances while biophosphonates seem to be the to promissory compounds able to prevent or reverse health problems of individuals exposed to uranium.

6. References

Abou-Donia, M., Dechkovskaia, A., Goldstein, L., Shah, D.,, Bullman, S., & Khan, W. (2002). Uranyl acetate-induced sensorimotor deficit and increased nitric oxide generation in the central nervous system in rats. *Pharmacol Biochem Behav.,* Vol. 72(4), pp. (881-90).

Adams, N., & Spoor, N. (1974). Kidney and bone retention functions in the human metabolism of uranium. *Phys. Med. Biol.* Vol. 19, pp. (460-471).

Aitken, M. (1999). Gulf War leaves legacy of cancer. *Br. Med. J.* Vol. 319, pp. (401).

Amabile, J., Leuraud, K., Vacquier, B., Caër-Lorho, S., Acker, A., & Laurier, D. (2009). Multifactorial study of the risk of lung cancer among French uranium miners: radon, smoking and silicosis. *Health Phys., Vol.,* 97(6), pp. (613-21).

ATSDR (1999), "Toxicological Profile for Uranium, Health Effects," available at http://www.atsdr.cdc.gov/toxprofiles/tp150-c2.pdf (last visited Nov. 29, 2008);

Barber, D., Hancock, S., McNally, A., Hinckley, J., Binder, E., Zimmerman, K., Ehrich, M., & Jortner, B. (2007). Neurological effects of acute uranium exposure with and without stress. *Neurotoxicology,* Vol. 28(6), pp. (1110-9).

Barbosa, F., Sertorio, J., Gerlach, R., & Tanus-Santos, J. (2006). Clinical evidence for lead-induced inhibition of nitric oxide formation. *Arch Toxicol.,*;80(12), pp. (811-6).

Bem, H., & Bou-Rabee, F. (2004). Environmental and health consequences of depleted uranium use in the 1991 Gulf War. *Environ. Int.,* Vol. 30, pp. (123–134).

Berry, J., Zhang, L., Galle, P., Ansoborlo, E., Henge-Napoli, M., & Donnadieu-Claraz, M. (1997). Role of alveolar macrophage lysosomes in metal detoxification. *Microsc. Res. Tech.,* Vol.36, pp. (313-323).

Bishayi, B., & Sengupta, M. (2006). Synergism in immunotoxicological effects due to repeated combined administration of arsenic and lead in mice. *Int. Immunopharmacol.,* Vol. 6, pp. (454-464).

Boice, J., Mumma, M., Schweitzer, S., &. Blot, W. (2003). Cancer mortality in a Texas county with prior uranium mining and milling activities, 1950-2001. *J Radiol Prot.* Vol. 23(3), pp. (247-262).

Boice, J., Mumma, M., & Blot, W. (2007). Cancer and noncancer mortality in populations living near uranium and vanadium mining and milling operations in Montrose County, Colorado, 1950-2000. *Radiat Res.,* Vol. 167(6), pp. (711-726).

Boice, J., Mumma. M., & Blot, W. (2010). Cancer incidence and mortality in populations living near uranium milling and mining operations in Grants, New Mexico, 1950-2004. *Radiat Res.,* Vol. 175(5), pp. (624-636).

Bozal, C., Martinez, A., Cabrini, R., & Ubios, A. (2005). Effect of ethane-1-hydroxy-1,1-bisphosphonate (EHBP) on endochondral ossification lesions induced by a lethal oral dose of uranyl nitrate. *Archives of Toxicology,* Vol. 79, pp. (475-481).

Briner, W., & Murray, J. (2005). Effects of short-term and long-term depleted uranium exposure on open-field behavior and brain lipid oxidation in rats. *Neurotoxicology and Teratology,*. vol. 27, pp. (135-44).

Cabrini, R., Gulielmotti, M., & Ubios, A. (1984). Prevention of the toxic effect of uranium on bone formation by tetracycline. *Acta Odontol Latinoam.,* 1(2), pp. (61-3).

Cooper, J., Stradling, G., Smith, H., & Ham, S. (1982). The behavior of uranium-233 oxide and uranyl-233 nitrate in rats. *Int J Radiat Biol Relat Stud Phys Chem Med.,* Vol. 41(4), pp. (421-433).

Chittenden, R., & Hutchinson, M. Some experiments on the physiologic al action of uranium salts, Tr. Conn. Acad. Arts and Sc., 1888-92, v i i i , 1.

Chittenden, R., & Hutchinson, M. The influence of uranium salts on the amylolytic action of Saliva and the proteolytic action of peps in an trypsin. Tr. Conn. Acad. Arts and Sc., 1885-88, v i i , 161.

de Rey, B., Lanfranchi, H., & Cabrini, R. (1983). Percutaneous absorption of uranium compounds. *Environ Res.*, Vol. 30, pp. (480-491).

de Rey, B., Lanfranchi, H., & Cabrini, R. (1984). Deposition pattern and toxicity of subcutaneously implanted uranium dioxide in rats. *Health Phys.*, Vol. 46, pp. (688-692).

Desmedt, L., Simaels, J., & Van Driessche, W. (1993). Ca(2+)-blockade, poorly selective cation channels in the apical membrane of amphibian epithelia. UO2(2+) reveals two channel types. *J. Gen. Physiol.*, Vol. 101, pp. (85-102).

Di Lella, L., Nannoni, F., Protano, G., & Riccobono, F. (2005). Uranium contents and 235U/238U atom ratios in soil and earthworms in western Kosovo after the 1999 war. *Sci. Total. Environ.*, Vol. 337, pp. (109–118).

Díaz Sylvester, P., López, R., Ubios, A., & Cabrini, R. (2002). Exposure to subcutaneously implanted uranium dioxide impairs bone formation. *Arch Environ Health.*, Vol. 57(4), pp. (320-5).

Domingo, J., Llobet, J., Tomás, J., & Corbella J. (1987). Acute toxicity of uranium in rats and mice. *Bull Environm Contam Toxicol.*, Vol. 39. pp. (168-174).

Domingo, J., Ortega, A., Llobet, J., & Corbella J. (1990). Effectiveness of chelation therapy with time after acute uranium intoxication. *Fundam Appl Toxicol.*, Vol. 14(1), pp. (88-95).

Domingo, J., Colomina, M., Llobet, J., Jones, M., Sing, P., & Campbell R. (1992). The action of chelating agents in experimental uranium intoxication in mice: variation with structure and time of administration. *Fundam and Appl Toxicol.*, Vol. 19, pp. (350-357).

Domingo, J. (2001). Reproductive and developmental toxicity of natural and depleted uranium: A review. *Reprod. Toxicol.*, Vol. 15, pp. (603–609).

Donaldson, K., Stone, V., Tran, C., Kreyling, W., & Borm, P. (2004). Nanotoxicology. *Occup Environ Med.*, Vol. 61, pp. (727-728).

Donnelly, G., & Holman, R. (1942). The stimulating influence of sodium citrate on cellular regeneration and repair in the kidney injured by uranium nitrate. *Journal of Pharmacology and Experimental Therapeutics*, Vol. 75, pp. (11-17).

Driscoll, K. (2000). TNFα and MIP-2: role in particle-induced inflammation and regulation by oxidative stress. *Toxicology Letters,* Volumes 112-113, pp. (177-183).

Driscoll, K., Carter. J., Hassenbein, D., & Howard, B. (1997). Cytokines and particle-induced inflammatory cell recruitment. *Environ Health Perspect.*,Vol. 105, (5), pp. (1159 - 1164).

Dublineau, I., Grison, S., Baudelin, C., Dudoignon, N., Souidi, M., Marquette, C., Paquet, F., Aigueperse, J., & Gourmelon P. (2005). Absorption of uranium through the entire gastrointestinal tract of the rat. *Int J Radiat Biol.*, Vol. 81(6), pp. (473-482).

Dublineau, I., Grison, S., Linard, C., Baudelin, C., Dudoignon, N., Souidi, M., Marquette, C., Paquet, F., Aigueperse, J., & Gourmelon, P. (2006). Short-term effects of depleted uranium on immune status in rat intestine. *J Toxicol Environ Health A.*, Vol. 69(17), pp. (1613-28).

Dublineau, I., Grandcolas, L., Grison, S., Baudelin, C., Paquet, F., Voisin, P., Aigueperse, J., & Gourmelon, P. (2007). Modifications of Inflammatory Pathways in Rat Intestine Following Chronic Ingestion of Depleted Uranium. Toxicological Sciences, Vol. 98(2), pp. (458–468).

Durbin, P., Kullgren, B., Xu, J., & Raymond, K. (1997). New agents for in vivo chelation of uranium (VI): efficacy and toxicity in mice of multidentate catecholate and hydroxypyridonate ligands. *Health Physics.*, Vol. 72, pp. (865-879).

Dursun, N., Arifoglu, C., Suer, C., & Keskinol, L. (2005). Blood pressure relationship to nitric oxide, lipid peroxidation, renal function, and renal blood flow in rats exposed to low lead levels. *Blood Trace Elem. Res.*, Vol. 104, pp. (141-149).

Ercal, N., Gurer-Orhan, H., & Aykin-Burns, N., (2001). Toxic metals and oxidative stress. Part 1. Mechanisms involved in metal-induced oxidative damage. *Curr. Top.Med. Chem.*, Vol. 1, pp. (529–539).

Evan, G., & Vousden, K. (2001). Proliferation, cell cycle and apoptosis in cancer. *Nature,* Vol. 411, pp. (342–348).

Fels, A., Cohn, Z (1986). The alveolar macrophage. *J Appl Physiol.,* Vol. 60(2), pp. (353-69). Review.

Fisher, D., Kathern, R., & Swint, M. (1991). Modified biokinetic model for uranium from analysis of acute exposure of UF6. *Health Phys.,* Vol. 60(3), pp. (335-342).

Fleisch, H., En: Biphosphonates in bone disease: from the laboratory to the patient. 2nd Edition. The Partenon Publishing Group, New York-London, 1995

Fulco, C., Liverman, C., & Sox, H. (2000). Gulf War and health, vol. 1. Washington, DC: National Academy Press.

Gazin, V., Kerdine, S., Grillon, G., Pallardy, M., & Raoul, H. (2004). Uranium induces TNF alpha secretion and MAPK activation in a rat alveolar macrophage cell line. *Toxicol. Appl. Pharmacol.* Vol. 194, pp. (49–59).

Gribovskaja, I., Brownlow, K., Dennis, S., Rosko, A., Marletta, M., & Stevens-Truss, R. (2005). Calcium-binding sites of calmodulin and electron transfer by inducible nitric oxide synthase. *Biochemistry,* Vol. 44, pp. (7593-7601).

Guglielmotti, M., Ubios, A., de Rey B., & Cabrini R. (1984). Effects of acute intoxication with uranyl nitrate on bone formation. *Experientia,* Vol. 40, pp. (474-476).

Guglielmotti, M., Ubios, A., & Cabrini, R. (1985). Alveolar wound healing alterations under uranyl nitrate intoxication. *J. Oral Pathol.* Vol. 14, pp. (565-572).

Gulgielmotti, M., Ubios, A., & Cabrini, R. (1987). Morphometric study of the effect of low dose of uranium on bone healing. *Acta Stereol.,* Vol. 6, pp. (357-366).

Guglielmotti, M., Ubios, A., Larumbe, J., & Cabrini R. (1989). Tetracycline in uranyl nitrate intoxication: Its action on renal damage and uranium retention in bone. *Health Phys.,* Vol. 57, pp. (403-405).

Gurer, H., & Ercal, N. (2000). Can antioxidants be beneficial in the treatment of lead poisoning? *Free Radic. Biol. Med.,* Vol. (29), pp. (927–945).

Haley, D. (1982). Morphologic changes in uranyl nitrate-induced acute renal failure in saline- and water-drinking rats. *Lab Invest.*, Vol. 46(2), pp. (196-208).

Harris BW et al. Experimental clearance of urannium dust from the body. In: Darries CN, Editors, (1961) Inhaled Particles and Vapuors. Pergamon

Harrison, J., & Stather, J. (1981). The gastrointestinal absorption of protactinium, uranium, and neptunium in the hamster. *Radiat Res.* Vol. 88(1), pp. (47-55).

Henge-Napoli, M., Archimbaud, M., Ansoborlo, E., Metivier, H., & Gourmelon, P. (1995). Efficacy of 3,4,3-LIHOPO for reducing the retention of uranium in rat after acute administration. *International Journal of Radiaiont Biology*, Vol. 68, pp. (389-393)

Henge-Napoli, M., Ansoborlo, E., Chazel, V., Houpert, P., Paquet, F., & Gourmelon, P. (1999). Efficacy of ethane-1-hydroxy-1,1 bisphosphonate (EHBP) for the decorporation of uranium after intramuscular contamination in rats. *International Journal of Radiation Biology*, vol.75, pp. (1473-1477).

Hoffman, D., Heinz, G, Sileo, L., Audet, D., Campbell, J., LeCaptain, L., & Obrecht, H. (2000). *J. Toxicol. Environ. Health A.*, Vol. 59, pp. (235–252).

Hunaiti, A., & Soud, M. (2000). Effect of lead concentration on the level of glutathione, glutathione S-transferase, reductase and peroxidase in human blood. *Sci. Total Environ.*, Vol. 248, pp. (45–50).

Hursh, J., Neuman, W., Toribara, T., Wilson, H., & Waterhouse C. Oral ingestion of uranium by man. (1969). *Health Phys.* Vol. 17(4), pp. (619-21).

Hursch, J., & Spoor, N. (1973). Data on man. In: Hodge HC, Stannard JN, Hursh JS, eds. Uranium, Plutonium Transplutonic elements. Handbook of Experimental Pharmacology, vol 36: 197-239. Berlin-Heidelberg: Springer-Verlag, 1973

ICRP. 1995. Age-dependent doses to members of the public from intake of radionuclides: Part 3, Ingestion dose coefficients. ICRP Publication 69. Oxford: Pergamon Press. International Commission for Radiation Protection.

ICRP. 1996. Age-dependent doses to members of the public from intake of radionuclides: Part 4, Inhalation dose coefficients. ICRP Publication 71. Oxford: Pergamon Press. International Commission for Radiation Protection.

Inoue, K., Takano, H., Yanagisawa, R., Sakurai, M., Ichinose, T., Sadakane, K., & Yoshikawa, T. (2005). Effects of nano particles on antigen-related airway inflammation in mice. *Respir. Res.*, Vol. 6, pp. (106–117).

Kalinich, J., Ramakrishnan, N., Villa, V., & McClain, D. (2002). Depleted uranium-uranyl chloride induces apoptosis in mouse macrophages. *Toxicology*, Vol. 179, pp. (105–114).

Kern, M., & Audesirk, G. (2000). Stimulatory and inhibitory effects of inorganic lead on calcineurin. *Toxicology*, Vol. 150, pp. (171-178).

Kirsch-Volders, M., Vanhauwaert, A., Eichenlaub-Ritter, U., & Decordier, I. (2003). Indirect mechanisms of genotoxicity. *Toxicol. Lett.* Vol. 63, pp. (140–141).

Kong, X., Liao, L., Lei, D, Huang, J., & Wen XD. (2000). Influence of lead on activity of nitric oxide synthase in neurons and vessel smooth muscle of small intestine in rats. *Hunan Yi Ke Da Xue Xue Bao*, Vol. 25(2), pp. (135-7).

Korashy, H., & El-Kadi, A. (2006). Transcriptional regulation of theNAD(P)H:quinone oxidoreductase 1 and glutathione S-transferase genes by mercury, lead, and copper. *Drug Metab Dispos.*, Vol. 34(1), pp. (152-65).

Kreyling, W., Semmler, M. & Möller, W. (2004). Dosimetry and toxicology of ultrafine particles. *J Aerosol Med.*, Vol. 17, pp. (140-152).

Kurttio, P., Auvinen, A., Salonen, L., Saha, H., Pekkanen, J., Mäkeläinen, I., Väisänen, S., Penttilä, I., & Komulainen, H. (2002). Renal effects of uranium in drinking water. *Environ Health Perspect.*, Vol. 110(4), pp. (337-342).

La Touche, Y., Willis, D., & Dawydiak O. (1987). Absorption and biokinetics of U in rats following an oral administration of uranyl nitrate solution. *Health Phys.*, Vol. 53(2), pp. (147-162).

Leach, L., Maynard, E., Hodge, H., Scott, J., Yuile, C., Sylvester, G., & Wilson, H. (1970). A five-year inhalation study with natural uranium dioxide (UO 2) dust. I. Retention and biologic effect in the monkey, dog and rat. *Health Phys.*, Vol. 18, pp. (599-612).

Leach, L., Gelein, R., Panner, B., et al. (1984). The acute toxicity of the hydrolysis products of uranium hexafluoride (UF6) when inhaled by the rat and guinea pig. Final report. ISS K/SUB-81-9039-3. DE84011539.

Leggett, R., & Harrison, J. (1995). Fractional absorption of ingested uranium in humans. *Health Phys.*, Vol. 68(4), pp. (484-498).

Lestaevel, P., Romero, E., Dhieux, B., Ben Soussan, H., Berradi, H., Dublineau, I., Voisin, P., & Gourmelon, P. (2009). Different pattern of brain pro-/anti-oxidant activity between depleted and enriched uranium in chronically exposed rats. *Toxicology*, Vol. 258(1), pp. (1-9).

Lin, R., Wu, L., Lee, C., & Lin-Shiau, S. (1993). Cytogenetic toxicity of uranyl nitrate in Chinese hamster ovary cells. *Mutat Res.* Vol. 319(3), pp. (197-203).

Linares, V., Belles, M., Albina, M., Sirvent, J., Sanchez, D., & Domingo, J. (2006). Assessment of the pro-oxidant activity of uranium in kidney and testis of rats. *Toxicol. Lett.*, Vol. 167, pp. (152-161).

Lohmann-Matthes, M., Steinmüller, C., & Franke-Ullmann G. (1994). Pulmonary macrophages. *Eur Respir J.* Vol. 7(9), pp. (1678-89). Review.

López, R., Díaz Sylvester, P., Ubios, A., & Cabrini, R. (2000). Percutaneous toxicity of uranyl nitrate: its effect in terms of exposure area and time. *Health Phys.*, Vol. 78(4) pp. (434-7).

Lu, S., & Zhao, F. (1990). Nephrotoxic limit and annual limit on intake for natural U. *Health Phys.*, Vol. 58(5) pp. (619-623).

MacNee, W., Li, X., Gilmour, P., & Donaldson, K.. (2000). Systemic effect of particulate air pollution. *Inhal Toxicol.*, Vol. 12 (3), pp.(233-244).

Macnider, W. (1916). The inhibition of the toxicity of uranium nitrate by sodium carbonate, and the protection of the kidney acutely nephropathic from uranium from the toxic action of an anesthetic by sodium carbonate. *J Exp Med 1*, Vol. 23(2), pp. (171-87).

Mao, Y., Desmeules, M., Schaubel, D., Bérubé, D., Dyck, R., Brûlé, D., & Thomas, B. (1995). Inorganic components of drinking water and microalbuminuria. *Environ Res.*, Vol. 71(2), pp. (135-140).

Martin, L., Krunkosky, T., Dye, J., Fischer, B., Jiang, N., Rochelle, L., Akley, N., Dreher, K., & Adler, K. (1997). The role of reactive oxygen and nitrogen species in the response of airway epithelium to particulates. Environ. *Health Perspect.* Vol. 105(5), pp. (1301–1307).

Martinez, A., Cabrini, R., & Ubios, A. (2000). Orally administered ethane-1-hydroxy-1,1-bisphosphonate reduces the lethal effect of oral uranium poisoning. *Health Physics.,* Vol. 78, pp. (668-671).

Martínez, A., Mandalunis, P., Bozal, C., Cabrini, R., & Ubios, A. (2003). Renal function in mice poisoned with oral uranium and treated with ethane-1-hydroxy-1,1-bisphosphonate (EHBP). *Health Phys.*, Vol. 85, pp. (343-347).

McDiarmid, M., Keogh, J., Hooper, F., McPhaul, K., Squibb, K., Kane, R., DiPino, R., Kabat, M., Kaup, B., Anderson, L., Hoover, D., Brown, L., Hamilton, M., Jacobson-Kram, D., Burrows, B., & Walsh, M. (2000). Health effects of depleted uranium on exposed Gulf War veterans. *Environ Res.*, Vol. 82(2), pp. (168-180).

McDiarmid, M., Hooper, F., Squibb, K., McPhaul, K., Engelhardt, S. M., Dipino, R., & Kabat, M. (2002). Health effects and biological monitoring results of Gulf War veterans exposed to depleted uranium. *Mil. Med.,* Vol. 167, pp. (123–124).

McDiarmid, M., Engelhardt, S., Oliver, M., Gucer, P., Wilson, P., Kane, R., Kabat, M., Kaup, B., Anderson, L., Hoover, D., Brown, L., Handwerger, B., Albertini, R., Jacobson-Kram, D., Thorne, C., & Squibb, K. (2004). Health effects of depleted uranium on exposed Gulf War veterans: A 10-year follow-up. *J Toxicol Environ Health A.,* Vol. 67(4), pp. (277-296).

McDiarmid, M., Engelhardt, S., Oliver, M., Gucer, P., Wilson, P., Kane, R., Cernich, A., Kaup, B., Anderson, L., Hoover, D., Brown, L., Albertini, R., Gudi, R., Jacobson-Kram, D., & Squibb, K., (2007). Health surveillance of Gulf War I veterans exposed to depleted uranium: Updating the cohort. *Health Phys.,* Vol. 93(1), pp. (60-73).

McDiarmid, M., Engelhardt, S., Dorsey, C., Oliver, M., Gucer, P., Wilson, P., Kane, R., Cernich, A., Kaup, B., Anderson, L., Hoover, D., Brown, L., Albertini, R., Gudi, R., & Squibb, K. (2009). Surveillance results of depleted uranium- exposed Gulf War I veterans: Sixteen years of follow-up. *J Toxicol Environ Health A.,* Vol. 72(1), pp. (14-29).

McNeilly, J., Jimenez, L., Clay, M., MacNee, W., Howe, A., Heal, M., Beverland, I., & Donaldson, K. (2005). Soluble transition metals in welding fumes cause inflammation via activation of NF-jB and AP-1. *Toxicol. Lett.,* Vol. 158, pp. (152-157).

Miller, A., Stewart, M., Brooks, K., Shi, L., & Page, N. (2002) Depleted uranium-catalyzed oxidative DNA damage: absence of significant alpha particle decay. *J Inorg Biochem.,* Vol. 91(1), pp. (246-52).

Miller, A., Brooks, K., Stewart, M., Anderson, B., Shi, L., McClain, D., & Page, N. (2003). Genomic instability in human osteoblast cells after exposure to depleted uranium: Delayed lethality and micronuclei formation. *J. Environ. Radioact.,* Vol. 64, pp. (247-259).

Miller, A., Stewart, M., & Rivas, R. (2010). Preconceptional paternal exposure to depleted uranium: transmission of genetic damage to offspring. *Health Phys.*, Vol. 99(3), pp. (371-9).

Monleau, M., Bussy, C., Lestaevel, P., Houpert, P., Paquet, F., & Chazel, V. (2005). Bioaccumulation and behavioural effects of depleted uranium in rats exposed to repeated inhalations. *Neuroscience Letters*, Vol. 390, pp. (31-6).

Monleau, M., De Meo, M., Frelon, S., Paquet, F., Donnadieu-Claraz, M., Duménil, G., & Chazel, V. (2006). Distribution and genotoxic effects after successive exposure to different uranium oxide particles inhaled by rats. *Inhal Toxicol.*, Vol. 18(11), pp. (885-894).

Muller, C., Ruzicka, L., & Bakstein, J. (1967). The sex ratio in the offspring of uranium miners *Acta Univ. Carolinae Med.*, Vol. 13, pp. (599–603).

Neuman, M., & Neuman, W. (1948). The deposition of uranium in bone; radioautographic studies. *J Biol Chem.*, Vol. 175(2), pp. (711-4).

Oberdörster, G., Ferin, J., & Lehnert BE. (1994). Correlation between particle size, in vivo particle persistence, and lung injury. *Environ Health Perspect.*, Vol. 102, pp. (173-179).

Oberdörster, G., Sharp, Z., Atudorei, V., Elder, A., Gelein, R., Lunts, A., Kreyling, W., & Cox, C. (2002) Extrapulmonary translocation of ultrafine carbon particles following whole-body inhalation exposure of rats. *J Toxicol Environ Health,* Vol. 65A, pp. (1531-1543).

Orcutt, J. (1949). The toxicology of compounds of uranium following application to the skin. In: Voegtlin C, Hodge HC, eds. Pharmacology and toxicology of uranium compounds. Vols 3 and 4. New York, NY: McGraw Hill Book Co., 377-414.

Orona, N. (2009). Tesis de Licenciatura en Biotecnología: Citotoxicidad del nitrato de uranilo sobre células del sistema fagocítico mononuclear. Universidad Nacional de San Martín-IIB.

Orona, N., & Tasat, D. DU (uranyl nitrate) - exposed rat alveolar macrophages encounter cell death through different intracellular signaling pathways. (umpublished).

Ortega, A., Domingo, J., Gomez, M., & Corbella, J. (1989). Treatment of experimental acute uranium poisoning by chelating agents. *Pharmacology and Toxicology,* Vol. 64, pp. (247-251).

Osburn, W., & Kensler, T. (2008). Nrf2 signaling: an adaptive response pathway for protection against environmental toxic insults. *Mutat Res.,* Vol. 659(1-2), pp. (31-9).

Patrick, L. (2006). Lead toxicity part II: the role of free radical damage and the use of antioxidants in the pathology and treatment of lead toxicity. *Altern. Med. Rev.* Vol. 11, pp. (114–127).

Pavlakis, N., Pollock, C., McLean, G., & Bartrop, R. (1996). Deliberate overdose of uranium: Toxicity and treatment. *Nephron.,* Vol. 72, pp. (313–317).

Peccorini, V., Ubios, A., Marzorati, M., & Cabrini, R.. (1990). Toxicidad del uranio. Estudio experimental de los productos del Complejo Minero Fabril de San Rafael (Mendoza), Gerencia de Investigaciones (CNEA).

Pellmar, T., Keyser, D., Emery, C., Hogan, J. (1999). Electrophysiological changes in hippocampal slices isolated from rats embedded with depleted uranium fragments. *Neurotoxicology*, Vol. 20(5), pp. (785-792).

Periyakaruppan, A., Kumar, F., Sarkar, S., Sharma, C., & Ramesh, G. (2007) Uranium induces oxidative stress in lung epithelial cells. *Archives of Toxicology*, Vol. 81(6), pp. (389-395).

Periyakaruppan, A., Sarkar, S., Ravichandran, P., Sadanandan, B., Sharma, C., Ramesh, V., Hall, J., Thomas, R., Wilson, B., & Ramesh, G. (2009). Uranium induces apoptosis in lung epithelial cells. *Arch Toxicol.*, Vol. J83(6), pp. (595-600).

Petitot, F., Frelon, S., Moreels, A., Claraz, M., Delissen. O., Tourlonias. E., Dhieux. B., Maubert.,C., & Paquet, F. (2007a). Incorporation and distribution of uranium in rats after a contamination on intact or wounded skin. *Health Phys.*, Vol. 92(5), pp. (464-474).

Petitot, F., Gautier, C., Moreels, A., Frelon, S., & Paquet, F. (2007b). Percutaneous penetration of uranium in rats after a contamination on intact or wounded skin. *Radiat Prot Dosimetry*. Vol. 127(1-4), (125-130).

Pinney, S., Freyberg, R., Levine, G., Brannen, D., Mark, L., Nasuta, J., Tebbe, C., Buckholz, J., & Wones, R. (2003). Health effects in community residents near a uranium plant at Fernald, Ohio, USA. *Int. Occup. Med. Environ. Health*, Vol. 16, pp. (139–153).

Prat, O., Berenguer, F., Malard, V., Tavan, E., Sage, N., Steinmetz, G., & Quemeneur, E. (2005). Transcriptomic and proteomic responses of human renal HEK293 cells to uranium toxicity. *Proteomics*, Vol. 5(1), pp. (297-306).

Prat, O., Bérenguer, F., Steinmetz, G., Ruat, S., Sage, N., & Quéméneur, E. (2010). Alterations in gene expression in cultured human cells after acute exposure to uranium salt: Involvement of a mineralization regulator. *Toxicol In Vitro*, Vol. 24(1), pp. (160-8).

Pujadas Bigi, M., Lemlich, L., Mandalunis, P., & Ubios, A. (2003). Exposure to oral uranyl nitrate delays tooth eruption and development. *Health Phys.*, Vol. 84(2), (163-9).

Pujadas Bigi, M., & Ubios, A. (2007). Catch-up of delayed tooth eruption associated with uranium intoxication. *Health Phys.*, Vol. 92(4), pp. (345-348).

Research Triangle Institute "Uranium: Draft for public comment", September 1997.

Saccomano, G., Auerbach, O., Kushner, M., Harley, N., Michels, R., Anderson, M., (1996) Bechtel, J. A comparison between the localization of lung tumors in uranium miners and non miners from 1947 to 1991. *Cancer*, Vol. 77(7), pp. (1278-1283).

Schins, R. & Borm, P. (1999). Mechanisms and mediators in coal dust induced toxicity: a review. *The Annals of Occupational Hygiene*, Vol. (43), pp.(7-33).

Schnelzer, M., Hammer, G., Kreuzer, M., Tschense, A., Grosche, B. (2010). Accounting for smoking in the radon-related lung cancer risk among German uranium miners: results of a nested case-control study. Health Phys. Vol. 98(1), pp. (20-8).

Simmons, M., & Murphy, S. (1993). Cytokines regulate L-arginine-dependent cyclic GMP production in rat glial cells. *Eur J Neurosci.*, Vol. 5(7), pp. (825-31).

Simons, T. (1993). Lead-calcium interactions in cellular lead toxicity. *Neurotoxicology*, Vol. 14, pp. (77-85).

Smith, D., & Black, S.(1975). Actinide concentrations in tissues from cattle grazing near the Rocky Flats Plant. Report: ISS NERC-LV-529-536.

Stearns, D., Yazzie, M., Bradley, A., Coryell V., Shelley, J., Ashby, A., Asplund, C., & Lantz, R. (2005). Uranyl acetate induces hprt mutations and uranium–DNA adducts in Chinese hamster ovary EM9 cells. *Mutagenesis*, Vol. 20(6), pp. (417-423).

Stoeger, T., Reinhard, C., Takenaka, S., Schroeppel, A., Karg, E., Ritter, B., Heyder, J., & Schulz, H. (2006). Instillation of six different ultrafine carbon particles indicates a surface area threshold dose for acute lung inflammation in mice. *Environ Health Perspect.*, Vol. 114(3), pp. (328-33).

Stohs, S. & Bagchi D. (1995). Oxidative mechanisms in the toxicity of metal ions. *Free Radic Biol Med* , Vol. 18(2): pp. (321-36).

Stokinger, H., Baxter, R., Dygert, H., et al. 1953. Toxicity following inhalation for 1 and 2 years. In: Voegtlin C, Hodge HC, eds. Pharmacology and toxicology of uranium compounds. New York, NY: McGraw-Hill, 1370-1776.

Stradling, G., Gray, S., Moody, J., & Ellender, M. (1991). Efficacy of tiron for enhancing the excretion of uranium from the rat. *Hum Exp Toxicol.,* Vol. 10(3), pp. (195-8).

Tannenbaum, A., Silverstone, H., & Koziol, J. (1951). Tracer studies of the distribution and excretion of uranium in mice, rats, and dogs. In: Tannenbaum A, ed. Toxicology of uranium compounds. New York, NY: McGraw-Hill, 128-181.

Tasat, D., & De Rey, B. (1987). *Citotoxic* effect of uranium dioxide on rat alveolar macrophages. *Environmental Research.*, Vol. 44, pp. (71-81).

Tasat, D., Orona, N., Mandalunis, P., Cabrini, R., & Ubios, A. (2007). Ultrastructural and metabolic changes in osteoblasts exposed to uranyl nitrate. *Arch Toxicol.,* Vol. 81(5), pp. (319-26).

Thiébault, C., Carrière, M., Milgram, S., Simon, A., Avoscan, L., & Gouget, B. (2007). Uranium induces apoptosis and is genotoxic to normal rat kidney (NRK-52E) proximal cells. *Toxicol Sci., Vol.* 98(2), pp. (479-87).

Thompson, J., & Nechay, B. (1981). Inhibition by metals of a canine renal calcium, magnesium-activated adenosinetriphosphatase. *J Toxicol Environ Health.*, Vol. 7(6), pp. (901-8).

Tirmarche, M., Baysson, H., & Telle-Lamberton, M. (2004). Uranium exposure and cancer risk: a review of epidemiological studies. *Rev Epidemiol Sante Publique., Vol.* 52(1), pp. (81-90).

Tissandie, E., Guéguen, Y., Lobaccaro, J., Paquet, F., Aigueperse, J., & Souidi M. (2006). Effects of depleted uranium after short-term exposure on vitamin D metabolism in rat. *Arch Toxicol.*, Vol. 80(8), pp (473-80).

Tissandié, E., Guéguen, Y., Lobaccaro, J., Grandcolas, L., Voisin, P., Aigueperse, J., Gourmelon, P., & Souidi, M. (2007). In vivo effects of chronic contamination with depleted uranium on vitamin D3 metabolism in rat. *Biochim Biophys Acta.*, Vol. 1770(2), pp. (266-72).

Ubios, A., Guglielmotti, M., Steimetz, T., & Cabrini, R. (1991). Uranium inhibits bone formation in physiologic alveolar bone modeling and remodeling. *Environ. Res.*, Vol. 54, pp. (17-23).

Ubios, A., Braun, E., & Cabrini R. (1994). Lethality due to uranium poisoning is prevented by ethane-1-hydroxy-1,1-bisphosphonate (EHBP). *Health Physics.*, Vol. 66, (540-544).

Ubios AM, Piloni MJ, Marzorati M, Cabrini RL. (1995). Bone growth is impaired by uranium intoxication. *Acta Odontol Latinoamer*, Vol 8, (3-8).

Ubios, A., Marzorati, M., & Cabrini, R. (1997). Skin alterations induced by long-term exposure to uranium and their effect on permeability. *Health Phys.*, Vol. 72(5), pp. (713-715).

Ubios, A., Braun, E., & Cabrini R. (1998). Effect of bisphosphonates on abnormal mandibular growth in rats intoxicated with uranium. *Health Physics.*, Vol. 75, pp. (610-613).

Uijt de Haag, P., Smetsers, R., Witlox, H., Kru¨s, H., & Eisenga, A.(2000). Evaluating the risk from depleted uranium after the Boeing 747-258F crash in Amsterdam, 1992. *J. Hazard. Mater.* Vol. A76, pp. 39–58.

UNSCEAR 1993 REPORT: SOURCES AND EFFECTS OF IONIZING RADIATION United Nations Scientific Committee on the Effects of Atomic Radiation UNSCEAR 1993 Report to the General Assembly, with Scientific Annexes.

USTUR. 2011. United States transuranium and uranium registries radiochemical analysis procedures manual. Pullman, WA: Washington State University, Nuclear Radiation Center, Manual # 01

Vaziri, N., & Ding, Y. (2001). Effect of lead on nitric oxide synthase expression in coronary endothelial cells: Role of superoxide. *Hypertension*, Vol. 37. pp. (223-226).

Voegtlin, C. & Hodge, H. (1949). Pharmacology and Toxicology of Uranium Compounds (First Ed.). New YorkLondon: McGraw Hill.

Walinder, G., Hammarstrom, L., & Billandelle, U. (1967). Incorporation of uranium. I. Distribution of intravenously and intraperitoneally injected uranium. *Br J Indust Med.*, Vol. 24, pp.(305-312).

Wan, B., Fleming, J., Schultz, T., & Sayler, G. (2006). In vitro immune toxicity of depleted uranium: effects on murine macrophages, CD4+ T cells, and gene expression profiles. *Environ Health Perspect.*, Vol. 114(1), pp. (85-91).

Williams, M., Tanner, J., & Hughes, P. (1974). Catch-up growth in male rats following growth retardation during the suckling period. *Pediatr Res.* Vol. 8, pp. (149-156).

WISE Uranium Project. www.wiseproject.org (1999)

Wrenn, M., Durbin, P., & Howard, B. (1985). Metabolism of ingested uranium and radium. *Health Phys.*, Vol. 48, pp. (601-633).

Xie, H., LaCerte, C., Douglas Thompson, W., & Wise, J. (2010). Depleted uranium induces neoplastic transformation in human lung epithelial cells. *Chemical Research in Toxicology*, Vol. 23 (2), pp. (373-378).

Zamora, M., Tracy, B., Zielinski, J., Meyerhof, D., & Moss, M. (1998). Chronic ingestion of uranium in drinking water: A study of kidney bioeffects in humans. *Toxicol Sci.*, Vol. 43(1), pp. (68-77).

Zamora, M., Zielinski, J., Meyerhof, D., & Tracy B. (2002). Gastrointestinal absorption of uranium in humans. *Health Phys.*, Vol. 83(1), pp. (35-45).

Zamora, M., Zielinski, J., Moodie, G., Falcomer, R., Hunt, W., & Capello, K. (2009). Uranium in drinking water: Renal effects of long-term ingestion by an aboriginal community. *Arch Environ Occup Health,* Vol. 64(4), pp. (228-241).

Zhu, Z., Yang, R., Dong, G., & Zhao ZY. (2005). Study on the neurotoxic effects of low-level lead exposure in rats. *J Zhejiang Univ Sci B., Vol* 6(7), pp. (686-92).

7

Photodynamic Therapy to Eradicate Tumor Cells

Ana Cláudia Pavarina[1], Ana Paula Dias Ribeiro[1], Lívia Nordi Dovigo[1], Cleverton Roberto de Andrade[2], Carlos Alberto de Souza Costa[2] and Carlos Eduardo Vergani[1]

[1]*Araraquara Dental School, UNESP- Univ Estadual Paulista, Department of Dental Materials and Prosthodontics*

[2]*Araraquara Dental School, UNESP- Univ Estadual Paulista, Department of Phisiology and Pathology*

Brazil

1. Introduction

The cell cycle is a collection of highly ordered processes involving numerous regulatory proteins that guide the cell through a specific sequence of events culminating in the duplication of the cell (Elledge, 1996). In general, the cell cycle can be altered to the advantage of many factors. Three basic cell cycle defects are mediated by misregulations of cyclin-dependent kinases (CDKs), unscheduled proliferation, genomic instability (GIN) and chromosomal instability (CIN). In the first case, mutations result in constitutive mitogenic signaling and defective responses to anti-mitogenic signals. The second, GIN, leads to additional mutations (Kastan & Bartek, 2004), and CIN is responsible for numerical changes in chromosomes (Lee et al., 1999). Moreover, data suggest that the mutations leading to tumorigenesis are even more numerous and heterogeneous than previously thought (Hudson et al., 2010). This accumulation of genetic mutations can arise by nucleotide substitutions, small insertions and deletions, chromosomal rearrangements and copy number changes that can affect protein-coding or regulatory components of genes. Cancer genomes usually acquire somatic epigenetic "marks" compared to non-neoplastic tissues from same organ (Hudson et al., 2010). In this context, a neoplasm (Greek, *neo* = new + *plasis* = growth) can be defined as an abnormal mass of tissue whose growth exceeds and is uncoordinated with that of the normal tissue, and persists in the same excessive manner after cessation of the stimuli that initiated the change.

Head and neck cancer is considered a worldwide problem due to its raising in developing countries (Lim et al., 2011). Approximately 90% of this type of cancer consists of oral squamous cell carcinoma (OSCC), which arises from the oral mucosal lining (Neville & Day, 2002). The risk factors related to OSCC includes tobacco and alcohol abuse, solar exposure, human papillomavirus, immunosuppression conditions, iron deficiency anemia in combination with dysphagia and esophageal webs, and tumor genesis that can occur as a result of genetic predisposition or epigenetic pathway involving DNA and/or histone modification (Neville & Day, 2002; Schweitzer & Somers, 2010). Although there have been many advances in the conventional treatment, the survival rate for patients with late stage of OSCC is the lowest of the major cancers, remaining at 50% over the last two decades (Neville & Day, 2002; Funk et al., 2002). The standard treatment for early OSCC includes surgery, radiation, chemotherapy or a combination of these procedures. However, the side effects of these treatments are severe and can result in structural defects leading to dysphagia, and also hyperpigmentation, scars and xerostomia (Hooper et al., 2004; Schweitzer & Somers, 2010). Therefore, alternative treatments have been proposed in order to reduce the toxicity and side effects from the conventional therapies.

Less invasive surgical modalities including the Photodynamic Therapy (PDT) are examples of new treatments that comprise the era of conservative surgery (Karakullukcu et al., 2011). In this context, the purposes of this chapter are: 1. discuss about PDT as a relatively new therapy for treating head and neck neoplasms including its advantages and disadvantages when compared to conventional treatments; 2. describe the pathophysiology of cancer cells that allows the photosensitizer to accumulate on these cells rather than normal tissue and mechanisms of cell/tumor death; 3. review the *in vivo* findings avaiable in the literature; and 4. present the anti-tumor effect observed *in vitro* when Hela cells were exposed to Curcumin, a natural photosensitizer.

2. Photodynamic therapy

PDT is a relative recent therapy, and excellent results have been reported after its application for the treatment of oral cancer and premalignant lesions (Allison et al., 2005; Yu et al., 2008), as well as bacterial and fungal infections (Teichert et al., 2002; Williams et al., 2006; Donnelly et al., 2008). The simplicity of PDT mechanism stimulated the interest for this therapy, which is characterized by the association of a photosensitizing agent (PS) and visible light with a wavelength compatible with the photosensitizer's absorption spectrum (Konopka & Goslinski, 2007; Buytaert et al., 2007). Photon absorption by the PS leads it to a triple state of excitation that may interact with the available oxygen, in two different ways. The reaction type 1 involves electron/hydrogen transfer directly from the PS or electron/hydrogen removal from a substrate molecule to form free radicals such as superoxide, hydroxyl radicals, and hydrogen peroxide. The reaction type 2 involves the production of the electronically excited and highly reactive state of oxygen known as singlet oxygen (Konopka & Goslinski, 2007). Both reactions can occur at the same time and the PS' characteristics and the substrate molecules are important components to define the ratio of each reaction. All these products originated from PDT may result in a cascade of oxidative events that cause direct cell death, destruction of tumor vascularization and activation of the host's immune response (Buytaert et al., 2007).

The success of PDT depends on several parameters such as the type and concentration of the PS, its localization during the irradiation, the pre-incubation period with the drug, type of light sources, light fluence and density, type of tumor and its level of oxygenation (Dolmas et al., 2003). Regarding the PS, the most commonly used in PDT for cancer treatment includes porphyrins, phthalocyanines, the 5-aminolevulinic acid and chlorine. These PSs have been preferentially selected due to their approval for clinical use. In the present, PDT has been approved for use in clinical treatment in the USA, EU, Canada, Russia and Japan (Bredell et al., 2010). The Food and Drug Administration (FDA) has approved the treatment of Barret's esophagus, obstructing esophageal and tracheobronchial carcinomas with

Photofrin, a hematoporphyrin derivate. Moreover, it has also approved the treatment for actinic keratosis with Levulan (aminolevulinic acid) and for macular degeneration with Verteporfin (benzoporphyrin derivate monoacide). In the EU, the same conditions have been approved adding the treatment for early head and neck cancer and palliative treatment for this type of cancer with Foscan, a tetrahydroxy-phenyl chloride (mTHPC) (Biel, 2007). This diversity of photosensitizers leads to differences in the PDT protocol, including differences in the dose and pre-incubation period with the photosensitive drug.

The pre-incubation period is extremely important in the photosensitizer toxicity and its intracellular localization. Hsieh et al. (2003) observed that when Photofrin remained in contact with tumor cells for one hour, the PS was retained in the plasma membrane; however, when this period was increased to 24 hours, the PS penetrated into the cells and caused cell damage even in the absence of light (Hsieh et al., 2003). Triesscheijn et al. (2004) found that a lower light dose was sufficient to reduce cell viability when a higher pre-incubation time was used for several cell types. Moreover, the PS localization defines the primary site of photodamage and also the type of cell death occurred after PDT. Generally, the PSs that accumulate in mitochondria or endoplasmic reticulum cause apoptosis, while those that accumulate in the plasma membrane or lysosomes predispose cells to necrosis (Buytaert et al., 2007). Also, PDT can induce authophagy in attempt to repair the photodamage and turn into a cell death signal if the initial response fails (Kessel & Oleinick, 2009). Usually the PSs do not accumulate in the nucleus, which results in a low mutagenic and genotoxic potential of PDT. A study evaluating DNA damage after PDT observed that the treatment with methylene blue associated with halogen light irradiation against a keratinocytes culture did not cause any DNA damage after 4 hours, even with a 90% percentage of cellular lysis (Zeina et al., 2003).

Light is an essential component of PDT reaction and it is able to interact and modify some cellular responses. According to Brancaleon & Moseley (2002), the choice of an adequate light source can be dictated by the location of the tumour, the light dose required and the type of the PS used. PDT has been traditionally performed using lasers (Light Amplification Stimulated Emission). Examples of this category of light are the argon, argon-pumped dye, metal vapour-pumped dye, solid state, and diode lasers. However, the use of these large and complex equipments can be limited due to the expensive technologies and maintenance required. Because of such disadvantages, lasers are not the only light sources that have been investigated for photodynamic purposes. Studies have reported the use of halogen light and LED (Light Emitting Diode). The LED is a category of light source that emits radiation in a wider range of the spectrum, but with a dominant wavelength. Moreover, the device has a lower cost and simpler technology compared to laser.

The main advantage of PDT is the limited tissue damage restricted to the illuminated and photosensitized area without long-term systemic side effects (Cooper et al., 2007; Karakullukcu et al., 2011). It also does not cause damage to normal structures like nerves, collagen fibers and large blood vessels, preserving the supporting components (Bredell et al., 2010). This means that when cancer cells are eliminated via apoptosis/necrosis the extracellular matrix remains forming a scaffold for the surrounding mucosal tissue to advance over, resulting in minimal scar formation. Moreover, this therapy can be repeated as often as needed without accumulative destructive effects and there is no interaction between current chemo- and radiotherapy protocols and PDT. The most common disadvantages are the photosensitivity that can lead to burn wounds, pain that can vary from mild to severe resolving within 2-3 weeks, and possible poor initial selectivity between tumor and normal tissue (Biel, 2007; Bredell et al., 2010). Although PDT presents some local side effects, it does not involve systemic or serious adverse events. Despite these advantages compared to the conventional treatments, the lack of clinical knowledge and especially the lack of established treatment protocols lead to the limited application of this treatment in the head and neck cancer field.

3. Characteristics of tumor cells and mechanisms of tumor destruction by PDT

The effectiveness of PDT treatment depends on the ability of the PS to selectively localize in tumor cells as opposed to normal cells. The selectivity of PDT to tumor cell is a result of some important characteristics of these cells such as greater proliferative rates, leaky vasculature, poor lymphatic drainage, high expression of low-density lipoprotein (LDL) receptors on tumor cells, to which the PS can bind, and low pH facilitating cellular uptake of the PS.

The antitumor effects of PDT against tumor cells include the direct cytotoxic effect, damage to the tumor vasculature, and induction of inflammatory reaction (systemic immunity) (Agostinis et al., 2011). Three PDT death pathways are described: apoptosis, autophagy and necrosis. The apoptosis mechanisms remain obscure and complex. Kessel & Luo (1998) described that apoptosis occurred when PS was accumulated in the mitochondria, but not when PS was located in the cytoplasmatic. Furre et al. (2006) showed that PDT can induce apoptosis through both cytochrone c-mediated caspase-dependent and AIF-induced caspase-dependent pathway. In addition, Baglo et al. (2011) described that PDT triggers a complex cellular response involving several biological pathways, but apparently involve both caspase dependent and independence apoptotic pathways. Specifically in head and neck cancer, PDT seems to active caspase-8 and caspase-9 pathway and their upstream NF-kB-JNK pathways (Chen et al., 2011). In summary, lysosomal membrane rupture and leakage of cathepsins involve mitochondria outer membrane permeabilization (MOMP) controlled by Bcl-2 family members. Autophagy is a complex cellular process involving dynamic membrane rearrangements and degradation of cell components through the lysosomal activity related to energy homeostasis, organelle turnover, and cancer death (Stromhaug & Klionsky, 2001). Buytaert et al. (2007) identified autophagy as a mode of cell death after PDT. While PDT treated murine embryonic fibroblasts WT (wild-type) and DKO-mtBAX (DKO reexpressing mtBAX) cells exhibited cellular shrinkage and membrane blebs characteristic of apoptotic cell death, the DKO (double-knockout, BAX and BAK) cells and DKO-SERCA (DKO overexpressing SERCA2) cells readily showed extensive cytosolic vacuolization which was remarkably prevented by class III phosphatidylinositol 3-kinase (PI3K), necessary for the sequestration of cytoplasmic material on autophagy process (Buytaert et al., 2007). The mechanisms and roles of autophagy following PDT could also differ markedly in different cell types, depending on a variety of cell characteristics, including their propensity to undergo apoptosis and in response to agents that produce different types or locations of damage and its lead to more resistant (organelle turnover) or death of cancer cells (Kessel & Oleinick, 2009). For necrosis, the PS is excited by light and then reacts with oxygen to generate reactive oxygen species (ROS) that ultimately lead to cell death. ROS can lead damage to proteins, nuclei acids and lipids, proteins change functions with attack to disulfide bonds and thiol groups. Another effect of ROS could be the modification of Ca^{2+} channels (Vanlangenakker et al., 2008). The increase of permeability of mitochondrial membrane is probably a dynamic process that can activate apoptosis or succumb to necrosis (Gramaglia et al., 2004). Mitochondria controls cytoplasmic concentration of Ca^{2+} by tuning the frequency of cytosolic Ca^{2+} waves, and this mechanism is potentially lethal and would rapidly activate a variety of enzymes in an uncontrolled fashion (Rasola & Bernardi, 2011). The fact that cancer cells can die through different mechanisms is a relevant clue in the choice and design of anticancer PDT (Panzarini et al., 2011).

The antivascular effects of PDT. Star et al. (1986) described direct evidence of endothelial damage from the histological sections of blood vessels showing destruction of the vessel wall. The effects observed were blanching (ischemia) of tumor, the circulation gradually slows down, leading to complete stasis (hour to day after irradiation), vasodilatation and hemorrhage (dose related) (Star et al., 1986). The directly cytotoxicity of endothelial cells were described with sublethal doses of PDT. Gomer et al. (1988) showed that bovine endothelial cells were significantly more sensitive than smooth muscle cells or fibroblast from the same species and conclude that endothelial cell photosensitivity may play a role in the vascular damage observed following PDT.

The inflammatory and immune response: The PDT oxidative stress and associated cell damage explain the strong acute inflammatory reaction. Moreover, the acute inflammation is characterized by increased expression of several pro-inflammatory cytokines including TNF-a, IL-1b, and IL-6, adhesion molecules E-selectin and ICAM-1, and rapid accumulation of leukocytes into the treated tumor bed (Brackett & Gollnick, 2011). It also involves the maturation and activation of dendritic cells, and ability to stimulate T cell activation (Brackett & Gollnick, 2011). Pre-clinical and clinical studies have demonstrated that PDT eliminates tumors by direct tumor cell death and indirectly by augmenting anti-tumor immunity. An in vitro study showed that PDT caused the suppression of several tumour-promoting factors in head and neck cancer such as MMP-2, MMP-9, uPA and VEGF. The study do not confirm whether PDT reduces the invasive potential of malignant head and neck cells, but it presents the suppression of factors responsible for tumor invasion which may be of therapeutic value (Sharwani et al., 2006).

4. *In vivo* findings for head and neck cancer treatmente with PDT

Several clinical studies have evaluated the potential of PDT for treating head and neck cancer using approved photosensitizers (Cooper et al., 2007; Biel, 2007; Schweitzer & Somers, 2010; Karakullukcu et al., 2011). In a study reported by Biel (2007), 276 patients with early head and neck squamous carcinoma were treated with Photofrin, a first generation PS. After 48 hours, the light exposition was performed with laser using different light fluencies and densities depending on the localization of the tumor (oral cavity, larynx) under local anesthesia. The cure rates with a single treatment for early laryngeal and oral cancers were 91% and 94%, respectively. A study using the same PS showed that PDT was able to provide local control in 80% of cases with diffuse aggressive mucosal carcinoma in situ of the oral cavity/oropharynx, especially in patients in which conventional treatment failed (Schweitzer & Somers, 2010).

The second generation of PSs was improved in order to give greater tumor selectivity and allow deeper light penetration into the tissue as a result of its longer excitation wavelength (Allison et al., 2004). Foscan, a clorin, was the only second-generation PS studied in clinical trials (Biel, 2007). A recent study analyzed 170 patients with 226 lesions that included primary (95) and non-primary (131) oral cavity and oropharynx neoplasms. It was observed an overall response rate of 90.7% with a complete response rate of 70.8% (Karakullukcu et al., 2011). This study also revealed the difference in response rates between different regions in the oral cavity. For example the tongue reacts significantly better than alveolar process, showing that some sub-sites present better characteristics for PDT as flat surface for a more homogenous light deliver. Cooper et al. (2007) evaluated 27 patients with secondary or multiple primary head and neck cancer treated with mTHPC intravenously. After 4 days, the tumor area was irradiated with a 652 nm laser under general anesthesia. After 24 hours from the treatment, necrosis and formation of slough at the tumor area was observed. In 28 of the 42 tumors, a complete remission was obtained which represents a complete response of 67%. The authors observed a great difference between the cure rates for stage I/in situ (85%) when compared to stage II/III (38%). These results emphasize the importance of an early diagnosis and periodic follow-up in order to discover new primary tumors in a curable stage.

The clinical studies using ALA mediated PDT showed that this drug has a limited depth accumulation that results in limited penetration of light reducing the application of this therapy. In the study of Fan et al. (1998), 18 patients that were diagnosed with dysplasia and malignant lesions in the oral cavity were treated with ALA mediated PDT, and only two of six patients with tumor obtained a complete response. Moreover, Sireon et al. (2001) observed that only a partial response was obtained after treating patients diagnosed with larynx carcinomas using ALA-PDT. Therefore, more studies with a greater number of patients using correct including criteria should be done in order to evaluate the potential of ALA-mediated PDT to treat head and neck cancer.

5. Evaluation of the antitumor potential of curcumin-mediated PDT

Recently, medicinal plants have become the focus of intense study regarding the prevention and/or treatment of several chronic diseases. Turmeric, the powdered rhizome of *Curcuma longa* L., has been used to treat a variety of inflammatory conditions and chronic diseases (Ammon & Wahl, 1991). Curcumin (CUR) is a naturally occurring, intensely yellow turmeric pigment that it is in worldwide use as a cooking spice, flavoring agent and colorant (Epstein et al., 2010). An increasing number of investigations have suggested that CUR exhibits potential therapeutic applications such as anti-inflamatory, antioxidant, antimicrobial, antifungal and anticancer properties (Epstein et al., 2010; Martins et al., 2009). Some studies also showed that CUR inhibits chemically induced carcinogenesis in the skin, forestomach, and colon when it is administered during initiation and/or postinitiation stages (Kawamori et al., 1999; Huang et al.,1994). Some studies have also propose that these effects may possibly be enhanced by combination with light (Bruzell et al., 2005; Dujic, et al., 2009), thus attracting researches to explore its use in several areas including photochemistry and photobiology. For this reason, an in vitro investigation was conducted to evaluate the association of CUR with LED light for the photoinactivation of a tumor cell line.

5.1 Photosensitizer and light source for PDT

Natural CUR (Fluka Co.) was obtained from Sigma Aldrich, St. Louis, MO, USA. A stock solution of CUR (200μM) was prepared in DMSO and then diluted in saline solution to obtain the concentrations to be tested (keeping the final concentration of DMSO at 10%). The 10% concentration of DMSO was selected since has been shown to have no effects on cells viability (data not shown). A light emitting diode (LED) based device, composed of eight royal blue LEDs (LXHL-PR09, Luxeon® III Emitter, Lumileds Lighting, San Jose, California, USA), was used to excite the CUR. The LED device provided a uniform emission from 440nm to 460nm, with maximum emission at 455nm. The irradiance delivered was of $22mW/cm^2$.

5.2 Hela culture, PDT treatments and analysis of cell morphology by SEM

The immortalized Hela cell line, purchased from Adolfo Lutz Institute (São Paulo, SP, Brazil), was selected to perform the anticancer evaluation because it is considered a classic prototype of a cell with high malignancy. The cell line was cultured in Eagle's minimum essential medium (Adolfo Lutz Institute, São Paulo, SP, Brazil.) supplemented with 10% bovine fetal serum (Gibco, Grand Island, NY, USA), with 100 IU/mL penicillin, 100μg/mL streptomycin and 2mmol/L glutamine (Gibco, Grand Island) in an humidified incubator with 5% CO_2 and 95% air at 37°C (Isotemp Fisher Scientific, Pittsburgh, PA, USA). The cells were sub-cultured every 3 days until an adequate number of cells were obtained for the study. After reaching approximately 80% density, the cells were trypsinized, seeded in sterile 24-well plates (30.000 cells/cm^2) and incubated for 72 hours.

After incubation, the culture medium was removed and cells were washed with phosphate buffer saline (PBS). Aliquots of 350 μL of CUR at final concentrations of 5, 10 and 20μM were transferred individually to wells in 24-well plates and were incubated in contact with the cells for 20 minutes at dark. After the incubation period, cells were irradiated for 4 minutes, corresponding to 5.28 J/cm^2 (C+L+). Additional wells containing cells exposed only to CUR (C+L-), or only to LED light (C-L+) were also evaluated. Negative control group was composed of cells not exposed to LED light or CUR (C-L-). Cell viability was evaluated by succinic dehydrogenase (SDH) enzyme, which is a measure of the mitochondrial respiration of the cell. In 10 wells, 900 μL of Eagle's medium associated with 100 μL of MTT solution (5mg/mL sterile PBS) (Sigma Chemical Co.) was added to the cells cultured in each well and incubated at 37°C for 4h. After this, the culture medium (Eagle's medium with the MTT solution) was aspirated and replaced by 600 μL of acidified isopropanol solution (0.04N HCl) to dissolve the blue crystals of formazan present in the cells. Cell metabolism was determined as being proportional to the absorbance measured at 570nm wavelength with the use of an ELISA plate reader (BIO-RAD, model 3550-UV, microplate reader, Hercules, CA, USA).

For Hela cell-line morphology analysis by Scanning Electron Microscopy (SEM), sterile cover glasses 12mm in diameter (Fisher Scientific) were placed on the bottom of the wells of all experimental and control groups immediately before the cell seeding. After the experimental conditions, the culture medium was removed and the viable cells that remained adhered to the glass substrate were fixed in 1mL of buffered 2.5% glutaraldehyde for 24hours and post-fixed with 1% osmium tetroxide for 1 hour. The cells adhered to the glass substrate were then dehydrated in a series of increasing ethanol concentrations (30, 50, 70, 95 and 100%) and immersed in 1,1,1,3,3,3-hexamethyldisilazane (HMDS; Acros Organics, Springfield, NJ, USA) for 90 minutes and stored in a desiccator for 24 hours. The cover glasses were then mounted on metal stubs, sputter-coated with gold and Hela cells morphology was examined by SEM (JEOL-JMS-T33A Scanning Microscope, Tokyo, Japan).

Each experimental treatment was repeated five times. The results obtained presented heterocedasticity. The Kruskall-Wallis and a post-hoc Dunn tests were used to detect differences in absorbance values (MTT) among investigated groups. Differences were considered statistically significant at $p < 0.05$.

5.3 Results

The box plot (Fig. 1) shows the median, 1st and 3rd quartiles, lowest and highest values obtained under each of the experimental and control conditions. Considering the negative control group (C-L-) as having 100% of Hela cell viability, a significant reduction in cell metabolism of 75.5, 81.6 and 87.3% was observed for curcumin concentrations of 5, 10, and 20 μM when the cells were irradiated by the blue LED.

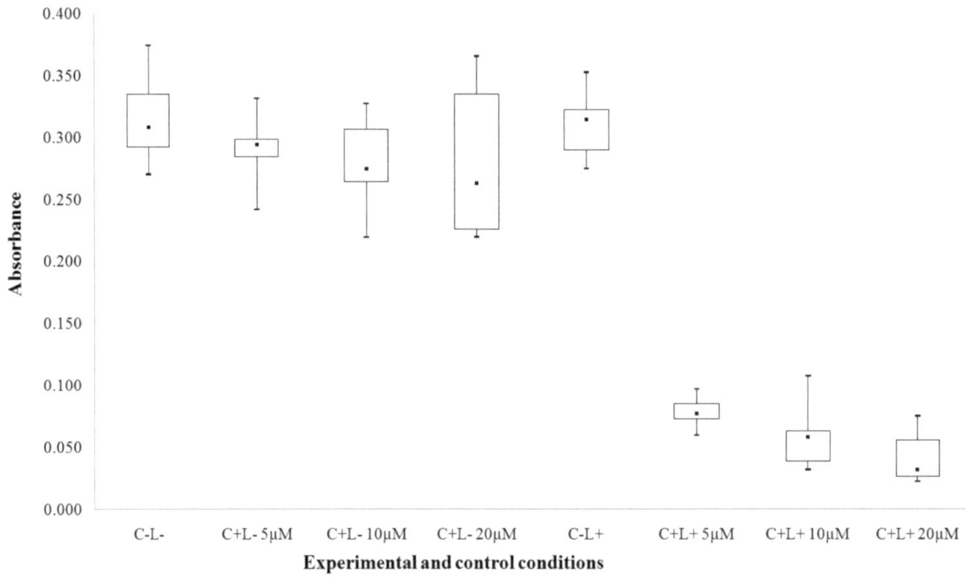

Fig. 1. Summary of absorbance values obtained after experiments with Hela cell culture.

In the C+L+ groups, there was statistically significant difference ($p<0.05$) between the CUR concentrations of 5 and 20 μM (Table 1). In comparison with the control group (C-L-), the dark toxicity of CUR was observed for the concentrations of 10 and 20 μM (reduction of 11.63 and 11.10% respectively), although no significant differences were detected among the three concentrations. The irradiation control group showed similar metabolism activity to the negative control group, showing that the light dose used in the present study did not alter the cell metabolism.

Experimental and control conditions	Mean Ranks	Post hoc test
C-L-	62.60	a
C+L- 5μM	54.50	abc
C+L- 10μM	48.70	c
C+L- 20μM	51.30	bc
C-L+	60.40	ab
C+L+ 5μM	23.15	d
C+L+ 10μM	14.55	de
C+L+ 20μM	8.80	e

Table 1. Post hoc multiple comparisons of mean ranks for the association of three curcumin concentrations with or without light against Hela cells. Significant differences ($p < 0.05$) among rows are indicated by different small letters.

Figure 2 (a-d) presents a panel of SEM micrographs of the Hela cell-line representative of the control and experimental groups. For the negative control group (no treatment) and the group treated with Curcumin (20 μM) alone, numerous Hela cells that remained adhered to the glass substrate exhibited wide cytoplasm and numerous fine cytoplasmic processes originated from the cell membrane (Figure 2a,b). In the groups submitted to PDT, there were a smaller number of Hela cells that remained adhered to the coverglass, which can explain the lower cell metabolism observed for the MTT assay. In PDT group using 5 μM of Curcurmin, it can be observed a small number of cells with ill -defined cytoplasmic membrane limits (Figure 2c). However, in PDT group using 20 μM of Curcumin, it is not observed any cell adhered to the glass substrate. Only rests of cytoplasmic membrane of dead cells were observed, suggesting necrosis death (Figure 2d).

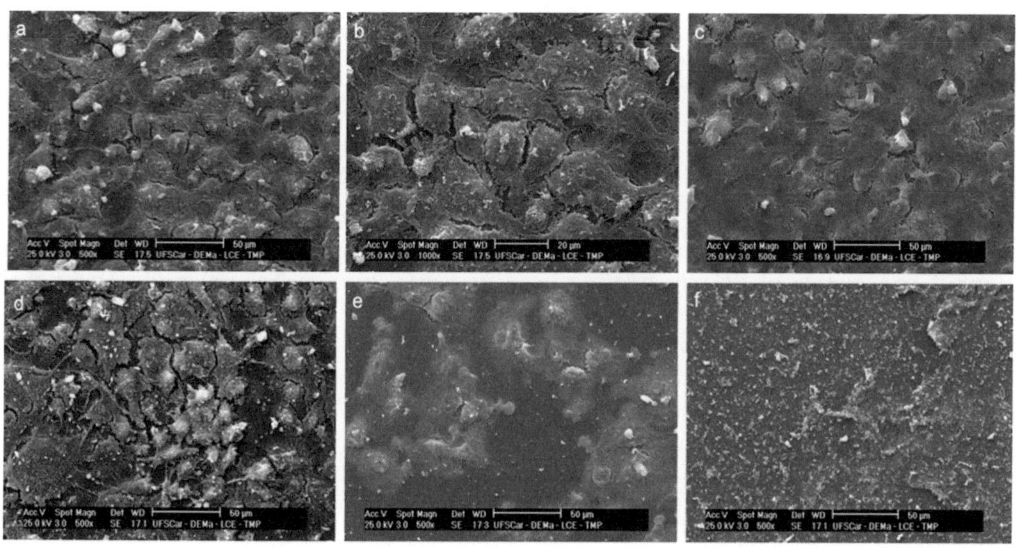

Fig. 2. a- Negative control group of Hela cell line, SEM, 500X; b- Negative control group of Hela cell line, SEM, 1000X; c- Hela cells exposed only to LED irradiation with a light fluence of 5.28 J/cm^2; d- Hela cells in contact with Curcumin (20μM), SEM, 500X; e- Group PDT with 5 μM of Curcumin, SEM, 500X; f- Group PDT with 20 μM of Curcumin, SEM, 500X.

5.4 Discussion

The association of CUR and light achieved a significant reduction in cell metabolism of 87.3%. On the other hand, the effect of CUR without illumination was also evaluated and the results showed a smaller degree of reduction in cell viability, when compared with the PDT results. Consistent with the data of the present study, Koon et al. (2006) reported that the cytotoxicity of CUR was enhanced by the irradiation using blue light with light fluence of 60 kJ/m^2, although dark cytotoxicity was also observed against a nasopharyngeal carcinoma cell line. The results of Park et al. (2007) also suggest the use of CUR as photosensitizer against skin cancer cells. The mode of action of CUR against tumor cells is likely to be induced by the apoptotic pathway. Curcumin has also been reported to selectively lead to apoptosis in various cancer cells without affecting normal and primary cells (Koon et al., 2006; Park et al., 2007; Weng et al., 2011). As regards the mode of action of photosensitized-CUR, evidences from the literature suggest that cell shrinkage, membrane bleeding and apoptosis can occur in different cell lines (Priyadarsini, 2009). In the present study, PDT-treated cells showed altered morphology, smaller size and smaller number of cells adhered to the cover glass. In addition, there were cell fragments and ill-defined cytoplasmic membrane limits in the cells subjected to PDT. This could be an indicator that the plasma membrane was the main target of the photosensitizer. As a rule, photosensitizer accumulation in the mitochondria or endoplasmic reticulum causes apoptosis, while its accumulation in the plasma membrane or lysosomes predisposes the cells to necrosis (Buytaert et al., 2007). Thus, it can be suggested that curcumin-mediated PDT promoted necrosis of Hela cells, although the type of cell death was not evaluated in the present investigation.

In conclusion, when CUR was irradiated with blue LED light, it proved to be an effective photosensitizing agent for the inactivation of tumor cells. However, some investigations have indicated that the main constraint to in vivo application is its low bioavailability, mainly due to its poor absorption in blood and fast metabolism. Thus, future investigations will be conducted to improve the solubility of CUR in an aqueous solution in an attempt to produce a formulation of CUR with adequate bioavailability and clinical efficacy for delivery to cancer cells.

6. Conclusion

According to the information discussed in this chapter, there is scientific evidence to support the efficacy of PDT in treating head and neck tumors. Several protocols of this treatment have been described and discussed, and the results from these studies have shown that PDT can be the option for those patients that have failed prior conventional treatments and an alternative to control multiple primary tumors without the morbidity of surgery or chemo/radiotherapy. PDT has the potential to be an important treatment modality for head and neck cancer due to its minimal side effects and reasonable local control. Therefore, more clinical studies are needed to establish treatment protocols with different photosensitizers as natural compounds in order to increase the application of this therapy in the community settings.

7. Acknowledgment

The authors would like to acknowledge the Center of Study in Optics and Photonics (CEPOF) at the Physics Institute of São Carlos (IFSC) of the University de São Paulo (USP) under supervision of Prof. Dr. Vanderlei S. Bagnato for developing the LED prototype specifically for this study. We also would like to acknowledge Prof. Dr. Iguatemy L. Brunetti for providing the Curcumin used in this study.

8. References

Agostinis, P.; Berg, K.; Cengel, K.; Foster, T.; Girotti, A.; Gollnick.; et al. (2011). Photodynamic therapy of cancer: An update. *CA: a cancer journal for clinicians*, Vol.61, No.4, (July-August), pp. 250-281, ISSN 0007-9235

Allison, R.; Cuenca, R.; Downie, G.; Camnitz, P.; Brodish, B. & Sibata, C. (2005). Clinical photodynamic therapy of head and neck cancers- a review of applications and outcomes. *Photodiagnosis and photodynamic therapy*, Vol.2, No.3, (September 2005), pp. 205-222, ISSN 1572-1000

Ammon, H. & Wahl, M. (1991). Pharmacology of Curcuma longa. *Planta Medica*, Vol.57, No.1, (February), pp. 1-7, ISSN 0032-0943

Baglo, Y.; Sousa, M.; Slupphaug, G.; Hagen, L.; Håvåg, S.; Helander, L.; et al. (2011). Photodynamic therapy with hexyl aminolevulinate induces carbonylation, posttranslational modifications and changed expression of proteins in cell survival and cell death pathways. *Photochemical & photobiological sciences*, Vol.10, No.7, (July 29), pp. 1137-1145, ISSN 1474-905X

Brackett, C & Gollnick, S. (2011). Photodynamic therapy enhancement of anti-tumor immunity. *Photochemical & photobiological sciences*, Vol.10, No.5, (May), pp. 649-652, ISSN 1474-905X

Biel, M. (2007). Photodynamic therapy treatment of early oral and laryngeal cancers. *Photochemistry and Photobiology*, Vol.83, No.5, (September/Octuber), pp. 1063-1068, ISSN 0031-8655

Brancaleon, L. & Moseley, H. (2002). Laser and non-laser light sources for Photodynamic Therapy. *Lasers in medical science*, Vol.17, No.3, pp. 173-186, ISSN 0268-8921

Bredell, M.; Besic, E.; Maake, C. & Walt, H. (2010). The application and challenges of clinical PD-PDT in the head and neck region: a short review. *Journal of photochemistry and photobiology B*, Vol.101, No.3, (December 2010), pp. 185-190, ISSN 1011-1344

Bruzell, E.; Morisbak E. & Tønnesen, H. (2005). Studies on curcumin and curcuminoids. XXIX. Photoinduced cytotoxicity of curcumin in selected aqueous preparations. *Photochemical & photobiological sciences*. Vol.4, No.7, (July), pp. 523-530, ISSN 1474-905X

Buytaert, E.; Dewaele, M. & Agostinis, P. (2007). Molecular effectors of multiple cell death pathways initiated by photodynamic therapy. *Biochimica et biophysica acta*, Vol.1776, No.1, (September), pp. 86-107, ISSN 0006-3002

Chen, H.; Liu, C.; Yang, H.; Chou, H.; Chiang, C. & Kuo, M. (2011). 5-aminolevulinic acid induce apoptosis via NF-kappaB/JNK pathway in human oral cancer Ca9-22 cells. *Journal of oral pathology & medicine*, Vol.40, No.6, (July), pp. 483-489, ISSN 0904-2512

Copper, M.; Triesscheijn, M.; Tan, I.; Ruevekamp, M. & Stewart, F. (2007). Photodynamic therapy in the treatment of multiple primary tumours in the head and neck, located to the oral cavity and oropharynx. *Clinical otolaryngology*, Vol.32, No.3, (June), pp. 185-189, ISSN 1749-4478

Dolmans, D.; Fukumura, D. & Jain, R. (2003). Photodynamic therapy for cancer. *Nature reviews. Cancer.*, Vol.3, No.5, (May 2003), pp. 380-387, ISSN 1474-175X

Donnelly, R.; McCarron, P.; Tunney, M. & Woolfson, A. (2007). Potential of photodynamic therapy in treatment of fungal infections of the mouth. Design and characterisation of a mucoadhesive patch containing toluidine blue O. *Journal of photochemistry and photobiology B*, Vol.86, No.1, (January 3), pp.59-69, ISSN 1011-1344

Dujic, J.; Kippenberger, S.; Ramirez-Bosca, A.; Diaz-Alperi, J.; Bereiter-Hahn, J.; Kaufmann, R.; Bernd, A. & Hofmann, M. (2009). Curcumin in combination with visible light inhibits tumor growth in a xenograft tumor model. *International journal of cancer*, Vol.124, No.6, (March), pp. 1422-1428, ISSN 0020-7136

Elledge, S. (1996). Cell Cycle Checkpoints: Preventing an Identity Crisis. *Science*, Vol.274, No.5293, (December 6), pp. 1664-1672. ISSN 0036-8075

Epstein, J.; Sanderson, I. & Macdonald, T. (2010). Curcumin as a therapeutic agent: the evidence from in vitro, animal and human studies. *The British journal of nutrition*, Vol.103, No.11, (June), pp. 1545-1557, ISSN 0007-1145

Fan, K.; Hopper, C.; Speight, P.; Buonaccorsi, G.; MacRobert, A. & Bown, S. (1996). Photodynamic therapy using 5-aminolevulinic acid for premalignant and malignant lesions of the oral cavity. *Cancer*, Vol.78, No.7, (October), pp. 1374-1383, ISSN 0008-543X

Funk, G.; Karnell, L.; Robinson, R.; Zhen, W.; Trask, D. & Hoffman, H. (2002). Presentation, treatment, and outcome of oral cavity cancer: a National Cancer Data Base report. *Head & Neck*, Vol.24, No.2, (February 2002), pp. 165–180, ISSN 1043-3074

Furre, I.; Møller, M.; Shahzidi, S.; Nesland, J. & Peng, Q. (2006). Involvement of both caspase-dependent and -independent pathways in apoptotic induction by hexaminolevulinate-mediated photodynamic therapy in human lymphoma cells. *Apoptosis*, Vol.11, No.11, (November), pp. 2031-2042, ISSN 1360-8185

Gomer, C.; Rucker, N. & Murphree, A. (1988). Differential cell photosensitivity following porphyrin photodynamic therapy. *Cancer Research*, Vol.48, No.16, (August), pp. 4539-4542, ISSN 0008-5472

Gramaglia , D.; Gentile, A.; Battaglia, M.; Ranzato, L.; Petronilli, V.; Fassetta, M.; Bernardi, P. & Rasola, A. (2004). Apoptosis to necrosis switching downstream of apoptosome formation requires inhibition of both glycolysis and oxidative phosphorylation in a BCL-X(L)- and PKB/AKT-independent fashion. *Cell death and differentiation*, Vol.11, No.3, (March), pp. 342-353, ISSN 1350-9047

Hopper, C.; Kubler, A.; Lewis, H.; Tan, I. & Putnam, G. (2004). mTHPC-mediated photodynamic therapy for early oral squamous cell carcinoma. *International Journal of Cancer*, Vol.111, No.1, (August 10), pp. 138–146, ISSN 0020-7136

Hsieh, Y.; Wu, C.; Chang, C. & Yu, J. (2003). Subcellular localization of Photofrin determines the death phenotype of human epidermoid carcinoma A431 cells triggered by photodynamic therapy: when plasma membranes are the main targets. *Journal of cellular physiology*, Vol.194, No.3, (March 2003), pp. 363-375, ISSN 0021-9541

Huang, M.; Lou, Y.; Ma, W.; Newmark, H.; Reuhl, K. & Conney, A. (1994). Inhibitory effect of dietary curcumin on forestomach, duodenal and colon carcinogenesis in mice. *Cancer Research*, Vol.54, No.22, (November), pp. 5841-5817, ISSN 0008-5472

Hudson, T.; Anderson, W. & Artez, A.(2010). International network of cancer genome projects. *Nature*, Vol.464, No.7291, (April 15), pp. 993-998, ISSN 0028-0836

Karakullukcu, B.; van Oudenaarde, K.; Copper, M.; Klop, W.; van Veen, R.; Wildeman, M. & Bing Tan, I. (2004). Photodynamic therapy of early stage oral cavity and oropharynx neoplasms: an outcome analysis of 170 patients. *European archives of otorhinolaryngology*, Vol.268, No.2, (February 2004), pp. 281-288, ISSN 0937-4477

Kastan, M. &. Bartek, J. (2004). Cell-cycle checkpoints and cancer. *Nature*, Vol.432, No.7015, (November 18), pp. 316-323, ISSN 0028-0836

Kawamori, T.; Lubet, R.; Steele, V.; Kelloff, G.; Kaskey, R.; Rao, C. & Reddy, B. (1999). Chemopreventive effect of curcumin, a naturally occurring anti-inflammatory agent, during the promotion/progression stages of colon cancer. *Cancer Research*, Vol.59, No.3, (February), pp. 597-601, ISSN 0008-5472

Kessel, D. & Luo, Y. (1998). Mitochondrial photodamage and PDT-induced apoptosis. *Journal of photochemistry and photobiology B*, Vol.42, No.2, (February), pp. 89-95, ISSN 1011-1344

Kessel, D. & Oleinick, N. (2009). Initiation of autophagy by photodynamic therapy. *Methods Enzymol*, Vol.453, pp. 1-16, ISSN 0076-6879

Konopka, K. & Goslinski, T. (2007) Photodynamic Therapy in Dentistry. *Journal of Dental Research*, Vol.86, No.8, (November 2007), pp. 694-707, ISSN 0022-0345

Koon, H.; Leung, A.; Yue, K. & Mak, N. (2006). Photodynamic effect of curcumin on NPC/CNE2 cells. *Journal of environmental pathology, toxicology and oncology*, Vol.25, No.1-2, pp. 205-215, ISSN 0731-8898

Lee, H.; Trainer, A.; Friedman, L.; Thistlethwaite, F.; Evans, M.; Ponder, B. & Venkitaraman A. (1999). Mitotic checkpoint inactivation fosters transformation in cells lacking the breast cancer susceptibility gene, Brca2. *Molecular Cell*, Vol.4, No.1, (July 1999), pp. 1-10, ISSN 1097-2765

Lim, S.; Lee, H. & Ho, A. (2011). A New Naturally-derived Photosensitiser and Its Phototoxicity on Head and Neck Cancer Cells. *Photochemistry and Photobiology*. (May 2), ISSN 0031-8655 doi: 10.1111/j.1751-1097.2011.00939.x.

Martins, C.; da Silva, D.; Neres, A.; Magalhães, T.; Watanabe, G.; Modolo, L.; Sabino, A.; de Fátima, . & de Resende, M. (2009). Curcumin as a promising antifungal of clinical interest. *The Journal of antimicrobial chemotherapy*, Vol.63, No.2, (February), pp. 337-339, ISSN 0305-7453

Neville B. & Day, T. (2002). Oral cancer and precancerous lesions. *CA: a cancer journal for clinicians*, Vol.52, No.4, (July/August 2002), pp. 195–215, ISSN 0007-9235

Panzarini, E.; Inguscio, V. & Dini, L. (2011). Timing the multiple cell death pathways initiated by Rose Bengal acetate photodynamic therapy. *Cell death & disease*, Vol.9, No.2, (June), pp. e169, ISSN 2041-4889

Park, K. & Lee, J. (2007). Photosensitizer effect of curcumin on UVB-irradiated HaCaT cells through activation of caspase pathways. *Oncology reports*, Vol.17, No.3, (March), pp. 537-540, ISSN 1021-335X

Priyadarsini, K. (2009). Photophysics, photochemistry and photobiology of curcumin: Studies from organic solutions, bio-mimetics and living cells. *Journal of photochemistry and photobiology. C*, Vol.10, No.2, (June), pp. 81-95, ISSN 1389-5567

Rasola, A. & Bernardi, P. (2011). Mitochondrial permeability transition in Ca(2+)-dependent apoptosis and necrosis. *Cell Calcium*, (May), ISSN 0143-4160. In press.

Schweitzer, V. & Somers, M. (2010). Photofrin-mediated photodynamic therapy for treatment of early stage (Tis-T2N0M0) SqCCa of oral cavity and oropharynx. *Lasers in Surgery and Medicine*, Vol.42, No.1, (January 2010), pp. 1-8, ISSN 0196-8092

Sharwani, A.; Jerjes, W.; Hopper, C.; Lewis, M.; El-Maaytah, M.; Khalil, H.; Macrobert, A.; Upile, T. & Salih. V. (2006). Photodynamic therapy down-regulates the invasion promoting factors in human oral cancer. *Archives of oral biology*. Vol.51, No.12, (December), pp. 1104-1111, ISSN 0003-9969

Sieron, A.; Namyslowski, G.; Misiolek, M.; Adamek, M. & Kawczyk-Krupka, A. (2001). Photodynamic therapy of premalignant lesions and local recurrence of laryngeal and hypopharyngeal cancers. *European archives of otorhinolaryngology*, Vol.258, No.7, (September), pp. 349-352, ISSN 0937-4477

Star, W.; Marijnissen, H.; van den Berg-Blok, A.; Versteeg, J.; Franken, K. & Reinhold, H. (1986). Destruction of rat mammary tumor and normal tissue microcirculation by hematoporphyrin derivative photoradiation observed in vivo in sandwich observation chambers. *Cancer research*, Vol.46, No.5, (May), pp. 2532-2540, ISSN 0008-5472

Stromhaug, P. & Klionsky, D. (2001). Approaching the molecular mechanism of autophagy. *Traffic*, Vol.2, No.8, (August), pp. 524-531, ISSN 1398-9219

Teichert, M.; Jones, M.; Usacheva, M. & Biel, M. (2002). Treatment of oral candidiasis with methylene blue-mediated photodynamic therapy in an immunodeficient murine model. *Oral Surg Oral Med Oral Pathol Oral Radiol Endod*, Vol.93, No.2 , (February 2002), pp. 155-160, ISSN 1079-2104

Triesscheijn, M.; Ruevekamp, M.; Aalders, M.; Baas, P. & Stewart, F. (2004). Comparative sensitivity of microvascular endothelial cells, fibroblasts and tumor cells after in vitro photodynamic therapy with meso-tetra-hydroxyphenyl-chlorin. *Photochemistry and Photobiology*, Vol.80, No.2, (September/October), pp. 236-241, ISSN 0031-8655

Vanlangenakker, N.; Vanden Berghe, T.; Krysko, D.; Festjens, N. & Vandenabeele, P. (2008). Molecular mechanisms and pathophysiology of necrotic cell death. *Current molecular medicine*, Vol.8, No.3, (May), pp. 207-220, ISSN 1566-5240

Weng, C.; Yung, B.; Weng, J. & Wu, M. (2010). Involvement of nucleophosmin/B23 in the cellular response to curcumin. *The Journal of nutritional biochemistry*, Vol.22, No.1, (January), pp. 46-52, ISSN 0955-2863

Williams, J.; Pearson, G. & Colles, M. (2006). Antibacterial action of photoactivated disinfection (PAD) used on endodontic bacteria in planktonic suspension and in artificial and human root canals. *Journal of Dentistry*, Vol.34, No.6, (July 2006), pp. 363-371, ISSN 0300-5712

Yu, C.; Chen, H.; Hung, H.; Cheng, S.; Tsai, T. & Chiang, C. (2008). Photodynamic Therapy outcome for oral verrucous hyperplasia depends on the clinical appearance, size, color, epithelial dysplasia, and surface keratin thickness of the lesion. *Oral Oncology*, Vol.44, No.6, (June 2008), pp. 595-600, ISSN 1368-8375

Zeina, B.; Greenman, J.; Corry, D. & Purcell, W. (2003). Antimicrobial photodynamic therapy: assessment of genotoxic effects on keratinocytes in vitro. *The British journal of dermatology.*, Vol.148, No.2, (February), pp. 229-232, ISSN 0007-0963

8

Wnt Signaling Network in Homo Sapiens

Bahar Nalbantoglu, Saliha Durmuş Tekir and Kutlu Ö. Ülgen
Department of Chemical Engineering, Boğaziçi University, Bebek-İstanbul, Turkey

1. Introduction

Signaling is a part of system communication among living cells by processing biological information that governs basic cellular activities and coordinates cell actions. The signals are transmitted by means of signaling molecules, and this process ends up with altering of gene transcription in the nucleus resulting in many cellular processes such as differentiation and proliferation. Systems biology research helps us understand the underlying structure of cell signaling networks and how changes in these networks affect the transmission and flow of information. Here we thoroughly investigate Wnt signaling, a major signaling system known to have role in regulating processes such as embryonic development and growth, cell differentiation, proliferation, migration and polarity.

There are two major types of Wnt signaling (Figure 1) classified as canonical (Wnt/β-catenin) and non-canonical (planar cell polarity (PCP) and the Wnt-calcium (Wnt/Ca^{2+})) pathways (Katoh, 2005; Nusse, 2005; DasGupta, 2005). In canonical Wnt signaling, the

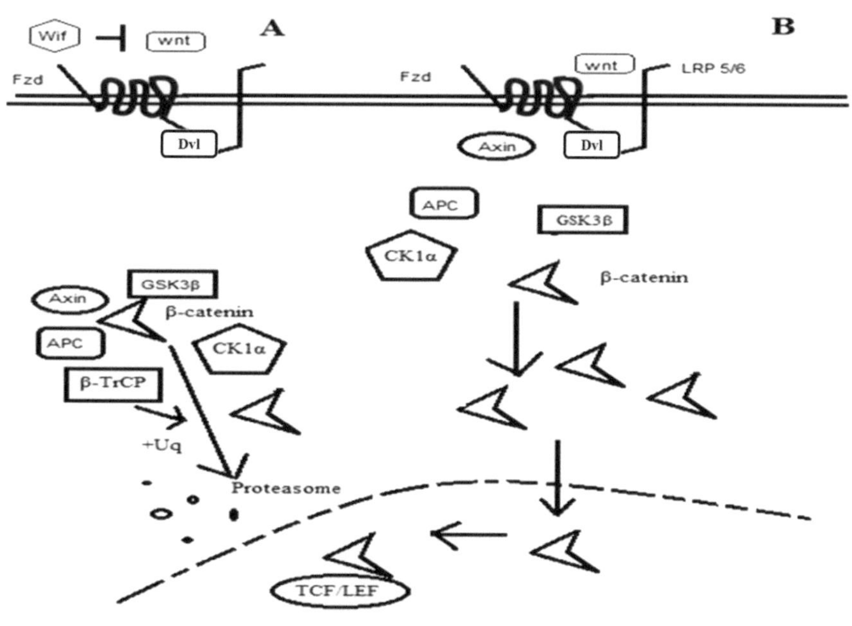

Fig. 1. Regulation of β-catenin stability by Wnt signaling.

absence of the Wnt ligand, the signal transduction is at non-operating state (off), the protein complex that involves APC and AXIN proteins facilitates the phosphorylation of β-catenin by glycogen synthase kinase (GSK3β). Then β-catenin binds to a protein called β–TrCP (beta-transducin repeat containing protein) that mediates the ubiquitination of CTNNB1 and the proteins tagged with ubiquitin are degraded by the proteosome that results in β-catenin destruction by the proteosomal degradation. When Wnt proteins bind to Frizzled receptors, this receptor activation works counter to destruction complex (APC-AXIN-GSK3β) by preventing β-catenin phosphorylation via Dishvelled protein. Then β–catenin accumulates and enters into the nucleus where it binds to a DNA binding protein of the TCF/LEF family and then activates new gene expression programs (Cadigan, 2008).

The illustration (Figure 1) depicts a cell in the absence (A) and in the presence (B) of Wnt protein. In unstimulated cells, β-catenin not complexed with the cadherin adhesion complex is phosphorylated by CKI and GSK3β, leading to β-TrCP dependent ubiquitination and proteosomal degradation. The presence of Wnt, on the hand, promotes LRP (lipoprotein receptor-related protein), cell-surface coreceptor, and Fz association, leading to recruitment of Dvl to the complex, followed by GSK3β and CKI casein kinase phosphorylation of LRP. This stabilizes recruitment of AXIN to the receptor, which in turn may disrupt the activity of the destruction complex, and hence allows accumulation of β-catenin and nuclear translocation (Cadigan, 2008). An overview of types of Wnt signaling in different species is illustrated in Figure 2.

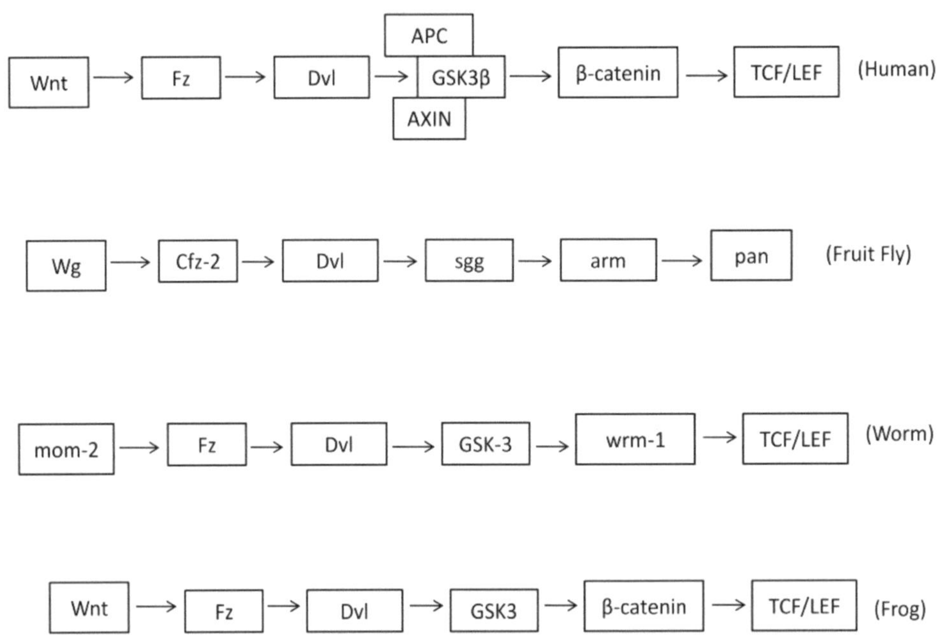

(a) Wnt/β-catenin

Fig. 2. (Continued)

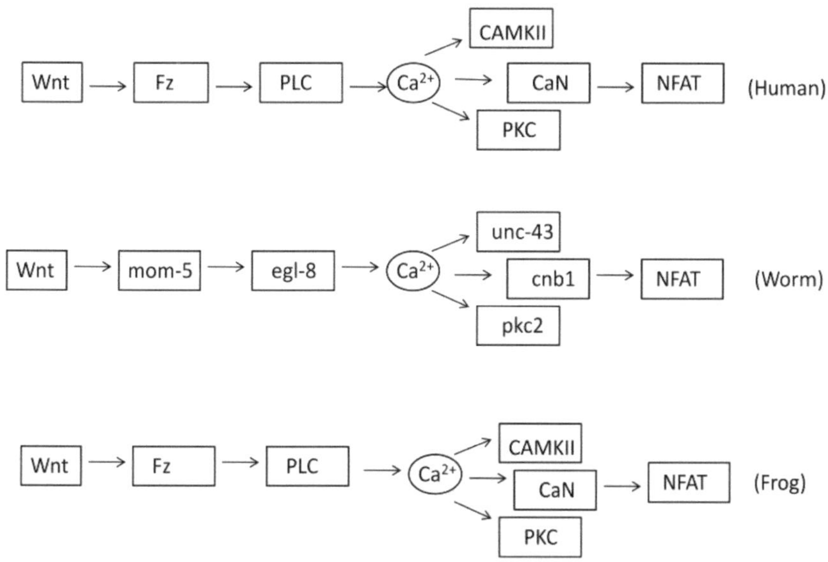

(b) Wnt/Ca^{2+}

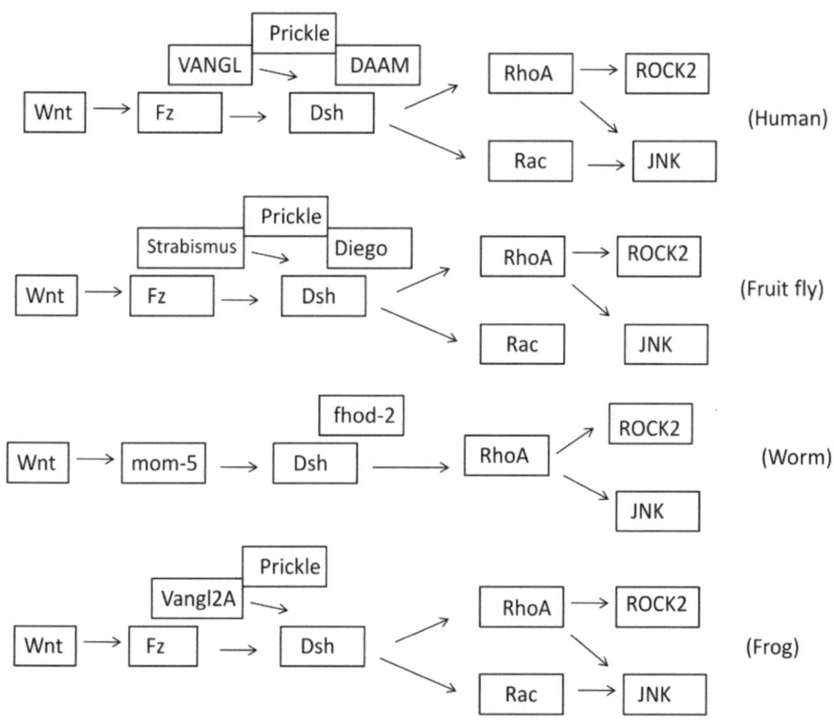

(c) Wnt/PCP

Fig. 2. An overview of Wnt signaling pathways:(a)Wnt/ β-catenin (b) Wnt/calcium (c) Wnt/Planar cell polarity in different species (KEGG: Kyoto Encyclopedia of Genes and Genomes; http://www.genome.jp/kegg/)

The number of non-canonical pathways is unclear (Veeman *et al.*, 2003; Eisenmann, 2005; Pandur, 2005). A Wnt signalling pathway that increases intracellular Ca^{2+} levels has been proposed as Wnt/Ca^{2+} pathway in vertebrates (Kuhl *et al.*, 2000). There is also a Wnt/JNK (Wnt/c-Jun N terminal kinase) pathway in vertebrates that controls cell polarity (Veeman *et al.*, 2003; Pandur 2005). A new Wnt signalling pathway called Wnt-5A/Ror2 signalling has been proposed in *X. leavis* (Schambony and Wedlich, 2007). In flies, the Wnt/PCP pathway (Figure 2c), which controls planar cell polarity, is the best-known non-canonical Wnt signalling pathway so far (Veeman *et al.*, 2003, Eisenmann, 2005). Processes such as P2/EMS signalling, T cell polarity, Z1/Z4 polarity and B cell polarity utilize non-canonical Wnt signalling in *C. elegans*. Furthermore, a non-canonical pathway, a PCP like pathway, has been proposed to control the polarity of B cells (Wu and Herman, 2006). Recently a novel non-canonical Wnt pathway, which is also β-catenin independent, in vertebrate and fly has been proposed to act in neuronal specification (Hingwing *et al.*, 2009).

Wnt-planar cell polarity (PCP) and Wnt-calcium(Ca^{2+}) pathways that are independent of β-catenin are considered the two most well-recognized non-canonical pathways in Wnt signaling. The activation of kinase C (PKC), Jun kinase (JNK), RHOA and nemo-like kinase (NLK) mediates the Wnt-planar cell polarity (PCP) pathway which is related to tissue polarity control and cell movement. The Wnt/PCP signaling pathway is mostly characterized in *D. Melanogaster.* Frizzled, Van Gogh, Stan/Flamingo, Dishevelled, Prickle and Diego proteins are the major proteins of Drosophilia Wnt/PCP signaling. Diego activates the pathway with binding to Dishevelled whereas Prickle prevents this association and negatively regulates the pathway. The tissue polarity is established by asymmetrical localization of FZD-Dsh-Diego-Stan complex and Vang-Prickle complex. The main proteins of the human Wnt/PCP pathway are mostly the homologs of the *Drosophilia* proteins. VANGL1, VANGL2 (Van Gogh homologs), DVL1, DVL2, DVL3 (Dishevelled homologs), PRICKLE1, PRICKLE2 (Prickle homologs) and ANKRD6 (Diego homolog) are the core proteins having essential roles in human Wnt/PCP signaling cascade. In Wnt/PCP signaling, Wnt ligands (WNT5A, WNT5B and WNT11) first bind to Frizzled receptors (FZD3, FZD6). Then, the association of Prickle, Vangl and Dishevelled proteins result in RHOA activation following the binding of Dishevelled proteins to Daam proteins which are implicated in actin-cytoskeleton re-organization. Moreover, by Dvl-dependent manner, the JNK cascade is activated by the association of dishevelled proteins with MAPK3. On the other hand, NLK signaling cascade is activated by Dvl-independent manner in which the Ca^{2+} release by FZD proteins activates MAP3K7 (TAK1). NLK protein is known to be an activator of Wnt/PCP signaling as well as the inhibitor of canonical Wnt pathway (Katoh, 2005; Katoh 2003; Wu and Maniatis, 1999).

Calcium has been implicated as an important messenger in Wnt pathway and recent studies showed that higher frequencies of calcium transients were associated with faster rates of outgrowth (Veeman *et al.*, 2003; Pandur 2005; Cadigan, 2008). WNT5a also activates the Wnt/Ca^{2+} signaling. The binding of Wnts to Frizzled receptors leads to an activation of heterotrimeric G-proteins and subsequent activation of phospholipase C by the G-protein beta/gamma dimer. This enzyme cleaves phosphatidylinositol-4,5-bisphosphate (PIP2) into inositol-1,4,5-trisphosphate (IP3) and diacylglycerol (DAG). IP3 is released from the membrane, and binds to the IP3 receptor which subsequently releases calcium ions from intracellular stores. This locally restricted calcium release subsequently activates calcium-sensitive proteins like protein kinase C (PKC), calcium-calmodulin dependent kinase II

(CamKII), and/or calcineurin (CaCN). IP3 is then degraded to inositol by means of specific phosphatases and recycled to PIP2 (Kuhl et al., 2000; Kohn and Moon, 2005; Kuhl, 2004; Slusarskia et al., 1997). Activated CaCN dephosphorylates NF-AT (nuclear factor of activated T-cells), which then translocates to nucleus to regulate gene expression. Studies performed for Xenopus showed that this signaling is active in the development of ventral cell fate which inhibits Wnt/β-catenin pathway (Saneyoshi et al., 2002; Veeman et al., 2003; Pandur, 2005). In addition to that, recent studies reported that there is a crosstalk between canonical and non-canonical Wnt signaling through GSK3b since it can phosphorylate NF-AT (Beals et al., 2003; Ohteki et al., 2000). The Wnt/PCP and Wnt/Calcium signaling pathways have overlaps since both of them are mediated by calcium release, and Dishevelled proteins play essential roles in activation of these cascades.

The deregulations and mutations in this signaling pathway cause several human diseases including lung, breast, colon and colorectal cancers. Recent works have demonstrated that Wnt-activated excessive β-catenin accumulation in nucleus plays an important role in tumour formation. On the other hand, one of the β-catenin independent signalling pathways; the Wnt/Ca^{2+} signalling pathway, is proposed to antagonize Wnt/β-catenin signalling pathway. Therefore, Wnt/Ca^{2+} cascade may have the ability to act as a tumour suppressor. However, still much more work has to be done on this subject as the mechanism of this antagonism is not well known. (Wang and Malbon, 2003; Giles *et al.*, 2003; Veeman *et al.*, 2003; Kohn and Moon, 2005; Slusarski and Pelegri, 2007; Maiese *et al.*, 2008).

Investigating Wnt signaling is therefore attractive for researchers to identify the suitable drug targets for therapeutic intervention in cancer treatment (Cadigan and Liu, 2006; Cong et al., 2004; Widelitz, 2005; Nusse, 2005). This signaling pathway has been studied and analyzed using different species such as worm (*Caenorhabditis elegans*), fly (*Drosophila melanogaster*), frog (*Xenopus leavis*) etc. in order to enucleate the governing mechanisms through essential components (DasGupta, 2005; Kuhl et al., 2000; Kohn and Moon, 2005). However, the whole Wnt mechanism is not well understood in human due to the restrictions on experiments with human, and more attention should be devoted to this organism.

In this study, the reconstruction of Wnt signaling network was computationally performed for *Homo Sapiens* via integration of interactome data and Gene Ontology annotations. The graph theoretical analysis was then performed for analyzing the topological properties of the network proteins. Moreover, the linear paths in which the signal is transferred from ligands (input) to transcription factors (output) were identified in order to follow signal transmittal as well as to identify the specific proteins of canonical and non-canonical Wnt pathways. Furthermore, the crosstalk analysis was applied in order to detect the significant bridging proteins in all these pathways. Finally, the proteins which might be targets of drugs against diseases involving Wnt pathways are indicated and these proteins are compared with the ones reported in literature in terms of topological properties and their roles in canonical and non-canonical Wnt pathways.

This work is aimed to give an insight on the role of Wnt signalling in maintaining the homeostasis as well as reacting to cellular stress. Understanding the molecular basis underlying the ability of Wnt proteins to perform antagonistic or similar signalling activities would eventually lead us to new ideas about how to suppress cancer cells in human metabolism.

2. Materials and methods

2.1 Network reconstruction by GO annotations

The reconstruction of Wnt signaling network in human was performed by integrating Gene Ontology (GO) annotations and PPI data. First of all, the core proteins that are known to have roles in Wnt signaling in *Homo Sapiens* were obtained from literature. Then an annotation collection table was prepared by using GO annotations (biological process, cellular component and molecular function) of the core proteins. All the GO annotations of all human proteins were downloaded from GO website (http://www.geneontology.org/GO.current.annotations.shtml) by October 2010. Next, the human proteins whose all three GO terms match with those in the annotation collection table were included into the network. Thereafter, the physical protein-protein interaction data among these human proteins were extracted from MINT (Ceol *et al.*, 2009) HPRD (Prasad *et al.*, 2009) and BioGRID (Stark *et al.*, 2010) by January 2011. Finally, the interaction partners of the proteins that passed the GO annotation filter were obtained and these protein pairs were used to reconstruct the network (Figure 3).

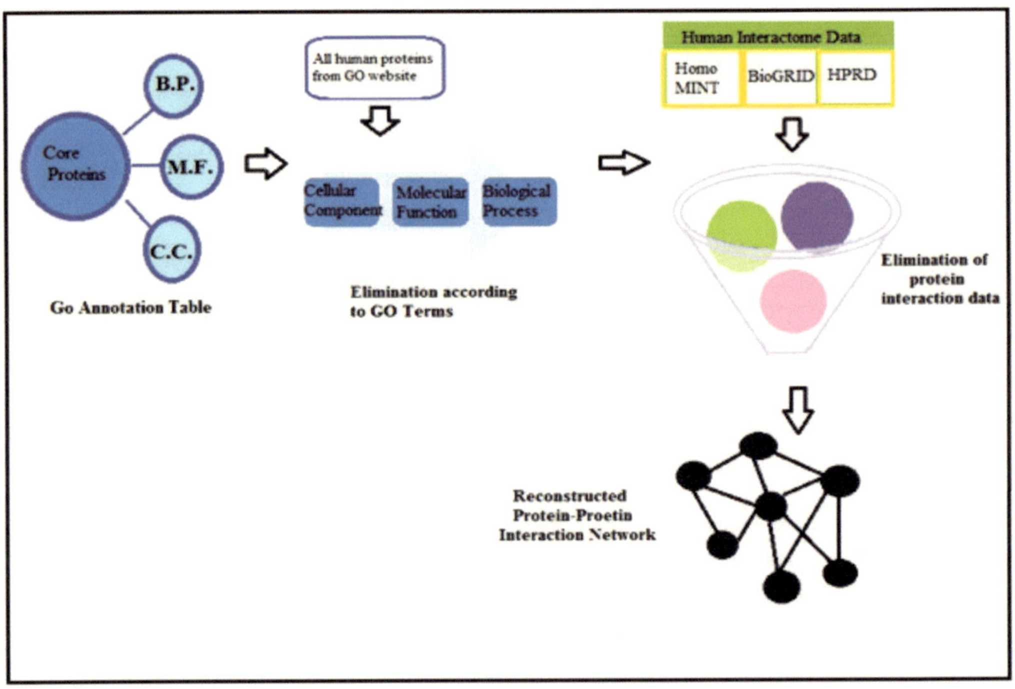

Fig. 3. The schematic representation of the algorithm used for the network reconstruction.

2.2 Graph theoretical analysis

Graph theoretical analysis enables better understanding of the structure of the complex networks and the distribution of the components via topological properties. The topology of the reconstructed network was determined by Network Analyzer Plug-in of Cytoscape (ver. 2.7.0). In the network analysis, the proteins are nodes whereas the interactions between the nodes are edges. The input of the calculations is the binary protein interactions and the output is the topological parameters such as degree distribution of the nodes, number of highly connected nodes (hubs), network diameter, mean path length, clustering coefficient, number of shortest paths and average shortest path length.

The network diameter is the maximum node eccentricity. The eccentricity here can be defined as the maximum non-infinite length of a shortest path from a node to another node in the network. The average shortest path length (or the characteristic path length) is the expected distance between two connected nodes. The betweenness centrality is another topological aspect and it is defined as the number of shortest paths going through a node. The proteins with high betweenness centrality values are called as bottleneck proteins. Another parameter, clustering coefficient of a node *N* can be defined as the ratio of the number of edges between the neighbors of *N* to the maximum number of edges that could possibly exist between the neighbors of *N*. The network clustering coefficient is the average of the clustering coefficients for all nodes in the network. This value is always between 0 and 1.

2.3 Module detection and analysis

In order to get an insight on the cellular role of the network, a Cytoscape plug-in, MCODE (1.31) was used to identify these highly interconnected regions in the present protein-protein interaction network. After the detection of the modules, the statistically overrepresented Gene Ontology functions of the proteins in the modules are obtained by the Biological Networks Gene Ontology tool (BINGO 2.44). BiNGO combines the relevant GO annotations and relates them upwards through the GO hierarchy. The output of this analysis gives the most significant GO annotations with the lowest P-value (Maere *et al.*, 2005).

2.4 Network decomposition analysis

In network decomposition analysis, the protein interaction networks are decomposed into linear paths in which ligands are inputs and transcription factors are outputs (Tekir et al., 2008). This analysis enables us to investigate the network communication route via linear paths. The linear paths of the reconstructed network were found by the NetSearch program (Steffen *et al.*, 2002). The participation frequencies of the proteins were calculated at a specified length for determining the essentiality of the proteins in the network. Next, crosstalk analysis, in which the network crosstalk value of a node is obtained by the difference in degree of the node in all considered networks and the maximum degree of this node in any individual pathway, was performed to detect the components that connect the pathways in signal transduction. A high path crosstalk value implies that a node is more important in the combined network than it was in the individual pathways (Zielinski *et al.*, 2009).

3. Results

In the present study, the protein-protein interaction sub-networks of Wnt signaling in *Homo sapiens* were reconstructed using the system biology approach in which many types of biological data and analysis techniques are integrated. Also, the topological analyses of these sub-networks were performed.

3.1 Reconstruction of Wnt signaling networks

In the first step, the proteins that are known to be involved in human Wnt signaling were identified through literature search. These selected Wnt proteins that belong to both canonical and non-canonical Wnt pathways in human were taken as the core proteins of the sub-networks to be constructed (Tables 1 and 2). The algorithm explained in section 2.1 was followed during the reconstruction process.

For canonical Wnt pathway 68 core proteins were identified. For PCP and Ca++ pathways these numbers were 33 and 32, respectively. For canonical pathway, 10592 physical protein interactions, which were obtained from MINT, BioGRID and HPRD databases, were accepted to the network. For PCP network 5928 protein-protein interactions and for Ca++ network 6080 protein-protein interactions were accepted to these sub-networks.

After removing the isolated smaller parts a network of 3251 nodes and 9304 edges was obtained for Wnt/β-catenin pathway; 1952 nodes and 5001 edges were obtained for PCP pathway; 2112 nodes and 5293 edges were obtained for Wnt/calcium pathway.

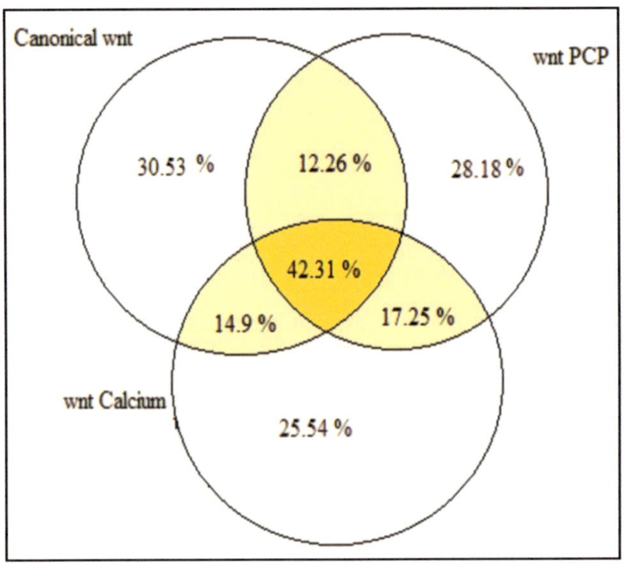

Fig. 4. Venn Schema Representation of Common Nodes.

Finally, the sub-networks were integrated and the whole Wnt network was obtained with 3489 nodes and 10092 edges. The common proteins of each sub-network were identified (Figure 4) and among them the proteins such as dishvelled (DVL1, DVL2, DVL3), AXIN (AXIN 1, AXIN 2), casein kinase (CSNK1A1, CSNK1D, CSNK1E, CSNK2A1, CSNK2B), β-catenin (CTNB1), frizzled (FZD1, FZD2, FZD4, FZD7, FZD8, FZD9), adenomatous polyposis coli protein (APC), glycogen synthase kinase-3 beta (GSK3β), mitogen-activated protein kinase (MAP3K2, MAP3K4, MAP3K7, MAP3K8), nuclear factor of activated T-cells (NFATC1, NFATC2), Smad family (SMAD 1, SMAD 2, SMAD 3, SMAD 4, SMAD 5, SMAD 6, SMAD 7, SMAD 9), transcription factor 7-like 2 (TCF7L2), cellular tumor antigen p53 (TP53), protein kinase C inhibitor protein 1 (YWHAZ) are found to be related to diseases like cancer, Alzheimer's, diabetes andosteoarthritis (Table 3).

General Name(Canonical)-Uniprot ID			
Wnt1	P04628	SENP2	Q9HC62
Wnt2	P09544	DKK1	O94907
Wnt2B	Q93097	NKD2	Q969F2
Wnt3	P56703	NKD1	Q969G9
Wnt3A	P56704	CXXC4	Q9H2H0
Wnt4	P56705	SKP1	P63208
Wnt7A	O00755	CUL1	Q13616
Wnt10B	O00744	NLK	Q9UBE8
FZD1	Q9UP38	RUVBL1	Q9Y265
FZD2	Q14332	SMAD4	Q13485
FZD4	Q9ULV1	SMAD3	P84022
FZD5	Q13467	CTBP1	Q13363
FZD7	O75084	CTBP2	P56545
LRP5	O75197	MAP3K7	O43318
LRP6	O75581	LEF1	Q9UJU2
DVL1	O14640	TCF7	P36402
DVL2	O14641	TCF7L1	Q9HCS4
DVL3	Q92997	BTRC	Q9Y297
FRAT1	Q92837	SIAH1	Q8IUQ4
FRAT2	O75474	EP300	Q09472
GSK3B	P49841	FBXW11	Q9UKB1
AXIN1	O15169	PSEN1	P49768
AXIN2	Q9Y2T1	WIF1	Q9Y5W5
APC2	O95996	PORCN	Q9H237
APC	P25054	CER1	O95813
PPP2CA	P67775	SFRP1	Q8N474
CSNK1A1	P48729	SFRP2	Q96HF1
CSNK1A1L	Q8N752	SFRP4	Q6FHJ7
CSNK1D	P48730	SFRP5	Q5T4F7
CSNK1E	P49674	SOX17	Q9H6I2
CSNK2A2	P19784	CHD8	Q9HCK8
CSNK2B	P67870	TBL1X	O60907
CTNNB1	P35222	CTNNBIP1	Q9NSA3

Table 1. Core proteins of canonical Wnt signaling pathway

General Name (PCP)-Uniprot ID		General Name (Ca++)-Uniprot ID	
Wnt5A	P41221	Wnt5A	P41221
Wnt5B	Q9H1J7	Wnt11	O96014
Wnt11	O96014	Wnt1	P04628
FZD3	Q9NPG1	PLCB1	Q9NQ66
FZD2	Q14332	PLCB2	Q00722
FZD6	O60353	PLCB3	Q01970
MAGI3	Q5TCQ9	PLCB4	Q15147
ROR1	Q01973	CAMK2A	Q9UQM7
ROR2	Q01974	CAMK2B	Q13554
PTK7	Q13308	CAMK2D	Q13557
VANGL1	Q8TAA9	CAMK2G	Q13555
VANGL2	Q9ULK5	CHP	Q99653
CELSR1	Q9NYQ6	PPP3CA	Q08209
CELSR2	Q9HCU4	PPP3CB	Q8N1F0
CELSR3	Q9NYQ7	PPP3CC	P48454
DVL1	O14640	PPP3R1	P63098
DVL2	O14641	PPP3R2	Q96LZ3
DVL3	Q92997	CHP2	O43745
PRINCKLE1	Q96MT3	PRKCA	P17252
PRINCKLE2	Q7Z3G6	PRKCB	P05771
NKD1	Q969G9	PRKCG	P05129
NKD2	Q969F2	NFAT5	O94916
ANKRD6	Q9Y2G4	NFATC1	O95644
DAAM1	Q9Y4D1	NFATC2	Q13469
DAAM2	Q86T65	NFATC3	Q12968
RHOA	P61586	NFATC4	Q14934
ROCK1	Q13464	FZD2	Q14332
ROCK2	O75116	FZD3	Q9NPG1
RAC1	P63000	FZD4	Q9ULV1
RAC2	P15153	FZD6	O60353
MAPK8	P45983	NLK	Q9UBE8
MAPK9	P45984		
MAPK10	P53779		

Table 2. Core proteins of non-canonical Wnt signaling pathway

1.1.1 Protein	1.1.2 Disease	1.1.3 References
β-catenin	Carcinogenesis, hepatocellular carcinomas Wilms' tumors	Klaus and Birchmeier,2008;Maiti *et al.*,2000
DVL	Lung cancer	Yang *et al.*,2010
FZDs	Gastric cancer,colorectal cancer& carcinogenesis	Kirikoshi, Sekihara and M. Katoh,2001;Ueno et al,2008
APC	Colorectal cancer, carcinogenesis	Klaus andBirchmeier, 2008; Ueno et al, 2008
KC1AL	Alzheimer Disease	Li,Yin and Kuret,2004
YWHAZ	Breast cancer, Obesity, Diabetes	Peng, Wang and Shan, 2009
sFRP(s)	colon cancer, mesothelioma, bladder cancer	Tan and Kelsey, 2009; Paul and Dey, 2008; Gehrke, Gandhirajan and Kreuzer, 2009.
GSK-3β	colorectal cancer	Ge and Wang, 2010
Smad3	Osteoarthritis	Valdes *et al*, 2010

Table 3. The common proteins found to be related to diseases.

3.2 Graph theoretical analysis

In order to gain insight on the characteristics of canonical and noncanonical pathways of Wnt signaling, the mean degree (average number of interactions per protein), clustering coefficients (normalized number of interactions between neighbors of each protein), mean path lengths, network diameters (longest path between any two nodes), power-law distribution exponents (γ), and centrality values were estimated using Network Analyzer. The degree distribution of each sub-network have scale-free topology and approximates a power law model ($P(k) \cong k^{-\gamma}$) with few nodes having high degree (hub proteins) and the others having low degree (Table 4). The network diameter value indicates the speed of signal flow. The diameters are 14, 13, and 15 for Wnt β-catenin, PCP and calcium signaling networks, respectively. The network diameter of the whole Wnt signaling network in which these three sub-networks are integrated, is found to be 15. The network diameter and the shortest path length distribution indicate small-world properties of the analyzed network. In addition to that, the average (mean) connectivity values are 5.72, 5.12 and 5.01 for β-catenin, PCP and calcium pathways. The topological properties of the present networks are consistent with many networks reported in literature (Table 4).

The hubs of the canonical pathway are obtained as KC1AL (Casein kinase I isoform alpha-like), YWHAZ (Protein kinase C inhibitor protein 1) and TBL1XR1 (F-box-like/WD repeat-containing protein). Casein kinase-1-alpha forms β-catenin destruction complex when connected to the proteins of APC, β-catenin and glycogen synthase kinase-3-beta (GSK3-β) (Faux *et al.*, 2008). KC1AL has interactions with the core proteins, AXIN1, AXIN2, CSNK1A1, CSNK1D and CSNK1E (String database). TBL1XR1, also a core protein of canonical Wnt signaling, is involved in signal transduction and cytoskeletal assembly and plays an essential role in transcription activation mediated by nuclear receptors and has effects on cytotypic differentiation. Besides, low levels of TBL1XR1 gene expression cause

Model	Number of Nodes	Number of Interactions	Power Law exponent(γ)	Mean Path Length	Network Diameter	Reference
Wnt/β-catenin (H. Sapiens)	3251	9304	1.78	4.46	14	Present work
Wnt/PCP (H. Sapiens)	1952	5001	1.80	4.61	13	Present work
Wnt/Ca^{+2} (H. Sapiens)	2112	5293	1.68	4.56	15	Present work
Wnt (whole) (H. Sapiens)	3489	10092	1.75	4.40	15	Present work
Wnt/β-catenin (D.melanogaster)	656	1253	1.78	4.80	13	Toku et al., 2010
Hedgehog (D.melanogaster)	568	975	1.75	4.80	14	Toku et al., 2010
EGFR (Oda et., 2005)	329	1795	1.86	4.70	11	Tekir et al., 2009
Signaling (S. cerevisiae)	1388	4640	1.76	6.81	9	Arga et al., 2007
DIP (C.elegans)	2638	4030	-	4.80	14	Wu et al., 2005
Sphingolipid (H. Sapiens)	3097	11064	1.68	4.10	13	Özbayraktar, 2011
Insulin_glucose transporting (H. Sapiens)	498	2887	1.53	2.9	5	Tekir et al., 2010
Ca-signaling (H. Sapiens)	1826	10078	1.49	3.57	11	Tiveci et al., 2011

Table 4. Graph theoretical properties of the protein interaction networks.The hubs of the Wnt/Ca^{2+} pathway are PRKCB (Protein kinase C beta type), PRKCA (Protein kinase C alpha type) and also YWHAZ (Protein kinase C inhibitor protein 1). Protein kinase C (PKC) is a family of serine- and threonine-specific protein kinases that can be activated by calcium and second messenger diacylglycerol. PKC family members phosphorylate a wide variety of protein targets and are known to be involved in diverse cellular signaling pathways. PRKCA also binds to RHOA which is another core protein in Wnt/PCP signaling. PRKCB, calcium-activated and phospholipid-dependent serine/threonine-protein kinase, is involved in various processes such as regulation of the B-cell receptor (BCR) signalosome, apoptosis and transcription regulation and it has an interation with the core protein, dishevelled 2 (DVL2) and the common hub protein YWHAZ. These hub proteins were also detected as the bottleneck proteins of the networks, due to their high betweenness centrality values. The topological properties of the hubs are listed in Table 5.

breast cancer (Kadota et al., 2009). YWHAZ (14-3-3 protein zeta/delta /Protein kinase C inhibitor protein 1), which is a member of highly conserved 14.3.3 proteins that are involved in many vital cellular processes such as metabolism, protein trafficking, signal transduction, apoptosis and cell cycle regulation, is a key component in both canonical and non-canonical Wnt signaling. In addition to its interaction with canonical pathway core protein of CSNK1A1, YWHAZ also has interactions with core proteins of NFATC2, NFATC4 and MAPK8 of non-canonical Wnt signaling. YWHAZ protein is the common hub and also a bottleneck protein in all reconstructed Wnt signaling sub-networks. YWHAZ contributes to chemotherapy resistance and recurrence of breast cancer (Ralhan et al., 2008).

Model	Uniprot ID (Name)	Betweenness Centrality	Closeness Centrality	Clustering Coefficient	Degree	Average Shortest Path Length
Wnt/Canonical	Q8N752 (KC1AL)	0.168	0.356	0.0060	241	2.817
	P63104 (YWHAZ)	0.124	0.350	0.0071	189	2.855
	Q9BZK7 (TBL1XR1)	0.052	0.289	0.0071	107	3.464
Wnt/PCP	P63104 (YWHAZ)	0.182	0.351	0.0094	133	2.850
Wnt/Ca^{2+}	P17252 (PRKCA)	0.160	0.353	0.0122	129	2.830
	P63104 (YWHAZ)	0.136	0.343	0.0099	125	2.917
	P05771 (PRKCB)	0.135	0.334	0.0074	149	2.997

Table 5. Topological properties of bottleneck proteins in human Wnt signaling.

3.3 Module detection and analysis

Scale-free networks are known to be composed of clustered regions and in biological networks these clustered regions correspond to molecular complexes named as modules (Bader and Houge, 2003). The canonical pathway was clustered into 75 complexes. Many of the proteins in the modules have roles in binding, catalytic activity and transcriptional regulation. The modules with significant molecular functions directly related to Wnt signaling were then detected by GO enrichment analysis. Some examples are as follows: The proteins in one module of Wnt/β-catenin (canonical) pathway were enriched in Wnt protein binding. NADH dehydrogenase (ubiquinone) activity was dominant in another module. In Wnt/Planar Cell Polarity (PCP) sub-network, a module showed potassium channel activity. The proteins of a module in Wnt/Ca^{2+}subnetwork were enriched in calcium ion binding.

The information obtained by module analysis such as finding of proteins behaving functionally similar in modules enabled us to confirm the present Wnt signaling network reconstructed using an integrated approach of interactomics and GO annotations.

3.4 Network decomposition analysis

The linear paths in the reconstructed Wnt signaling network as a whole and those in each canonical and noncanonical Wnt pathway were determined via NetSearch algorithm (Steffen et al. 2002) in order to examine the signal transmittal steps. In this algorithm, the membrane (ligand) proteins were set as input whereas the transcription factors were set as output components (Table 6) of Wnt signaling network in *Homo Sapiens.*

In the Wnt signaling network as a whole, the shortest path length is found to be 4, which includes 5 proteins connected by 4 linear interactions for two linear paths from Wnt3A to LEF1 (Table 7). The path length is increased in order to cover all the proteins in the network. However, a maximum number of 12 steps that has 1 086 956 linear paths in which only 59 (50%) of 118 core proteins and 1244 (34%) of 3676 proteins are covered, is achieved due to computer capacity. The linear paths were found to reach to LEF1 (Q9UJU2) in canonical subnetwork and NFATC1 (O95644), NFATC2 (Q13469), NFATC3 (Q12968) in noncanonical subnetwork.

Input Protein (Uniprot_ID)	Protein Name	Output Protein (Uniprot_ID)	Protein Name
P04628	Wnt1	O94916	NFAT5
P09544	Wnt2	O95644	NFATC1
Q93097	Wnt2B	Q13469	NFATC2
P56703	Wnt3	Q12968	NFATC3
P56704	Wnt3A	Q14934	NFATC4
P56705	Wnt4	Q9UJU2	LEF1
P41221	Wnt5A	P36402	TCF7
Q9H1J7	Wnt5B	Q9HCS4	TF7L1
O00755	Wnt7A		
O00744	Wnt10B		
O96014	Wnt11		

Table 6. Input and output proteins of the linear paths.

Path Length	Input Protein				Output Protein
4	*P56704* (Wnt 3A)	*Q07954* (LRP1)	*P12757* (SKIL)	*Q13485* (SMAD4)	*Q9UJU2* (LEF1)
	P56704 (Wnt 3A)	*Q07954* (LRP1)	*P12757* (SKIL)	*Q15796* (SMAD2)	*Q9UJU2* (LEF1)

Table 7. The linear paths at path length of 5.

3.4.1 Canonical vs non-canonical Wnt pathways

Network decomposition analysis was performed for canonical and non-canonical Wnt pathways separately. A maximum number of 12 steps that has 815627 linear paths, in which only 33 of 68 core proteins (42%) and 1115 of 3251 proteins (32%) are participated, can be obtained for canonical pathway. The number of linear paths at 12 steps is found to be 29082 for non-canonical pathway, in which 546 of 2547 nodes and only one core protein of 60 core proteins are covered. It has 1098373 linear paths at 14 steps, and 27 of 60 core proteins (48%) and 817 of 2547 proteins (34%) are covered. This result seems to be logical since the diameter of the non-canonical pathway is found to be larger than that of canonical pathway, which implies that the signal transfer is slower in non-canonical pathway. A minimum number of 4 steps (5 proteins) was necessary to reach the end transcription factor in canonical pathway whereas the signal has to pass at least 7 proteins in case of non-canonical pathways such as PCP or Wnt/Ca2+ signaling. In general, the information flow preferring short routes is faster in canonical pathways.

3.4.2 Participation of proteins in linear paths

For identification of the significant proteins in the whole Wnt network, the percentages of each protein contributing to linear paths were calculated (Table 8) and the proteins having participation percentages higher than 20 are discussed below. T cell specific transcription factor 1-alpha (LEF1) has the highest percentage since it is one of the output proteins. WNT7A and WNT1 are the input proteins. These three proteins (WNT7A, WNT1 and LEF1)

Uniprot ID	Protein Name	Recommended Name	Canonical/ Noncanonical	Participation in linear paths (%)	Degree
Q9UJU2	LEF1	T cell-specific transcription factor 1-alpha	Canonical	56.19	17
O00755	WNT7A	Protein Wnt-7A	Canonical	51.91	2
O00144	FZD9	Frizzled-9	Canonical/PCP/Ca^{2+}	51.91	4
Q99750	MDFI	MyoDfamilyinhibitor	Canonical/PCP/Ca^{2+}	50.94	50
P04628	WNT1	Proto-oncogene Wnt-1	Canonical/Ca^{2+}	47.20	10
Q9HD26	GOPC	Golgi-associated PDZ andcoiled-coil motif-containing protein	Canonical/PCP	46.89	18
Q9H461	FZD8	Frizzled-8	Canonical/PCP/Ca^{2+}	46.87	4
P33992	MCM5	DNA replicationlicensingfactor MCM5	Canonical/Ca^{2+}	42.94	6
Q14566	MCM6	DNA replicationlicensingfactor MCM6	Canonical/Ca^{2+}	38.83	28
Q15797	SMAD1	SMAD familymember 1	Canonical/PCP/Ca^{2+}	29.23	60
P28070	PSB4	Proteasomesubunit beta type-4	Canonical	28.75	19

Table 8. Proteins with the highest participation percentages in Wnt signaling pathway.

are also the core proteins of the canonical Wnt signaling sub-network and they bind to essential proteins, which are common to many paths in the network. Frizzled 9 (FZD9), which is a receptor for Wnt proteins, is common to all three sub-networks of Wnt signaling. It leads to the activation of dishevelled proteins, inhibition of GSK-3β kinase, nuclear accumulation of β-catenin and activation of Wnt target genes. It was hypothesized that FZD9 may be involved in transduction and intercellular transmission of polarity information during tissue morphogenesis and/or in differentiated tissues (www.uniprot.org). Another protein common to all three Wnt sub-networks is MyoD family inhibitor protein (MDFI), which regulates the transcriptional activity of TCF7L1/TCF3 by direct interaction to it, and it prevents TCF7L1/TCF3 from binding to DNA. The DNA replication licensing factor proteins (MCM5 and MCM6) have interaction with each other and MCM5 also binds to MDFI and β-catenin, which is an essential protein for Wnt signaling pathway. Besides that, SMAD1-OAZ1-PSMB4 complex mediates the degradation of the CREBBP/EP300 repressor SNIP1.

When the proteins with low participation percentages in linear paths are evaluated according to the criteria of low betweenness and high closeness centrality values, four proteins (LRSAM1, MLTK, MARK1 and miyosin 9) seem to be important for consideration as putative drug targets (either by activation or inhibition) and need further examination (Table 9).

	Protein _ID	**Name**	**Protein _ID**	**Name**	**Protein _ID**	**Name**	**Protein _ID**	**Name**
Input Protein	O00755	Wnt7A	P04628	Wnt1	P04628	Wnt1	P04628	Wnt1
	O00144	FZD9	Q9H461	FZD8	Q9H461	FZD8	Q9H461	FZD8
	Q99750	MDFI	Q9HD26	GOPC	Q9HD26	GOPC	Q9HD26	GOPC
	Q12906	ILF3	P13569	CFTR	P13569	CFTR	P13569	CFTR
	Q8N752	KC1AL	P08670	VIME	P08670	VIME	P08670	VIME
	Q9UQM7	CAMK2A	O43353	RIPK2	Q12873	CHD3	O43353	RIPK2
	Q13554	CAMK2B	P05771	PRKCB	Q14974	IMB1	P05771	PRKCB
	P48443	RXRG	**Q9P0L2**	**MARK1**	Q00722	PLCB2	**Q9NYL2**	**ZAK**
	Q6UWE0	**LRSAM1**	P31947	SFN	Q96QT4	TRPM7	P31947	SFN
	Q99816	TS101	P63104	YWHAZ	**P35579**	**MHY9**	P63104	YWHAZ
	Q13464	ROCK1	P30291	WEE1	P19838	NFKB1	P30291	WEE1
	Q15796	SMAD2	P84022	SMAD3	P17252	PRKCA	P84022	SMAD3
Output Protein	Q9UJU2	LEF1	Q9UJU2	LEF1	O95644	NFAC1	Q9UJU2	LEF1
Path Length	12		12		12		12	

Table 9. Linear paths of lowest participant proteins.

LRSAM1 (leucine rich repeat and sterile alpha motif containing1), also called RIFLE and TAL (TSG101-associated ligase), is an E3 type ubiquitin ligase. TSG101 itself is a tumor suppressor gene, which has a role in maturation of human immunodeficiency virus, and LRSAM1 is implicated in its metabolism directly by polyubiquitination (Guernsey et al., 2010). The functional disruption of TSG101 led both to cellular transformation and to tumors that metastasized spontaneously in nude mice (Li and Cohen, 1996). In addition to that, although genomic alterations in TSG101 are rare in human cancer, functional inactivation of the gene enhances metastatic growth of murine fibroblasts (Li and Cohen 1996). Another protein is ZAK (MLTK - Q9NYL2) which inhibits human lung cancer cell growth via ERK and JNK activation in an AP-1-dependent manner (Yang et al., 2010). Also, overexpression of ZAK results in apoptosis (OMIM).

Another protein is serine/threonine-protein kinase MARK1. Cellular studies showed that overexpression of MARK1 resulted in shorter dendrite length and decreased transport speed. MARK1 overexpression in individuals with autism may underlie subtle changes in synaptic plasticity linked to dendritic trafficking (Maussion et al., 2008; OMIM). The last protein is miyosin9. Fechtner syndrome, which is an autosomal dominant disorder characterized by the triad of thrombocytopenia, giant platelets, and Dohle body-like inclusions in peripheral blood leukocytes, with the additional features of nephritis, hearing loss, and eye abnormalities, mostly cataracts, is caused by heterozygous mutation in the gene encoding nonmuscle myosin heavy chain-9 (MYH9; 160775) on chromosome 22q11 (Peterson et al., 1985; OMIM). ZAK and MARK1 both bind to SFN which has interaction with YWHAZ. YWHAZ is found to be hub and bottleneck protein in these reconstructed canonical and noncanonical Wnt pathways due to its high degree and betweenness centrality value, respectively. YWHAZ also has a low participation percentage of 0.95 in linear paths. YWHAZ is found to be a key mediator protein in various diseases involving various types of cancers, heart diseases, obesity, diabetes and autism (Nguyen and Jordá, 2010). Key mediators are proteins that bind to significant proteins (mostly hubs) and so they can be chosen as the drug targets.

3.4.3 Specific proteins in linear paths

The proteins in the linear paths ending at transcription factors specific to canonical and noncanonical pathways were further examined in detail. The proteins, which participate in the linear paths leading to one transcription factor only, are called specific proteins of that particular pathway.

286 specific proteins were obtained where 262 of them belong to canonical (transcription factor LEF1) and 24 of them belong to non-canonical pathway (transcription factor NFATC). They were then investigated according to their topological properties such as lower betweenness centrality, higher closeness centrality and higher clustering coefficient than the average for drug target identification. As a result, 51 proteins (48 canonical, 3 noncanonical) meet these criteria. Among 51 proteins 4 proteins in canonical pathway seem to be important since they are either related to important diseases or connected to significant proteins in the network. These proteins are Myc proto-oncogene protein (MYC), TGF-beta receptor type-2 (TGFR2), cyclin-dependent kinase inhibitor 3 (CDKN3) and F-box-like/WD repeat-containing protein TBL1X (canonical).

MYC is a protein that participates in the regulation of gene transcription. The mutations and overexpressions seen in MYC resulted in cell proliferation and consequently formation of cancer. The translocations such as t (8:14) are the reasons of the development of Burkitt's lymphoma. Soucek *et al.*, 2008 demonstrated that the temporary inhibition of MYC selectively killed lung cancer cells in mouse, making it a potential drug target in cancer (Gearhart *et al.*, 2007; Soucek *et al.*, 2008). TGFR2 is the receptor protein of TGF-beta and also known to be involved in tumor suppression. It forms receptor complexes with serine/threonine protein kinases and has role in activation of SMAD transcriptional regulators. The mutations and defects seen in this protein are associated with Lynch sendrome, Loeys-Deitz aortic aneurysm syndrome, Osler-Weber-Rendu syndrome, hereditary non-polyposis colorectal cancer type 6 (HNPCC6) and esophageal cancer (Tanaka *et al.* , 2000; Lu *et al.*, 1998).

TBL1X is a protein that plays an essential role in transcription activation mediated by nuclear receptors. Besides, it is a component of E3 ubiquitin ligase complex containing UBE2D1, SIAH1, CACYBP/SIP, SKP1, APC and TBL1X proteins. It has interactions with essential proteins of Wnt signaling such as APC and β-catenin and it is also a core protein of reconstructed canonical Wnt signaling pathway (Matsuzawa and Reed, 2001). CDKN3 is a member of cyclin-dependent kinases (CDKs) which have roles in regulating cell cycle, transcription, mRNA processing, and differentiation of nerve cells (Gyuris *et al.*, 1993). The overexpression and defects seen in this protein leads to prostate cancer and hepatocellular carcinoma (HCC) (Yeh *et al.*, 2003; Lee *et al.*, 2000).

These specific proteins except TBL1X are related to cancer and they are suitable for drug target applications according to their topological properties. Hence, they need more attention with further experimental investigation.

3.4.4 Crosstalk of proteins in Wnt sub-networks

Signaling networks are communicating systems and they interact with each other rather than behaving in isolation. If a node has a high network crosstalk value, which is defined as the difference in degree of the node in all considered networks and the maximum degree of this node in any individual pathway, it means that this component is a branch node connecting two or more pathways. The network crosstalk analysis indicated 239 proteins that are found to be common among Wnt sub-networks.

One of the highest crosstalk values belongs to YWHAZ protein (Table 10). This is rational since this protein was obtained as the hub and bottleneck protein of all canonical and non-canonical Wnt pathways. Besides, DVL2 has a significant crosstalk value. Dishevelled proteins also have high participation in the subnetworks since they interact with the core proteins such as frizzled receptors and GSK3β in Wnt/β-catenin sub-network, and with frizzled receptors and DAAM1 in Wnt/PCP sub-network. Smad proteins also have considerable crosstalk value since they have interactions with AXIN, beta-catenin and LEF1 proteins. PRKCA, which was found as hub and core protein in Wnt/calcium sub-network, has a non-zero crosstalk value. AXIN protein is also a significant protein that has participation in β-catenin destruction complex with APC, GSK3β and CKIα. Detecting these connector proteins by network crosstalk analysis is a promoter step for further experimental studies towards cancer treatment. However, further elaboration on the crosstalk mechanism is difficult due to the fact that the reconstructed networks are undirected.

Proteins		Network crosstalk values
YWHAZ	Hub-Core protein (all sub-networks)	11
DVL2	Core protein (β-catenin and Wnt/ PCP sub-networks)	11
CAMK2A	Core protein (Wnt/Ca^{2+} sub-network)	4
SMAD3-4	Core proteins (β-catenin subnetwork)	4
GSK3B	Core protein (β-catenin sub-network)	2
PRKCA	Hub-Core protein (Wnt/ Ca^{2+} subnetwork)	2
RAC1	Core protein (Wnt/PCP sub-network)	2
NFATC2	Core protein (Wnt/Ca^{2+} subnetwork)	1
AXIN1	Core protein (β-catenin sub-network)	1

Table 10. Proteins and network crosstalk values

4. Discussion

4.1 Wnt signaling in maintaining homeostasis and managing cellular stress

Homeostasis, balance of cellular processes, is an important phenomenon since cells are the factories that maintain the intracellular environment and keep the conditions stable. Therefore, it is essential for cells to maintain homeostasis for the organism to remain healthy. Wnt signaling, being related to embryonic development, generation of cell polarity and specification of cell death, is highly effective in maintaining homeostasis in adults (Peifer and Polakis, 2000). In canonical Wnt pathway, for example, the stabilization of β-catenin plays an essential role in cellular homeostasis. In the absence of Wnt ligands, a destruction complex is formed by AXIN, APC, GSK-3β and β-catenin, that results in β-catenin phosphorylation by GSK-3β followed by ubiquitination and degradation that keeps β-catenin level low in cytoplasm. Wnt ligands, on the other hand, enhance the β-catenin accumulation via inhibition of GSK-3β by dishevelled proteins and free β-catenin is transferred into the nucleus where it interacts with transcription factors. Therefore, AXIN, APC and GSK-3β proteins are significant players for homeostasis.

The mutations seen in AXIN result in hepatocellular carcinoma, which implies that, it has a multi-objective position in tumorigenesis and embryonic axis formation. It is also reported that the main role of AXIN, beside controlling β-catenin level, is to down-regulate cell growth and help sustain cellular homeostasis (Zhang et al., 2001). AXIN is known to be is a "switch" protein for JNK and Wnt signaling pathways. It binds to MEKK1 and activates JNK signaling. MEKK1 is related to microtubule cytoskeletal stress and apoptosis. During JNK activation, AXIN-MEKK1-APC-β-catenin complex transduces the cytoskeletal stress signals for apoptosis (Yujiri et al., 1999; Zhang et al., 2001).

4.2 Wnt/Ca^{2+}-Wnt/β-catenin antagonistic mechanism in *H. Sapiens*

The non-canonical Wnt signalling pathways do not signal through β-catenin and they can antagonize the functions of canonical Wnt pathway (Mc Donald and Silver, 2009). Wnt5a is known to activate non-canonical signalling via cGMP(cylic guanosine-3'5'-monophosphate) that actives protein kinase G. This leads to an increase in the cellular concentration of Ca^{2+}

and this Ca^{2+} increase triggers activation of calcium sensitive proteins. Wnt5a also inhibits the activation of canonical signalling via activation of NFAT which is mediated by activation of PLC (phospolipase C). PLC increases the calcium level that results in activation of CaCN (calcineurin) which activates NFAT.

Wnt/Ca^{2+} signalling pathway can inhibit Wnt/β-catenin pathway in two different ways: CACN-NFAT branch and CAMKII-TAK1-NLK branch (Figure 5). CACN-NFAT branch for inhibiting β-catenin function is mediated by PLC activation, which involves the β/γ subunits of heterotrimeric G-proteins leaving its α unit behind. PLC activation generates diacylglycerol (DAG) and inositol-1,4,5-trisphosphate (IP3) which eventually increases Ca^{+2} concentration in the cell. The calcium increase sets off the CaCN activation that results in dephosphorylation of NFAT (nuclear factor of activated T-cells). NFAT then translocates to nucleus to regulate gene expression. This CaCN-NFAT activated way inhibiting the canonical Wnt signalling pathway is covered in our reconstructed network (Saneyoshi et al., 2002; Veeman et al., 2003; Pandur, 2005). Moreover, the reconstructed network successfully

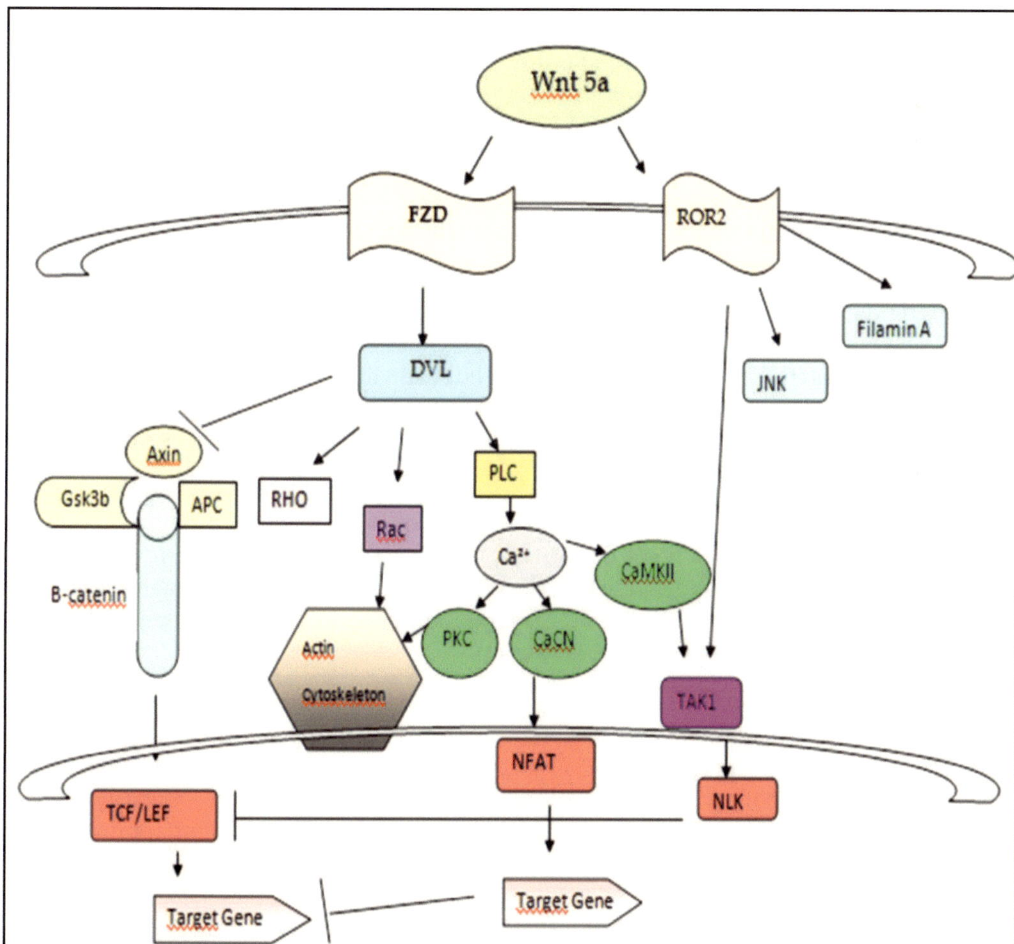

Fig. 5. General representation of Wnt signaling pathway in vertebrates.

covers CAMKII-TAK1-NLK branch, which is known to inhibit Wnt/β-catenin signalling pathway. As it is mentioned above, the PLC activation results in calcium release. The increase seen in Ca^{2+} level may trigger activation of another calcium sensitive protein; Ca^{2+}-calmodulin-dependent protein kinase II (CamKII) which further activates TGF-β activated kinase 1 (TAK1). TAK1 then stimulates nemo-like kinase (NLK), which has a role in TCF phosphorylation. Afterwards, the phosphorylation of TCF inhibits TCF/β–catenin complex (Kuhl et al., 2000; Pandur, 2005).

Besides stimulating non-canonical signaling and inhibiting canonical signalling through CamkII mentioned above, Wnt-5a can also inhibit the activation of canonical signalling through ROR2 signalling pathway that stimulates TAK1-NLK pathway in turn. ROR2 receptor also actives the actin binding protein Filamin A and JNK pathway (Mc Donald and Silver, 2009).

As a consequence it can be said that Wnt5a exhibits tumor suppressor activity through inhibiting the activation of canonical Wnt signalling. Recent research showed that, in HTC116 and HT-29 colon cancer cell lines, the activation of β-catenin-mediated transcription is reduced by Wnt5a (Macleod et al., 2007; Ying et al., 2008). Additionally, the reconstructed network provides a chance to investigate the antagonism between Wnt/Ca^{2+} and Wnt/β-catenin signalling pathways. Although the static nature of the network cannot directly explain the interaction characteristics between these pathways, a dynamic model can enlighten the antagonism between Wnt/Ca^{2+} and Wnt/β-catenin pathways.

4.3 Potential drug targets in the reconstructed Wnt signaling networks

Wnt signaling pathways regulate many cellular processes such as proliferation, migration and differentiation in embryonic development and maintenance of homeostasis in matured tissues. The deregulations and mutations in Wnt signaling pathway are known to result in cancer. Unfortunately, there is no selective inhibitor for the deficiencies in Wnt signaling. That is why targeting key components, such as SFRPs, WIF-1, DKK-1, APC, AXIN2, ICAT, LEF1 and β-catenin, of the Wnt signaling seems to be reasonable in cancer treatment (Aguilera et al., 2007).

The topological parameters such as centrality values or participation percentages in linear paths are important criteria in identification of putative target proteins. Therefore, the nodes that have lower average shortest path length, higher clustering coefficient, higher closeness centrality, lower betweenness centrality and higher participation percentages than the average values are further investigated (Table 11).

In our reconstructed networks, FZD9, WNT7A and LEF 1 proteins are found to be essential due to their high participation percentages in linear path analysis. Albers et al. (2011) show that the Wnt receptor Frizzled-9 (FZD-9) can be a new potential target for the treatment of osteoporosis by promoting bone formation. Also, it is known that the re-expression of WNT7A and signaling through FZD9 are associated with increased differentiation and used in the lung cancer treatment (Winn et al., 2005). Frizzled proteins are the receptors for Wnt ligand, and they are structurally similar to G protein-coupled receptors (GPCRs) which are targets of more than 50% of chemically applicable drugs (Yanaga and Sasaguri, 2007). So targeting frizzled proteins seems to be logical in cancer treatment. In addition to that, β-catenin has a connectivity value of 40 and participation percentage of 4.36%. β-catenin is

Uniprot ID	Name	Average Shortest Path Length	Betwenness Centrality	Closeness Centrality	Clustering Coefficient	Degree	Participation Percentage
P35222	β-catenin	**3.42**	1.43×10^{-2}	**0.292**	1.92×10^{-2}	40	**4.36**
Q9UJU2	LEF1	**3.58**	4.52×10^{-3}	**0.279**	4.41×10^{-2}	17	**56.2**
P49841	GSK3b	**3.42**	2.97×10^{-3}	**0.292**	8.57×10^{-2}	15	2.08×10^{-1}
P25054	APC	**3.66**	**7.45×10^{-4}**	**0.273**	**1.67×10^{-1}**	4	5.47×10^{-2}
Q8N474	SFRP1	6.49	**6.93×10^{-4}**	0.154	0	4	-
O94907	DKK-1	**4.22**	**5.94×10^{-4}**	0.237	0	2	-
Q9Y2T1	AXIN2	4.63	**5.73×10^{-4}**	0.216	**3.33×10^{-1}**	3	1.54×10^{-3}
Q14332	fzd2	4.57	**2.42×10^{-4}**	0.219	0	2	-
Q9NSA3	ICAT	4.41	**3.36×10^{-5}**	0.227	0	2	1.30×10^{-3}
Q9Y5W5	WIF-1	6.49	0	0.154	0	1	-
Q6UWE0	LRSAM1	**4.18**	8.73×10^{-4}	**0.239**	0	5	8.10×10^{-5}
Q9P0L2	MARK1	**3.90**	4.78×10^{-5}	**0.256**	0	2	8.10×10^{-5}
Q9NYL2	ZAK	**3.90**	6.21×10^{-4}	**0.256**	0	3	8.10×10^{-5}
P35579	MHY9	**3.58**	5.61×10^{-4}	**0.280**	1.11×10^{-1}	10	8.10×10^{-5}
O00144	FZD9	4.80	4.71×10^{-3}	0.208	0	4	**51.9**
O00755	WNT7A	5.57	**4.27×10^{-5}**	0.179	0	2	**51.9**
Q8N752	KC1AL	**2.83**	1.40×10^{-1}	0.354	6.09×10^{-3}	241	4.81
P63104	YWHAZ	**2.83**	1.18×10^{-1}	**0.353**	6.97×10^{-3}	200	9.46×10^{-1}
Q9BZK7	TBL1XR1	**3.46**	4.62×10^{-2}	**0.289**	7.14×10^{-3}	107	1.60×10^{-1}
Q16667	CNKD3	**4.31**	0	**0.232**	**1**	2	8.10×10^{-5}
P37173	TGFR2	**4.31**	**1.57×10^{-4}**	**0.232**	**6.0×10^{-1}**	5	4.69×10^{-2}
P01106	MYC	**4.14**	**1.53×10^{-4}**	**0.242**	**2.0×10^{-1}**	5	2.00×10^{-3}
Average		4.40	9.76×10^{-4}	0.232	1.13×10^{-1}	5.8	$7.7.78 \times 10^{-1}$

Table 11. The topological values of the target proteins.

encoded by an oncogene and has functions in the maintenance of epithelial cell layers by regulating cell growth and adhesion between cells. β-catenin also anchors the actin cytoskeleton (Peifer and Polakis, 2000; Zhang et al., 2001). Luu et al. (2004) suggested that targeting β-catenin could be a rational approach in cancer treatment.

In the present Wnt network, there are two essential proteins (AXIN2 and APC) that have higher clustering coefficient values than the average and it is known that essential proteins tend to be more cliquish within the interaction network (Yu *et al.*, 2004; Estrada E., 2006). Ranking proteins according to their centrality measures can additionally be useful in selecting possible drug targets. Consequently, GSK3β and APC can be seen as potential drug targets in Wnt signaling for having higher closeness centrality value than the average. APC is related with colorectal cancer and APC-activating mutations are very common in colorectal cancer (Estrada E., 2006; Garber, 2009; Yanaga and Sasaguri, 2007).

Moreover, the betweenness centrality and bridging centrality (nodes between modules and connecting clusters defined by the ratio of the number of interactions of a neighboring node over the number of remaining edges) are also effective in identifying the drug targets due to their position in communication (Hopkins, 2008, Hwang et al., 2008). In order to prevent side effects and high lethality, the essential nodes with lower betweenness centrality values are chosen as drug targets on the purpose of not affecting the neighbors of the targeted protein. It is seen that APC, DKK1, AXIN2, FZD2, Wnt7A, ICAT and WIF1 are consistent with this fact (Table 11). SFRP1 protein needs special attention since its loss causes breast cancer (Klopocki et al., 2004).

It is further seen that the nodes which have low participation percentages as well as low degrees (LRSAM1, MARK1, ZAK, MHY9), the nodes which are defined as specific proteins (CNKD3, TGFR2 and MYC) and the nodes which are detected as hub proteins (YWHAZ, TBL1XR1, KC1AL) have the quality of conformance since they have lower average shortest path length and higher closeness centrality values than the average. These proteins can also be suggested as potential drug targets and more attention should be given through experimental analysis.

The gene expression data (microarray data) belonging to these proteins are within reach via several database sources. However, due to the disease heterogeneity, the expression level of a gene /protein can be up-regulated as well as down-regulated in cancer and the expression type may also differ according to the cancer type. Hence, it is difficult to obtain a right answer for the expression level of a gene/protein in diseases like cancer.

5. Conclusion

Recently, the evolutionarily conserved signaling pathways which are involved in embryonic development are on the march for many researches since the deregulations seen in the mechanism of these pathways results in several diseases, especially in cancer. Hence, interaction networks have begun to be appreciated because it may be useful to understand the general principles of biological systems by means of systems biology. Wnt signaling is a major signaling pathway which has important roles in embryonic development of many species. Hence, in this study, Wnt signaling pathway is investigated with the aim of getting an insight on the role of Wnt signalling in maintaining homeostasis as well as managing cellular stress, understanding the molecular basis underlying the ability of Wnt proteins to perform antagonistic or similar signalling activities and identifying the suitable drug targets for therapeutic intervention in cancer treatment.

The reconstruction of Wnt signaling network was performed for *Homo sapiens* via integration of interactome data and Gene Ontology annotations. The reconstruction process was applied to both canonical (Wnt/β-catenin) and non-canonical Wnt signaling pathways (Wnt/planar cell polarity; Wnt/calcium). The reconstructed whole Wnt signaling network contains 3489 nodes and 10092 interactions. AXIN, APC and GSK-3β proteins are found to be significant players for homeostasis. Moreover, AXIN-MEKK1-APC-β-catenin complex is important in transducing the cytoskeletal stress signals leading to apoptosis.

The ligand Wnt5a has dual role; it activates non-canonical signalling and also inhibits the activation of canonical signalling through a calcium dependent mechanism. This antagonism between noncanonical Wnt/Ca^{2+} and canonical Wnt/β-catenin signalling pathways is successfully covered in our reconstructed network. CNKD3, TGFR2 and MYC, which are the specific proteins in linear paths leading to specific transcription factors in canonical pathway, are proposed as potential drug targets for cancer. The reconstructed large-scale protein-protein interaction network of Wnt signaling in *H. sapiens* will allow system biologist to see the global picture and guide them in designing experiments. For further research, experimental and clinical studies can be carried out for the validation of the proposed drug targets leading to design novel drugs.

6. Acknowledgments

The financial support for this research was provided by the Research Funds of Boğaziçi University and TÜBİTAK through projects 5554D and 110M428, respectively. The scholarship for Saliha Durmuş Tekir, sponsored by TÜBİTAK, is gratefully acknowledged.

7. References

Aguilera, O., Muñoz, A., Esteller, M. and Fraga,M. F., 2007, "Epigenetic alterations of the Wnt/beta-catenin pathway in human disease", *Endocr Metab Immune Disord Drug Targets*, Vol. 7(1), pp. 13-21.

Apweiler, R., Bairoch, A., Wu, C. H., Barker, W. C., Boeckmann, B., Ferro, S., Gasteiger, E., Huang, H., Lopez, R., Magrane, M., Martin, M. J., Natale, D. A., O'Donovan, C., Redaschi, N., Yeh, L. S., 2004, "UniProt: the Universal Protein knowledgebase", *Nucleic Acids Res.*, ; Vol. 32, pp. D115-9.

Arga, K. Y., Önsan, Z.I., Kırdar, B., Ülgen, K. Ö. and Nielsen, J.,2007, "Understanding Signaling in Yeast: Insights From Network Analysis", *Biotechnology and Bioengineering*, Vol. 97, No. 5, pp. 1246-1258.

Beals, C. R., Sheridan, C. M., Turck, C. W. et al., 1997, "Nuclear export of NF-ATc enhanced by glycogen synthase kinase 3", *Science*, Vol. 275, pp. 1930-1934.

Bateman, A., Coin, L., Durbin, R., Finn, R. D., Hollich, V., Griffiths-Jones, S., Khanna, A., Marshall, M., Moxon, S., Sonnhammer, E. L., Studholme, D. J., Yeats, C., Eddy, S. R., 2004, "The Pfam protein families database", Nucleic Acids Research, Vol. 32, pp. D138–D141.

Cadigan, K. M. and Liu, Y. I. 2006, "Wnt signaling: complexity at the surface", *J. Cell. Sci.*, Vol. 119, pp. 395-402.

Cadigan, K. M. ,2008, "Wnt-β-catenin signaling" , *Curr. Biol.*, Vol. 18: R943-947.

Cong, F., Schweizer, L.and Varmus, H., 2004, "Wnt signals across the plasma membrane to activate the beta-catenin pathway by forming oligomers containing its receptors, Frizzled and LRP", *Development*, Vol. 131, pp. 5103-5115.

DasGupta, R., 2005, "Functional Genomic Analysis of the Wnt-Wingless Signaling Pathway", *Science*, Vol. 308, Issue 826.

Eisenmann, D. M., 2005, "Wnt signaling", WormBook, pp. 1-17.

Estrada E., 2006, "Virtual identification of essential proteins within the protein interaction network of yeast", *Proteomics*, Vol. 6, pp. 35-40.

Garber, K., 2009, "Drugging the Wnt Pathway: Problems And Progress", *Journal of the National Cancer Institute*, Vol. 101, Issue 8, pp. 548-550.

Ge, X. and Wang, X., 2010, "Role of Wnt canonical pathway in hematological malignancies", *Journal of Hematology & Oncology*, Vol. 3, Issue 33.

Gearhart, J., Pashos, E. E., Prasad, M. K., 2007, "Pluripotency Redeux -- advances in stem-cell research", *N Engl J Med*, Vol. 357(15), pp. 1469-72

Gehrke, I., Gandhirajan, R. K. and Kreuzer, K. ,2009, "Targeting the Wnt/β-catenin/TCF/LEF1 axis in solid and haematological cancers: Multiplicity of therapeutic options", *Eur J Cancer*, Vol. 45, Issue 16, pp. 2759-2767.

Giles, R. H., van Es, J. H., and Clevers, H., 2003, "Caught up in a Wnt storm: Wnt signaling in cancer", *Biochim Biophys Acta*, Vol. 1653, pp. 1-24.

Guernsey, D. L., Jiang, H. , Bedard, K. et al., 2010, " Mutation in the Gene Encoding Ubiquitin Ligase LRSAM1 in Patients with Charcot-Marie-Tooth Disease", *PLoS Genet* ., Vol. 6, No. 8.

Gyuris, J., Golemis, E., Chertkov, H., Brent, R., 1993, "Cdi1, a human G1 and S phase protein phosphatase that associates with Cdk2", *Cell*, Vol. 75, pp. 791-803.

Hingwing, K., Lee, S., Nykilchuk, L., Walston, T., Hardin, J., Hawkins, N., 2009, "CWN-1 functions with DSH-2 to regulate C. elegans asymmetric neuroblast division in a β-catenin independent Wnt pathway", *Dev Biol*, Vol. 328, pp. 246-256.

Hopkins A. L., 2008, "Network pharmacology: the next paradigm in drug discovery", *Nature Chemical Biology*, Vol. 4, pp. 682 - 690.

Hwang, W.C., Zhang, A. and Ramanathan, M, 2008, "Identification of information flow modulating drug targets: a novel bridging paradigm for drug discovery", *Clinical Pharmacology & Therapeutics* , Vol. 84, pp. 563-572.

Kadota, M., Sato, M., Duncan,B., Ooshima, A. et al. , 2009, "Identification of novel gene amplifications in breast cancer and coexistence of gene amplification with an activating mutation of PIK3CA", *Cancer Res.*, Vol. 69, Issue 18, pp. 7357–7365.

Klaus, A., and Birchmeier, W.,2008, "Wnt signaling and its impact on development and cancer", *Nat Rev Cancer*, Vol. 8, pp. 378-398.

Kanehisa, M. and Goto, S., 2000, "KEGG: kyoto encyclopedia of genes and genomes", Nucleic Acids Res., Vol. 28, No. 1, pp. 27-30.

Katoh, M., 2005, "WNT/PCP signaling pathway and human cancer", *Oncol Rep.*, Vol. 14, Issue 6, pp. 1583-8.

Katoh M and Katoh, M., 2003, " Identification and characterization of human PRICKLE1 and PRICKLE2 genes as well as mouse Prickle1 and Prickle2 genes homologous to Drosophila tissue polarity gene prickle", *Int J Mol Med*, Vol. 11,pp. 249-256.

Klaus, A., and Birchmeier, W., 2008, "Wnt signaling and its impact on development and cancer", *Nat Rev Cancer*, Vol. 8, pp. 378-398.

Klopocki, E., Kristiansen, G., Wild, P. J, Klaman, I., Castanos-Velez, E., Singer, G., Stöhr, R., Simon, R., Sauter, G., Leibiger, H., Essers, L., Weber, B., Hermann, K., Rosenthal, A., Hartmann, A., Dahl, E., 2004, "Loss of SFRP1 is associated with breast cancer progression and poor prognosis in early stage tumors", Int J Oncol. , Vol. 25(3), pp. 641-9.Kohn, A. D., and Moon, R. T., 2005, "Wnt and calcium signaling: beta-cateninindependent pathways", *Cell Calcium*, Vol. 38, pp. 439-446.

Kirikoshi, H., Sekihara, H. and Katoh, M., 2001, "Expression profiles of 10 members of Frizzled gene family in human gastric cancer", *Int J Oncol.*, Vol. 19, Issue 4 pp. 767-71.

Kuhl, M., 2004, "The Wnt/Calcium Pathway: Biochemical Mediators, Tools and Future requirements", *Frontiers in Bioscience*, Vol. 9, pp. 967-974.

Kuhl, M., Sheldahl, L. C., Park, M., Miller, J. R., and Moon, R. T., 2000, "The Wnt/Ca2+ pathway: a new vertebrate Wnt signaling pathway takes shape", *Trends Genet,* Vol. 16, pp. 279-283.

Li, G., Yin, H. and Kuret,,J., 2004, "Casein Kinase 1 Delta Phosphorylates Tau and Disrupts Its Binding to Microtubules", *Journal of Biological Chemistry,* Vol. 279, pp. 15938-15945.

Li L. and Cohen, S.N. , 1996, "Tsg101: A novel tumor susceptibility gene isolated by controlled homozygous functional knockout of allelic loci in mammalian cells", *Cell,* Vol. 85, pp. 319-329.

Lu, S.-L., Kawabata, M., Imamura, T., Akiyama, Y., Nomizu,T., Miyazono,K., Yuasa, Y., 1998, "HNPCC associated with germline mutation in the TGF-beta type II receptor gene", *Nature Genet,,* Vol. 19, pp. 17-18

Luu, H. H., Zhang,R., Haydon, R.C. et al., 2004, "Wnt/beta-catenin signalling pathway as a novel cancer drug target", *Curr Cancer Drug Targets,* Vol. 4, pp. 653–671.

Macleod, R. J.,Hayes,M., Pacheo, I., 2007, "Wnt5a secretion stimulated by the extracellular calcium-sensing receptor inhibits defective Wnt signalling in colon cancer cells", *Am J Physiol Gastrintest Liver Physiol,* Vol 293, G403-411.

Maere, S., Heymans, K., Kuiper, M., 2005, "BiNGO: a Cytoscape plugin to assess overrepresentation of gene ontology categories in biological networks", *Bioinformatics,* Vol. 21, pp. 3448-3449.

Maiese, K., Li, F., Chong, Z. Z., and Shang, Y. C., 2008, "The Wnt signaling pathway: aging gracefully as a protectionist?", *Pharmacol Ther,* Vol. 118, pp. 58-81.

Maiti, S., Alam, R., Amos, C. I., Huff, V., 2000, "Frequent Association of beta-Catenin and WNT1 Mutations in Wilms Tumors", *Cancer Res* , Vol. 60, pp. 6288-6292.

Matsuzawa, S. I., Reed, J. C., 2001, "Siah-1, SIP, and Ebi collaborate in a novel pathway for beta-catenin degradation linked to p53 responses", *Mol Cell.,* Vol. 7(5), pp. 915-26

Maussion, G., et al., 2008, "Convergent evidence identifying MAP/microtubule affinity-regulating kinase 1 (MARK1) as a susceptibility gene for autism", *Hum. Mol. Genet.,* Vol. 17, pp. 2541-2551.

Mcdonald S and Silver,A., 2009, "Wnt5a: opposing roles in cancer", *British J. Cancer,* Vol. 101, pp.209-14.

Nguyen, T. P. and Jordán,F., 2010, "A quantitative approach to study indirect effects among disease proteins in the human protein interaction network", *BMC Syst Biol.,* Vol. 4, Issue 103.

Nusse, R., 2005, " Wnt signaling in disease and in development", *Cell Research,* Vol. 15, pp. 28-32.

Otheki, T. et al., 2000, "Negative regulation of T-cell proliferation and interleukin 2production by the serine threonine kinase Gsk-3", *J. Exp. Med.,* Vol. 192, pp. 99-104.

Özbayraktar, B. K.,2011, "From Yeast to Human: Unraveling Sphingolipid Metabolism through Macroscopic and Microscopic Analyses", PhD Thesis, Boğaziçi Üniversitesi , İstanbul.

Pandur, P., 2005, "Recent discoveries in vertebrate non-canonical Wnt signaling: Towards a Wnt signaling network", *Advances in Dev Biol,* Vol. 14, pp. 91-106.

Paul S. and Dey, A., 2008, "Wnt Signaling and Cancer Development: Therapeutic Implication", *Neoplasma,* Vol. 55, pp. 165-176.

Peifer,M. and Polakis ,P., 2000," Wnt signaling in oncogenesis and embryogenesis-a look outside the nucleus", *Science*, Vol. 287,1606-1609.

Peng, Z., Wang, H. and Shan, C., 2009, "Expression of ubiquitin and cullin-1 and its clinicopathological significance in benign and malignant lesions of the lung", *Zhong Nan Da Xue Xue Bao Yi Xue Ban.*, Vol. 34, Issue 3, pp. 204-9.

Peterson, L. C., Rao, K. V. and Crosson, J. T. et al., 1985, " Fechtner syndrome: a variant of Alport's syndrome with leukocyte inclusions and macrothrombocytopenia", *Blood*, Vol. 65, pp. 397-406.

Prasad, T. S. K., Goel, R., Kandasamy, K. et al., 2009, "Human Protein Reference Database--2009 update", Nucleic Acids Res., Vol. 37, pp. D767-72.

Ralhan, R., DeSouza, L. V., Matta, A., Tripathi, S. C, Ghanny, S., Gupta, S. D., Bahadur, S. and Siu, K. W. M., 2008, "Discovery and Verification of Head-and-neck Cancer Biomarkers by Differential Protein Expression Analysis Using iTRAQ Labeling, Multidimensional Liquid Chromatography, and Tandem Mass Spectrometry", *Molecular and Cellular Proteomics*, Vol. 7, pp. 1162-1173.

Saneyoshi, T., Kume, S., Amasaki, Y., and Mikoshiba, K., 2002, "The Wnt/calcium pathway activates NF-AT and promotes ventral cell fate in Xenopus embryos", *Nature*, Vol. 417, pp. 295-299.

Schambony, A., and Wedlich, D., 2007, "Wnt-5A/Ror2 regulate expression of XPAPC through an alternative noncanonical signaling pathway", *Dev Cell*, Vol. 12, pp. 779-792.

Slusarski, D. C., and Pelegri, F., 2007, "Calcium signaling in vertebrate embryonic patterning and morphogenesis", Dev Biol, Vol. 307, pp. 1-13.

Slusarskia, D. C., Yang-Snyderb , J., Busaa, W. B. and Moon, R. T. , 1997, "Modulation of Embryonic Intracellular Ca2+ Signaling by Wnt-5A", *Developmental Biology*, vol. 182, pp. 114-120.

Soucek, L., Whitfield, J., Martins, C. P., Finch, A. J., Murphy, D. J., Sodir, N. M., Karnezis, A. N., Swigart, L. B., Nasi, S., Evan, G. I., 2008, "Modelling Myc inhibition as a cancer therapy", *Nature*,Vol. 455 (7213), pp. 679–83.

Stark, C., Breitkreutz, B. J., Chatr-Aryamontri, A. *et al.*, 2010, "The BioGRID Interaction Database: 2011 update", *Nucleic Acids Res.*, Vol. 39, D698-704.

Steffen, M. , Petti, A., Aach, J. , D'haeseleer, P. and Church, G., 2002, "Automated modelling of signal transduction networks", *BMC Bioinformatics*, Vol. 3, No. 34, pp. 1471-2105.

Tan, S. G. and Kelsey, K. T., 2009, "Epigenetic Inactivation of SFRP1 and SFRP2 Genes as Biomarkers of Invasive Bladder Cancer", Honors Thesis, Brown University.

Tanaka, S., Mori, M., Mafune, K., Ohno, S., Sugimachi, K., 2000, "A dominant negative mutation of transforming growth factor-beta receptor type II gene in microsatellite stable oesophageal carcinoma", *Brit. J. Cancer*, Vol. 82, pp. 1557-1560

Tekir, S. D., Ümit,P., Toku, A. E. and Ülgen, K. Ö., 2010, "Reconstruction of Protein-Protein Interaction Network of Insulin Signaling in Homo Sapiens", *Journal of Biomedicine and Biotechnology*, Vol. 2010, Article ID 690925.

Tekir, S. D., Arga, K. Y. and Ülgen, K. Ö., 2009, "Drug targets for tumorigenesis: Insights from structural analysis of EGFR signaling network", *Journal of Biomedical Informatics*, Vol. 42, pp. 228–236.

Tiveci, S., 2011, "Ca Signaling in Multiple Species", PhD Thesis, Bogaziçi Üniversitesi , Istanbul.

Toku, A. E., 2010, "Investigation of Crosstalk Between Wnt and Hedgehog Signaling Pathways in fly", MSc Thesis, Bogaziçi Üniversitesi, Istanbul.

Ueno, K., Hiura, M., Suehiro, Y., Hazama, S., Hirata, H., Oka, M., Imai, K., Dahiya, R., Hinoda, Y., 2008, "Frizzled-7 as a Potential Therapeutic Target in Colorectal Cancer", *Neoplasia*, Vol. 10, pp. 697–705.

Valdes, A. M., Spector, T. D., Tamm, A., Kisand, K., Doherty, S. A., Dennison, E. M., Mangino, M., Tamm, A., Kerna, I., Hart, D. J., Wheeler, M., Cooper, C., Lories, R. J., Arden, N. K. and Doherty, M., 2010, "Genetic variation in the SMAD3 gene is associated with hip and knee osteoarthritis", *Arthritis Rheum.*, Vol. 66 Issue 8, pp. 2347-52.

Veeman, M. T., Axelrod, J. D., and Moon, R. T., 2003, "A second canon: Functions and mechanisms of beta-catenin-independent Wnt signaling", *Dev Cell*, Vol. 5, pp. 367-377.

Wang, H. Y., Malbon, C. C., 2003, "Wnt signaling, Ca2+, and cyclic GMP: visualizing Frizzled functions", *Science*, Vol. 300, pp. 1529-1530

Widelitz, R., 2005, "Wnt signaling through canonical and non-canonical pathways: recent progress", *Growth Factors*, Vol. 23, pp. 111-116.

Wu, M., and Herman, M. A., 2006, "A novel noncanonical Wnt pathway is involved in the regulation of the asymmetric B cell division in C. Elegans", *Dev Biol*, Vol. 293, pp. 316-29.

Wu, Q. and Maniatis, T., 1999, " A striking organization of a large family of human neural cadherin-like cell adhesion genes", *Cell*, Vol. 97, pp. 779-790.

Wu, X. R., Zhu, Y., Li, Y. 2005, "Analyzing Protein Interaction Networks via Random Graph Model", *International Journal of Information technology*, Vol. 12, pp. 1251-1260.

Yanaga F. T. and Sasaguri, T., 2007, "The Wnt/β-Catenin Signaling Pathway as a Target in Drug Discovery", *J Pharmacol Sci.*, Vol. 104, pp. 293 - 302.

Yang, Z. Q., Zhao, Y., Liu, Y., Zhang, J. Y., Zhang, S., Jiang, G. Y., Zhang, P. X., Yang, L. H., Liu, D., Li, Q. C., Wang, E. H., 2010, "Downregulation of HDPR1 is associated with poor prognosis and affects expression levels of p120-catenin and β-catenin in nonsmall cell lung cancer", *Molecular Carcinogenesis*, Vol. 49, Issue 5, pp. 508–519.

Yeh, C. T., Lu, S. C., Chen, T. C., Peng, C. Y., Liaw, Y. F., 2003, "Aberrant transcripts of the cyclin-dependent kinase-associated protein phosphatase in hepatocellular carcinoma", *Cancer Res.*, Vol. 60, pp. 4697-4700.

Ying, J., Li, H., Yu, J., Ng, K. M., Poon, F. F., Wong, S. C., Chan, A. T., Sung,J. J, Tao, Q.,2008, "Wnt5a exhibits tumor-suppressive activity through antagonizing the Wnt/beta-catenin signaling, and is frequently methylated in colorectal cancer", *Clin Cancer Res.*, Vol. 14, pp. 55-61.

Yu, H., Greenbaum, D., Xin, L. H. et al., 2004," Genomic analysis of essentiality within protein networks", *Trends Genet*, Vol. 20,pp. 227-231.

Yujiri, T., Fanger, G. R. et al., 1999, "MEK kinase 1 (MEKK1) transduces c-Jun NH2-terminal kinase activation in response to changes in the microtubule cytoskeleton", *J. Biol. Chem.*, Vol. 274, pp. 12605-12610.

Zanzoni, A., Montecchi-Palazzi, L., Quondam, M., Ausiello, G., Helmer-Citterich, M., Cesareni, G., 2002, "MINT: a Molecular INTeraction database", FEBS Lett., Vol. 513, No. 1, pp. 135-40.

Zhang, Y. et al., 2001, "Differential Molecular Assemblies Underlie the Dual Function of AXIN in Modulating the Wnt and JNK Pathways", *Journal of Biological Chemistry*, Vol. 276, No. 34, pp. 32152-32159.

Zielinski R., Przytycki,P. F., Zheng, J., Zhang, D., Przytycka, T. M. and Capala, J., 2009, "The crosstalk between EGF, IGF, and Insulin cell signaling pathways- computational and experimental analysis", *BMC Bioinformatics*, Vol. 3, No. 88.

9

Imaging Cellular Metabolism

Athanasios Bubulya and Paula A. Bubulya
Wright State University, Dayton, Ohio
USA

1. Introduction

Imaging tools that aid in identifying the precise location of diseased cells within a patient's tissues, and that measure the physiological status of these cells, have clear impact for medical scientists in a wide range of specialties ranging from clinical oncology to cardiology to neurology. Furthermore, laboratory scientists can utilize imaging methods to gain insight into the subcellular localization and kinetics of key branches for major biosynthetic pathways. Our predominant focus for discussion in this chapter is imaging of living cellular systems, and we also discuss how dynamics of key cellular pathways can be revealed from observations in fixed cells. We discuss some of the recent advances that continue to lead scientists toward imaging metabolic pathways for understanding and diagnosing human disease. The goal of this chapter is to provide the novice researcher with an overview of a variety of approaches for imaging cellular metabolism both for medical and research purposes.

Understanding the metabolic differences between normal cells and cancer cells is a major objective of biomedical research. Cancer cells exhibit increased metabolic rates for pathways needed to support uncontrolled cellular proliferation, and this has been exploited for treatment of tumors as well as for diagnostic purposes (Locasale et al., 2009). The problem for understanding and treating cancer is to learn not only what makes each specific type of cancer unique, but also to learn the commonly aberrant pathways in cancers such as higher glucose metabolism or altered membrane biosynthesis pathways that can be exploited as targets for developing diagnostic tools and anticancer therapies. There is continued hope for novel therapies not only in the well-documented oncogene/tumor suppressor-related cellular signaling pathways, but also in areas of renewed interest, such as unique regulation of stress response pathways by cancer cells (reviewed in Luo et al., 2009). Regardless of the pathway targeted, cancer therapy would ideally leave normal cells unaffected while specifically interfering with altered metabolic pathways of cancer cells.

2. Exploiting aberrant metabolism to image cancer cells

Increased glucose metabolism in cancer cells, initially observed by Warburg over 90 years ago, has fueled the development of labeled metabolites to differentially label cancer cells from surrounding normal tissues (Warburg, O., 1956). To detect the labeled malignant tissue, the use of minimally invasive in vivo imaging techniques has increased rapidly over the last two decades. Here we briefly describe various types of metabolites and the imaging methods used to observe cancer cells in vivo. Of particular interest is the use of imaging techniques in conjunction with labeled metabolic markers to determine location, size and response to drug treatment of cancerous tissue in vivo. The current non-invasive imaging techniques include but are not limited to positron emission tomography (PET), magnetic resonance spectroscopy (MRS), magnetic resonance imaging (MRI), single-photon emission computed tomography (SPECT) and computed tomography (CT). These techniques have been extensively reviewed elsewhere and the reader is encouraged to refer to those recent articles (Plathow & Weber, 2008; Condeelis & Weissleder, 2010).

2.1 In vivo imaging probes

According to the Molecular Imaging and Contrast Database (MICAD), there are currently 1107 imaging probes and contrasting agents currently used for in vivo studies (Chopra et al., 2011). The most commonly used metabolic marker is the synthetic glucose analog, 2-deoxy-2-(^{18}F)fluoro-D-glucose (FDG) (Ido et al., 1978). FDG has wide clinical use because malignant cells display high glucose metabolic rates as compared to normal tissue (Warburg, 1956, Sokoloff et al. 1977). In the cell, FDG becomes phosphorylated by hexokinase, it can not be further metabolized, and it is in essence trapped within the cell. FDG uptake is therefore indicative of the rates of both glucose uptake and glucose phosphorylation. FDG uptake is monitored using PET and this method of probe detection has become the choice of clinicians for identifying tumors in vivo and has been used to successfully detect head and neck, prostate, breast, lung and liver cancer (Ben-Haim & Ell, 2009). Additional metabolic processes have been targeted for clinical use, and PET metabolic markers have been developed that detect apoptosis, angiogenesis, hypoxia, cell proliferation and amino acid metabolism. Metabolic markers currently being used include [^{11}C]Methionine, L-[3-^{18}F]-□-methyltyrosine [^{18}F]FMT, [^{11}C]Thymidine, ^{18}F-3'-fluoro-3'-deoxy-thymidine [^{18}F]FLT, [^{11}C]Choline [^{18}F]Choline, [^{11}C]Acetate, ^{68}Ga-NOTA-RGD [^{18}F]Galacto-RGD, ^{18}F-fluoromisonidazole [^{18}F]FMISO, [^{18}F]FAZA, ^{64}Cu-ATSM, ^{99m}Tc-Annexin-V and [^{124}I]Annexin-V (reviewed in Lee, 2010). Here we highlight PET markers that are associated with cell proliferation, amino acid metabolism and lipid metabolism.

Collectively [^{11}C]Methionine, [^{18}F]FMT, [^{11}C]Thymidine and [^{18}F]FLT are used as indicators of cell proliferation. These compounds can further be classified into amino acid and nucleic acid analogs. The radiolabeled amino acid [^{11}C]methionine, is easily taken up by tumor cells due to their increased protein synthesis, and has been very useful for identifying malignancies in the central nervous system (Comar et al., 1976, reviewed in Nanni et al., 2010). Furthermore, the amount of [^{11}C]methionine present in malignant tissue can be used to determine tumor grade (reviewed in Nanni et al., 2010). For example, [^{11}C] methionine, was recently used to grade the aggressiveness of glioblastoma in patients with grade IV gliomas (Kawai et al., 2011). Additionally, [^{11}C]Methionine/ PET is a promising tool for determining active tumor regions providing valuable information for chemotherapy (Tsien et al., 2011). Deng et al. (2011) recently synthesized S-^{11}C-methyl-L-cysteine as a PET tracer and suggest that it shows improved distinction between malignant tissue and inflammatory response as compared with [^{11}C]Methionine.

[^{18}F]FMT as a metabolic indicator is useful for detecting tumors and monitoring response to therapy. Biochemically, FMT is a tyrosine analog that is not incorporated into nascent proteins; however, it does reflect amino acid uptake that is increased in cancer cells

(Ishiwata et al., 2004). In patients with non-small cell lung carcinoma, FMT uptake in primary adenocarcinoma was suggested to be an indicator of poor prognosis (Kaira et al., 2009). FMT has also been useful in identifying bone lesions (Ishiwata, et al., 2004). The D isomer of FMT was used to monitor squamous cell carcinoma in a mouse model system. Irradiated mice showed a decrease in uptake D-FMT after radiation in contrast to FDG and ^{11}C Met (Murayama et al., 2009). These results suggest that D-FMT may be a good indicator of early tumor response to treatment. The nucleotide analogs [^{11}C] thymidine and [^{18}F]FLT have also been used to detect malignancies. FLT is a thymidine analog that can be phosphorylated by thymidine kinase 1 which is elevated in proliferating cells (Rasey, et al., 2002). Phosphorylation traps FLT, resulting in its accumulation in cells. While FLT undergoes phosphorylation it is not clear if it is incorporated in DNA. Like FMT, FLT has been most useful in monitoring tumor response to therapy (reviewed in Barwick et al., 2009). In fact, FLT was similar to FMT in its ability to monitor early tumor response to radiation (Murayama, et al., 2009).

Both [^{11}C]Choline and [^{18}F]Choline, are indicators of phospholipid metabolism in cells. Choline is transported into cells where it is metabolized to phosphocholine by choline kinase, an enzyme that is frequently upregulated in tumor cells (Ramirez de Molina, et al. 2002 a,b). Phosphocholine is negatively charged and remains trapped within the cells. Labeled choline has been used extensively in prostate cancer studies. For example, an increase of ^{11}C choline in prostatic malignancies has been recently been shown to be an indicator of aggressiveness in prostate cancer patients (Piert et al., 2009). Several recent reviews discuss imaging in prostate cancer and the reader is directed to these reviews for further reading (Edmonds et al., 2009; Jadvar, 2009; Zaheer et al., 2009). Labeled choline, including ^{18}F-fluoroethylcholine in animal models, has also been used to detect hepatocellular carcinoma and brain tumors (Talbot et al., 2006; Kubota et al., 2006; Kolthammer et al., 2011).

Acetate is taken up by cells, converted to acetyl CoA and ultimately incorporated into the cell membrane (Howard & Howard, 1975). [^{11}C]Acetate has recently been used to detect increased glial tumor metabolism (Liu et al., 2006; Tsuchida et al., 2008). 2-^{18}F-fluoroacetate has been successful in detecting glioblastoma in a mouse model system (Marik et al., 2009). Like labeled choline, labeled acetate is also a useful tracer for prostate cancer. A review has recently been published by Jadvar (2011) comparing the use of labeled acetate versus FDG or labeled choline and discussing the utility of these imaging probes in prostate cancer detection.

2.2 Imaging of molecular complexes in living cells

Our knowledge about the function of molecular complexes and their subcellular localization has rapidly expanded with the development of a wide array of encoded fluorescent probes that have enabled direct observation of metabolic pathways in living cells (reviewed in Zhang et al., 2002). Several of these probes utilize fluorescence resonance energy transfer (FRET) to detect interaction between molecules or to assess cellular levels of metabolites and monitor their compartmentalization. Among such FRET probes that sense metabolites, "cameleon" measures intracellular calcium (Miyawaki et al., 1997), and other probes have been designed to non-destructively sense phosphorylation by specific kinases such as protein tyrosine kinases (Ting et al., 2001) or protein kinase A (Zhang et al., 2001).

Fluorescent nanosensors have also been developed for concentration-dependent sensing of maltose in yeast (Fehr et al., 2002), as well as for detecting glucose uptake and subcellular compartmentalization (Fehr et al., 2003; Fehr et al., 2005) or ribose uptake (Lager et al., 2003) in mammalian cells. Furthermore, bimolecular fluorescence complementation (BiFC) can detect interaction between protein partners in living cells. Individual proteins are fused with non-fluorescent fragments of green fluorescent protein (GFP; or one of its variants). If the two fusion proteins interact, this brings the GFP fragments in close enough proximity to reconstitute fluorescence, and imaging reveals subcellular localization of the complex (Hu et al., 2002). Multicolor BiFC allows detection of multiple complexes simultaneously in living cells, and it can be used to measure the efficiency of complex formation between a protein of interest and each of its known partners (Hu & Kerppola, 2003). One example for how BiFC can be used to monitor cell physiological readout for cancer pathways in single cells was recently demonstrated by visualizing activation of caspase-2, the initiator caspase for mitochondrial apoptosis pathway (Bouchier-Hayes et al., 2009). This is interesting in light of the evidence that caspase-2 is a tumor suppressor. Caspase -/- mouse embryonic fibroblasts (MEFs) resisted apoptosis, and they showed increased proliferation as well as tumor formation most likely compounded by lost function of p53 (Ho et al., 2009).

3. Nuclear organization and gene expression

3.1 Nuclear organelles

Organization of nuclear compartments can reflect metabolic status in mammalian cells. This is exemplified by numerous accounts of altered nuclear structure, nuclear organelles and nuclear biochemistry observed in a wide range of diseases. Tying altered gene expression to changes in nuclear organization is an area of intense current research (reviewed in Rajapaske and Groudine, 2011). Among the best examples of such observable alteration is seen in the perinucleolar compartment (PNC), a nuclear organelle found adjacent to the nucleolus that contains RNA binding proteins and is enriched with transcripts synthesized by RNA polymerase III (Huang et al., 1998). Metabolism of RNA polymerase III transcripts is suggested as a primary factor regulating PNC size (Wang et al., 2003). Variation in PNC size is medically relevant because the size of the PNC has been directly correlated with disease staging. Analysis of clinically staged breast cancer tissue samples showed that presence of PNCs increases with disease progression, such that metastatic tumors have the largest and most abundant PNCs (Kamath et al., 2005). Because PNC prevalence correlated with metastatic potential and malignancy in other solid tumors (Norton et al., 2008), PNC status can therefore be used as a simple and relatively low-cost prognostic marker for tumor progression. Further study is needed to determine if the PNC changes occur in other cancers, or if the changes are a result or a cause of cellular transformation. Regardless, defining the primary functions of the PNC and understanding the biochemical pathways housed in this nuclear organelle could lead to developing very specific tools for knocking out breast tumors and other types of tumors. Along these lines, an automated high-throughput imaging screen performed in living cells expressing fluorescently tagged PNC component polypyrimidine tract binding protein (PTB-GFP) identified compounds that disassemble the PNC (Norton et al., 2009). This work shows promise not only for cancer drug development, but also for the general assessment and screening of compounds that effect nuclear structural changes, as well as for scientists to determine where compounds interfere with cellular biochemistry in order to better understand the metabolic pathways that regulate metastasis.

Despite decades of intense research, new players involved in protein coding gene expression are still being identified and characterized. Many of the factors localize to specific nuclear organelles such as Cajal bodies and nuclear speckles (also called SC35 domains or interchromatin granule clusters; reviewed in Spector and Lamond, 2010) Nuclear speckles are storage sites for pre-mRNA processing factors from which factors are exchanged with the nucleoplasm and recruited to nascent transcripts for co-transcriptional pre-mRNA processing (reviewed in Spector & Lamond, 2010). The organization of nuclear speckles reflects as well as impacts the global status of pre-mRNA synthesis and the efficiency of pre-mRNA processing. Inhibition of RNA polymerase II by alpha-amanitin supports the recruitment model, as this treatment causes RNA processing factors to remain in enlarged rounded nuclear speckles speckle rounding (Lamond & Spector, 2003). Disassembly of nuclear speckles distributes components throughout the nucleoplasm and alters pre-mRNA processing (Sacco-Bubulya & Spector, 2002). Purification of nuclear organelles has identified many of the nuclear proteins whose components are used for synthesis and processing of RNA (Mintz et al., 1999; Saitoh et al., 2004; Andersen et al., 2004) and is beginning to reveal functions for RNAs (Prasanth et al., 2005; Tripathi et al., 2010). Polyadenylated RNA was previously shown to be enriched in nuclear speckles by fluorescence in situ hybridization methods (Visa et al., 1993; Huang et al., 1994). As these new players are identified, synthetic gene reporter systems will be incredibly useful tools for determining the kinetics of their assembly at transcription sites, as well as to pin down their specific functions in gene expression.

Visualizing biosynthetic pathways in the nucleus has relied on a variety of experimental approaches. Labeling cellular structures or contents (e.g. lipids, mitochondria, DNA) non-immunologically with fluorescent molecules, or by using immunocytochemical approaches, has been described extensively elsewhere (Spector & Goldman, 2006). Incorporation of nucleotide analogs into nascent strands is a common way to label nucleic acids during their synthesis, and can be used to visualize entire chromosomes, individual DNA replication foci, or transcription factories. Nucleotide analogs can be radioactive, enzymatic, fluorescent or in some other way tagged for detection. Radioactive nucleotide incorporation relies on detection by autoradiography which has the disadvantage of requiring long exposure times of several months, offers low resolution, and is not typically the preferred labeling method. Recent technical advances using nucleotide incorporation approaches or various molecular tagging methods, as well as advances in super-resolution imaging systems (Huang et al., 2009), are certain to continue rapidly expanding our knowledge about localization and kinetics of nuclear pathways.

3.2 Transcription and RNA processing

BromoUTP incorporation is a widespread tool for to labeling transcription sites in situ. Cells are gently permeabilized to allow uptake of nucleotide analog, followed by incubation in a transcription buffer cocktail that promotes elongation of nascent transcripts (Haukenes et al., 1997). A short pulse of labeling in mammalian cells (~5-8 minutes at 37 degrees Celsius for HeLa cells) is sufficient to globally label nascent RNA, which can be subsequently detected in transcription foci throughout the nucleus corresponding to RNAs synthesized predominantly by RNA polymerases II and III (Sharma et al., 2010; Sacco-Bubulya and Spector, 2002) or specifically in nucleoli corresponding to ribosomal RNAs synthesized by RNA polymerase I (Dundr et al., 2002). Fluorouridine is also useful for labeling transcription sites in situ (readers should note that this is halogenated, not fluorescent), and has the advantage that it directly enters the cells without the need for detergent permeabilization (Boisvert et al., 2000). In addition, permeabilizing cells using different detergents and various permeabilization times can affect labeling efficiency, for example, in nucleolar transcription foci versus transcription foci in the nucleoplasm, and nucleotide analogs appear to be incorporated with different degrees of efficiency in distinct nuclear compartments in mammalian cells (P. Bubulya, unpublished observations). RNAs synthesized by distinct RNA polymerases I, II or III have been shown in dedicated nuclear transcription "factories" by sophisticated twists on BrUTP incorporation methods (Pombo et al., 1999). Regardless, in all these studies the cells are not alive; following pulse labeling, removal of excess nucleotide, and fixation of the cells, nucleotide incorporation is detected by subsequent labeling of the halogenated nucleotide. Studies in living cells have revealed information regarding the kinetics of transcription and processing machinery assembly and transcript elongation rates for RNA polymerases I and II (Dundr et al., 2002; Kimura et al., 2002) as well as the cell cycle regulated assembly of the nuclear organelles that house these machineries.

The synthesis and processing of ribosomal RNAs for ribosome production is the most robust gene expression pathway in mammalian cells. Although products of all RNA polymerases are required for final assembly of ribosomes, ribosome biogenesis occurs in nucleoli (reviewed in Leary & Huang, 2001, Hernandez-Verdun et al., 2002). Activation of the rRNA genes evades global inhibition of transcription on condensed chromosomes during mitosis. Imaging methods have demonstrated the involvement of pre-rRNAs in the onset of nucleolar organization at daughter cell chromatin and that different chromatin regions containing rRNA genes come together during nucleolar maturation (Hernandez-Verdun et al., 2002). Ribosomal components are delivered to newly forming nucleolar bodies as they are needed for pre-rRNA processing, and individual components show rapid exchange in and out of these bodies during assembly (Dundr et al., 2000). During interphase, individual components of RNA polymerase I enter nucleoli independent of others, and different incorporation efficiency and nucleolar residence time for independent components suggests sequential assembly of complexes with each round of transcription (Dundr et al., 2002).

Reporter gene constructs that have been engineered and stably integrated into cellular genomes to visualize RNA polymerase II transcription in living cells made it possible to study the position and activity of transcription sites in real time, in the context of chromatin structure and global nuclear organization. Tandem arrays of DNA containing repeated binding sites for fluorescently-tagged DNA binding proteins can be stably integrated into chromatin where they become assembled into higher order chromatin structure. These arrays have been used to visualize chromatin dynamics, and they have been coupled to reporter genes downstream to visualize the kinetics of chromatin unfolding and transcript synthesis in living cells (reviewed in McNally, 2009 and in Rafalska-Metcalf & Janicki, 2007).

In one system, integrated tandem arrays of mouse mammary tumor virus driving a *ras* reporter (called MMTV-LTR-ras-BPV) created a specific locus containing 800-1200 binding sites for gluococorticoid receptor (GR). Following controlled expression of a GFP-tagged-GR (GFP-GR) and hormone addition, GR nuclear translocation as well as GR-GFP recruitment to the locus were observed in living cells (McNally et al., 2000). Photobleaching techniques determined that GFP-GR undergoes its rapid exchange on locus chromatin (McNally et al., 2000). Reporter transcripts were detected by RNA fluorescence in situ hybridization (RNA-FISH) in the vicinity of GR-GFP accumulation on the reporter locus. Since GR-GFP reflected the underlying DNA binding sites on chromatin, the extent of locus chromatin decondensation could in turn be observed and correlated with the amount of transcription at that site (Muller et al., 2001).

Other systems have been developed to allow visualization of chromatin independent of transcription activity, such that both the transcriptional inactive and transcriptionally active chromatin can be observed, and the transitions between these two states can be studied. One such system consists of a tandem array of synthetic reporter sequences in baby hamster kidney (BHK) cells containing lac operator repeats proximal to an inducible reporter gene encoding cyan fluorescent protein that is targeted to peroxisomes via a serine-lysine-leucine tag (SKL; Tsukamoto et al., 2000). Expression of fluorescently-tagged lac repressor protein in these cells labels the locus chromatin, and inducing transcription allows live observation of the transition from the inactive to active state, revealing continued chromatin decondensation at the locus over the time course of imaging (Tsukamoto et al., 2000). While reporter transcripts could be detected in that system by RNA-FISH, this method does not allow real time imaging of transcripts. The same basic reporter system was then further developed such that transcripts would contain MS2 bacteriophage viral replicase translational operator sequences (MS2 stem loops). These sequences are recognized by the MS2 coat protein that can be fluorescently tagged and expressed in the reporter cells for imaging transcription in real time (Janicki et al., 2004). This latter system, referred to as U2OS 2-6-3, enables visualization of DNA, RNA and protein product for a single reporter locus in living cells (Janicki et al., 2004). For that reason, it is a very powerful tool for understanding all the steps of RNA polymerase II-mediated gene expression in any phase of the cell cycle. For example, the transition from condensed to decondensed chromatin, the switch from heterochromatic to euchromatic modifications, the exchange of histones H3/H3.3, the recruitment of transcription factors, the kinetics of reporter transcript production, and the recruitment of pre-mRNA processing machinery were all observed in living cells (Janicki et al., 2004). A further advantage of these live cell reporter systems for transcription is that the lac repressor can be fused both with a fluorescent protein AND a second protein/domain that sequesters it to a particular subnuclear compartment. For example, fusing lac repressor to the fluorescent protein mCherry and to lamin B1, and placing the expression of this triple fusion protein under the control of an inducible promoter, allowed for inducible tethering of a reporter locus to the nuclear periphery (Kumaran and Spector, 2008). Despite that the peripheral regions of the nucleus are generally thought of as heterochromatic and transcriptionally silent, this reporter locus retained its transcriptional activity upon relocation to the nuclear lamina. This tethering system opens many possibilities for gaining both spatiotemporal and kinetic information about gene expression events occurring on a gene locus during its nuclear repositioning, as well as a novel way to test how nuclear compartments in diseased cells might demonstrate misregulated gene expression pathways (Kumaran and Spector, 2008).

Many continuing studies are using such systems to learn more about transcription regulation and how it relates to nuclear dynamics or RNA trafficking. The involvement of proteins and RNAs in pre-mRNA synthesis and processing, or on chromatin condensation/decondensation during gene regulation can be systematically evaluated either by RNAi-mediated knockdown or overexpression studies. For example, depletion of 7SK RNA resulted in upregulation of transcription on a modified version of the U2OS 2-6-3 gene reporter system (Prasanth et al., 2010). Furthermore, the transcripts themselves can be observed directly in living cells. Individual mRNP particles can be tracked to gain information about mRNP movement through the nucleoplasm by using the U2OS 2-6-3 cells (Shav-Tal et al., 2004). Photoactivation studies showed the single mRNPs freely diffused in all directions as they moved away from the transcription site. In addition, general inhibitors of cellular metabolism surprisingly caused a decreased mobility of mRNP particles, mostly likely explained by overall restructuring of nuclear organization under these conditions, that was restored when energy levels were reset to normal (Shav-Tal et al., 2004). At least one new system has been recently developed for examining the trafficking of mRNA (and visualizing its protein product) that is certain to shed light on the spatial and temporal kinetics of synthesis, export, and cytoplasmic transport of mRNA to its cellular location where it is ultimately translated into protein (Ben-Ari, et al., 2010). Our capability for RNA imaging is on the verge of complete transformation due to a newly developed tool for tagging and imaging RNAs in living cells (Paige et al., 2011). This new "RNA version" of green fluorescent protein, called Spinach, was successfully used to image the dynamics of 5S ribosomal RNA in living cells. Spinach as well as other RNA-tagging molecules in a range of excitation and emission spectra will undoubtedly open new avenues for discovery with potential applications in RNA-RNA and RNA-protein FRET (Paige et al., 2011).

Pre-mRNA splicing regulation can be studied by using minigene reporters that typically contain only a small segment of genomic DNA from a given gene subcloned into a mammalian expression vector. A well-characterized beta-tropomyosin (BTM) minigene (Helfman et al., 1988; Huang & Spector, 1996) was stably integrated into HeLa cells (Sacco-Bubulya and Spector, 2002; Sharma et al., 2011). The BTM minigene is useful for monitoring constitutive as well as alternative splicing of minigene transcripts. One advantage of expressing transcripts from reporter minigenes is that a comparison can be made between precursors versus spliced transcripts produced at the reporter locus that can be visualized by RNA-FISH (Sacco-Bubulya and Spector, 2002; Sharma et al., 2011). Altered splicing of the reporter transcript can be detected after various treatments, for example, RNAi-mediated depletion or overexpression of specific splicing factors. The extent of reporter transcript processing can be monitored in situ. Sharma et al. (2011) recently demonstrated that HeLa cells treated with siRNA duplexes targeting the splicing factor Son showed increased skipping of exon 6 in BTM transcripts both by quantitative PCR and by RNA-FISH. In addition, a genome-wide screen identified human transcription and splicing targets for Son that include chromatin modifiers and cell cycle regulators (Sharma et al., 2011).

Many of the above mentioned reporter gene loci have been instrumental in studying the onset of pre-mRNA synthesis and processing in post-mitotic nuclei. Somewhat similar to what was observed during nucleolar reassembly following mitosis, sequential nuclear entry of pre-mRNA synthesis and processing factors also occurs in an ordered sequence according to the timing for when factors are needed (Prasanth et al., 2003). Interestingly, transcription factors and RNA polymerase II are detected in daughter nuclei first, followed by RNA processing factors. Only after all components are available in the nucleus for coupled RNA synthesis and processing are the elongating RNA polymerase II and exon junctions in reporter mRNAs detected (Prasanth et al., 2003). Recently, live cell imaging demonstrated that specific factors regulate post-mitotic reinitiation of transcription (Zhao et al., 2011). Active transcription sites are remembered, or "bookmarked", by histone post-translational modification, specifically by histone H4 lysine 5 acetylation. This mark persists on transcriptionally inactive mitotic chromatin. Bromodomain protein 4 (BRD4) recognizes this mark to increase kinetics of RNA polymerase II transcription on a reporter locus following mitosis as compared to interphase cells (Zhao et al., 2001).

4. DNA replication

5-Bromodeoxyuridine (BrdU) incorporation is a common way for observing DNA replication in situ, and it is useful for general labeling of nuclear replication foci. Typically a pulse of BrdU is administered to live synchronized cells, and localization of the BrdU is observed after preserving cells by fixation and immunofluorescence labeling with antibodies against BrdU to show progression of replication in different subnuclear regions over time (Nakamura et al., 1986). This approach proved to be substantially better than radioactive nucleotide incorporation for replication labeling due to limited spatial resolution of autoradiography. Perhaps continued development of more sensitive and faster methods to incorporate nucleotide analogs will open new avenues for DNA replication studies. One such approach employs 5-ethenyl-2′-deoxyuridine incorporation followed by detection using "click" chemistry. This approach is useful in that it can be done in live cells or tissues to label cells in S-phase and also for direct imaging and high-throughput applications, although the cytotoxicity of the reaction prevents long-term cell survival (Salic & Mitchison, 2008). All these types of approaches have been used widely to label replicating DNA.

The overall DNA replication pattern changes progressively through S-phase as different nuclear regions undergo DNA replication at different times, and this generally correlates with chromatin status such that euchromatin replicates earlier in S-phase than heterochromatin (O'Keefe et al., 1992). Labeling of replication foci has been done in a variety of cell types and with multiple different labeling methods that commonly rely on labeled nucleotide analog such as bromo- or biotinylated-dUTP, and allow for comparison of replication foci to other cellular structures (Nakayasu & Berezney, 1989). Similarly in live cell analysis by GFP-PCNA labeling, one can see that replication factories in live cells show limited subnuclear movement, and each individual replication site arises and recedes independently from others (Leonhardt, 2000). The global position of chromosome organization in interphase nuclei and through mitosis has been analyzed after tagged-dUTP incorporation as well as "chromosome painting" methods. Such approaches have shown that the gene density of chromosomes may position them differently in nuclear space. Gene-dense chromosomes more frequently appear in the nuclear interior, while gene-poor chromosomes are found at the nuclear periphery (Bolzer et al., 2005; Spector 2003). In addition, the timing of replication for specific gene loci has been determined by coupling fluorescence in-situ hybridization methods with the nucleotide analog incorporations. For example, alpha-satellite DNA at centromeres replicates at mid-S-phase in a variety of human cell types (O'Keefe et al., 1992). Also, active rDNA replication foci labeled by EdU pulse became activated early in S-phase, while inactive rDNA repeats replicated later (Dimitrova 2011). Early replicated rDNA repeats were then repositioned to the nucleolar interior where they coincided with RNA polymerase I transcription factor UBF, indicating that these replicated regions are transcriptionally active rDNA genes (Dimitrova 2011).

After completion of a number of cell cycles following tagged-dUTP incorporation, the number of labeled chromosome territories eventually decreases by successive distribution to daughter nuclei such that individual chromatin territories are labeled and can be compared relative to other objects or gene loci. Fluorescently tagged nucleotide (e.g. Cy5-dUTP) is also useful for imaging DNA strand condensation into chromosomes and to monitor chromosome organization through mitosis (Manders et al., 1999). Chromosome painting techniques in fixed cells have supported the model that individual chromosomes occupy discrete nuclear regions during interphase (Cremer et al., 1993). A significant advance in understanding chromatin dynamics was made with histone H2B-green fluorescent protein (GFP), which incorporates into mononucleosomes without disturbing the cell cycle (Kanda et al., 1998). H2B-GFP labeled chromatin was originally used to follow the position and segregation of double minute chromosomes during mitosis (Kanda et al., 1998). Widespread use of H2B-GFP in many subsequent studies has given more global insight into the organization and dynamics of chromatin. H2B-GFP observation in living HeLa cells has shown that chromosomes maintain nuclear position or "neighborhoods" through interphase, but become repositioned to new neighborhoods following mitosis (Walter, 2003). In other studies, mitotic chromosomes were demonstrated to have remarkably ordered positioning suggesting that chromosome position is transmissible (Gerlich et al., 2003). Photobleaching experiments showed clearly that a region of YFP-H2B chromatin located at the extreme end of an interphase nucleus would become distributed to many subnuclear regions distributed throughout post-mitotic daughter nuclei (Walter et al., 2003). Because spindle tension draws sister chromatids of adjacently situated chromosomes at the metaphase plate to roughly the same regions of the daughter cell, there is relative symmetry of chromosome arrangement in daughter nuclei. However, the pattern of chromosome territories in daughters is not similar to the pattern of the mother cell, and overall chromosome territory pattern appears to rearrange significantly with each cell cycle (Walter et al., 2003).

Temporal control of replication timing must be reinitiated with each cell cycle, and many questions regarding how the timing and organization is controlled remain unanswered (reviewed in Lucas & Feng, 2003). It will be particularly interesting to follow the activity of specific replication proteins through the cell cycle, or to know what specific nuclear structures/complexes might support dynamic assembly of replication foci. Basic research in this regard will be crucial for us to understand the mechanisms that underlie observed differences in replication of specific gene loci in diseased cells versus normal cells. For example, malignant cells demonstrated less synchrony in replication of homologous loci, having early and late replicating alleles rather than synchronously replicating alleles found in normal cells (Amiel et al., 1998). Also, the spatial organization of DNA replication foci within the nucleus during early S-phase, and the association of these replication foci with intranuclear lamin A/C was altered in immortalized cells (Kennedy et al., 2000). An understanding of how replication of a gene locus is tied to its transcriptional activity may also shed light on why replication organization and timing is altered in disease. In order to address such questions, it will be important to analyze cell cycle dependent events in large numbers of cells. A very promising new technique for measuring cell cycle dependent growth was demonstrated recently, using spatial light interference microscopy (SLIM) coupled with a fluorescence marker for S-phase to analyze cell cycle phase within a cell population (Mir et al., 2011). The applications for this technique to a range of cell types, as well as to microscopy systems that utilize multi-channel fluorescence imaging, open endless possibilities for developing variations on this method to image cellular metabolism in the context of cell growth even within a complex cellular population.

5. Conclusion

The rapid progress recently made toward developing metabolic tracer molecules shows great promise for new applications in clinical diagnostics. Further characterization of novel imaging probes is needed to understand how they can be used to image and identify malignant tissues. Rapidly screening novel tracer molecules for efficacy in identifying tumors in cell culture systems, animal models and clinical trials is a crucial ongoing challenge aimed toward building a battery of tools for imaging cancer metabolism in patients. Feeding into clinical studies is a vast amount of knowledge gained from basic research characterizing metabolic pathways in single cells. This information has potential for wide use for diagnostic imaging, but awaits further research and development into translational medicine that will utilize novel biomarkers and imaging technologies. Finally, continued development of super-resolution imaging platforms for both basic research and clinical use are certain to have a major impact on our understanding of molecular complexes, especially with regard to colocalization of specific protein-protein, protein-RNA or protein-DNA complexes within the overall context of cellular architecture.

6. References

Amiel, A., Litmanovitch T., Lishner M., Mor A., Gaber E., Tangi I., Fejgin M. & Avivi, L. 1998. Temporal differences in replication timing of homologous loci in malignant cells derived from CML and lymphoma patients. Genes Chromosomes Cancer 22: 225–231.

Andersen, J.S., Lam Y.W. Leung A.K.L., Ong S._E., Lyon C., Lamond A.I., & Mann M. 2004. Nuclolar proteome dynamics. Nature. 433:77-83.

Barwick, T., Bencherif B., Mountz J.M. & Avril N. 2009. Molecular PET and PET/CT imaging of tumour cell proliferation using F-^{18} fluoro-L-thymidine: a comprehensive evaluation. Nucl. Med. Commun. 30: 908-17.

Ben-Ari, Y., Brody Y., Kinor N., Mor A., Tsukamoto T., Spector D.L., Singer R.H. & Shav-Tal Y. 2010. The life of an mRNA in space and time. J Cell Sci. 123: 1761-1774.

Ben-Haim, S. & Ell P. 2009. ^{18}F-FDG PET and PET/CT in the evaluation of cancer treatment response. J. Nucl. Med. 50: 88-99.

Bolzer A., Kreth G., Solovei I., Koehler D., Saracoglu K., Fauth C., Muller S., Eils R., Cremer C., Speicher M.R. & Cremer, T. (2005) Three-Dimensional Maps of All Chromosomes in Human Male Fibroblast Nuclei and Prometaphase Rosettes. PLoS Biol 3(5): 157.

Boisvert, F.M., Hedzel M.J. & Bazett-Jones, D.P.. 2000. Promyelocytic leukemia (PML) nuclear bodies are protein structures that do not accumulate RNA. J. Cell Biol. 148: 283-292.

Bouchier-Hayes, L., Oberst A., McStay G.P., Connell S., Tait S.W.G., Dillon C.P., Flanagan J.M., Beere H.M. & Green D.R. 2009. Characterization of cytoplasmic caspase-2 activation by induced proximity. Mol. Cell. 25: 830-840.

Chopra, A., Shan L., Eckelman W.C., Leung K., Latterner M., Bryant S.H. & Menkens A. 2011. Molecular Imaging and Contrast Agent Database (MICAD): Evolution and Progress. [Internet]. Mol. Imaging Biol. Oct 12.

Comar, D., Cartron J.C., Maziere M. & Marazano, C. 1976. Labelling and metabolism of methionine-methyl-^{11}C. Eur. J. Nucl. Med. 1: 11-14.

Condeelis, J. &Weissleder R. 2010. In vivo imaging in cancer. Cold Spring Harb. Perspect. Biol. 2:a003848.

Cremer, T., Kurz A., Zirbel R., Dietzel S., Rinke B., Schrock E., Speicher M.R., Mathieu U., Jauch A., Emmerich P., Scherthan H., Reid T., Cremer C. & Lichter, P. 1993. Role of chromosome territories in the functional compartmentalization of the cell nucleus. Cold Spring Harbor Symp. Quant. Biol. 58: 777–792.

Deng, H., Tang X., Wang H., Tang G., Wen F., Shi X., Yi C., Wu K. & Meng Q. 2011. S-^{11}C-methyl-L-cysteine: a new amino acid PET tracer for cancer imaging. J. Nucl. Med. 52: 287-93.

Dimitrova, D.S. 2011. DNA replication initiation patterns and spatial dynamics of the human ribosomal RNA gene loci. J. Cell Sci. 124: 2743-2752.

Dundr, M., Misteli T. & Olson, M.O.J. 2000. The dynamics of postmitotic reassembly of the nucleolus. J. Cell Biol. 150:433-446.

Dundr, M., Hoffmann-Rohrer U., Hu Q., Grummt I., Rothblum L.I., Phair R.D. & Misteli T. 2002. A kinetic framework for a mammalian RNA polymerase in vivo. Science. 298: 1623-1626.

Emonds, K.M., Swinnen J.V., Mortelmans L. & Mottaghy F.M. 2009. Molecular imaging of prostate cancer. Methods. 48: 193-9.

Fehr, M., Frommer W.B. & Lalonde S. 2002. Visualization of maltose uptake in living yeast cells by fluorescent nanosensors. Proc. Natl. Acad. Sci. 99:9846-9851.

Fehr, M., Lalonde S., Lager I., Wolff M. W. & Frommer W. B. 2003. In vivo imaging of the dynamics of glucose uptake in the cytosol of COS-7 cells by fluorescent nanosensors. J. Biol. Chem. 278: 19127-19133.

Fehr, M., Takanaga H., Ehrhardt D.W. & Frommer W.B. 2005. Transport across the endoplasmic reticulum membrane by genetically encoded fluorescence resonance energy transfer nanosensors. Mol. Cell. Biol. 25: 11102-11112.

Gerlich, D., Beaudouin J., Kalbfuss B., Daigle N., Eils R. & Ellenberg J. 2003. Global chromosome positions are transmitted through mitosis in mammalian cells. Cell. 112: 751-764.

Haukenes, G., Szilvay A.M., Brokstad K.A., Kanestrom A. & Kalland K.H. 1997. Labeling of RNA transcripts of eukayotic cells in culture with BrUTP using liposome transfection reagent (DOTAP). Biotechniques. 22: 308-312.

Helfman, D.M., Ricci W.M. & Finn L.A. 1988. Alternative splicing of tropomyosin pre-mRNAs in vitro and in vivo. Genes & Dev. 2: 1627-1638.

Hernandez-Verdun, D., Roussel P. & Gebranes-Younes J. 2002. Emerging concepts of nucleolar assembly. J Cell Sci. 115:2265-2270.

Ho, L.H., Taylor R., Dorstyn L., Cakourous D., Bouillet P. & Kumar S. 2009. A tumor suppressor function for caspase-2. Proc. Nat. Acad. Sci. 106: 5336–5341.

Howard, B.V. & Howard W.J. 1975. Lipids in normal and tumor cells in culture. Prog. Biochem. Pharmacol. 10: 135–66.

Hu, C.-D., Chinenov Y. & Kerppola T.K. 2002. Visualization of interactions among bZIP and Rel family proteins in living cells using bimolecular fluorescence complementation. Mol. Cell 9: 789-798.

Hu, C.-D. & Kerppola T.K. 2003. Simultaneous visualization of multiple protein interactions in living cells using multicolor fluorescence complementation analysis. Nat. Biotech. 21: 539-545.

Huang, S., Deerinck T.J., Ellisman M.H., & Spector D.L. 1994. In vivo analysis of the stability and transport of nuclear poly(A)+ RNA. J. Cell Biol. 126: 877–899.

Huang, S. & Spector D. L. 1996. Intron-dependent recruitment of pre-mRNA splicing factors to sites of transcription. J. Cell Biol. 133: 719-732.

Huang, S., Deerinck T.J., Ellisman M.H. & Spector D.L. 1998. The perinucleolar compartment and transcription. J. Cell Biol. 143: 35-47.

Huang, B., Bates M. & Zhuang X. 2009. Super-resolution fluorescence microscopy. Ann. Rev. Biochem. 78: 993-1016.

Ido, T., Wan C-N., Casella, V., Fowler J. S., Wolf A. P., Reivich M. & Kuhl D. E. 1978. Labeled 2-deoxy-D-glucose analogs. 18F-labeled 2-deoxy-2-fluoro-D-glucose, 2-deoxy-2-fluoro-D-mannose and 14C-2-deoxy-2-fluoro-D-glucose. J. Labelled Compd. Radiopharmac. 14: 175–183.

Ishiwata, K., Kawamura K., Wang W.F., Furumoto S., Kubota K., Pascali C., Bogni A. & Iwata R. 2004. Evaluation of O-[^{11}C]methyl-L-tyrosine and O-[^{18}F]fluoromethyl-L-tyrosine as tumor imaging tracers by PET. Nucl. Med. Biol. 2: 191-8.

Jadvar, H. 2009. Molecular imaging of prostate cancer: a concise synopsis. Mol. Imaging. 8: 56-64.

Jadvar, H. 2011. Prostate cancer: PET with ^{18}F-FDG, ^{18}F- or ^{11}C-acetate, and ^{18}F- or ^{11}C-choline. J. Nucl. Med. 52: 81-9.

Janicki, S.M., Tsukamoto T., Salghetti S.E., Tansey W.P., Sachidanandam R., Prasanth K.V., Ried T., Shav-Tal Y., Bertrand E., Singer R.H. & Spector D.L. 2004. From Silencing to Gene Expression: Real-Time Analysis in Single Cells. Cell. 116: 683-698.

Kaira, K., Oriuchi N., Shimizu K., Tominaga H., Yanagitani N., Sunaga N., Ishizuka T., Kanai Y., Mori M. & Endo K. 2009. ^{18}F-FMT uptake seen within primary cancer on PET helps predict outcome of non-small cell lung cancer. J. Nucl. Med. 50: 1770-6.

Kamath, R.V., Thor A.D., Wang C., Edgerton S.M., Slusarczyk A., Leary D.J., Wang J., Wiley E.L., Jovanovic B., Wu Q., Nayar R., Kovarik P., Shi F., & Huang S. 2005. Perinucleolar comparment prevalence has an independent prognostic value for breast cancer. Cancer Res. 65: 246-253.

Kanda, T., Sullivan K.F. & G. M. Wahl. 1998. Histone-GFP fusion protein enables sensitive analysis of chromosome dynamics in living mammalian cells. Curr. Biol. 8: 377-385.

Kawai, N., Maeda Y., Kudomi N., Miyake K., Okada M., Yamamoto Y., Nishiyama Y. & Tamiya T. 2011. Correlation of biological aggressiveness assessed by ^{11}C-methionine PET and hypoxic burden assessed by 18F-fluoromisonidazole PET in newly diagnosed glioblastoma. Eur. J. Nucl. Med. Mol. Imaging. 38: 441-50.

Kennedy, B.K., Barbie D.A., Classon M., Dyson N. & Harlow E. 2000. Nuclear organization of DNA replication in primary mammalian cells. Genes Dev. 14:2855-2868.

Kimura, H., Sugaya K. & Cook P.R. 2002. The transcription cycle of RNA polymerase II in living cells. J. Cell Biol. 159: 777-782.

Kolthammer, J.A., Corn D.J., Tenley N., Wu C., Tian H., Wang Y. & Lee Z. 2011. PET imaging of hepatocellular carcinoma with ^{18}F-fluoroethylcholine and ^{11}C-choline. Eur. J. Nucl. Med. Mol. Imaging. 38: 1248-56.

Kubota, K., Furumoto S., Iwata R., Fukuda H., Kawamura K. & Ishiwata K. 2006. Comparison of ^{18}F-fluoromethylcholine and 2-deoxy-D-glucose in the distribution of tumor and inflammation. Ann. Nucl. Med. 20: 527–533.

Kumaran, R.I. & Spector D.L. 2008. A genetic locus targeted to the nuclear periphery in living cells maintains its transcriptional competence. J. Cell Biol. 180: 51-65.

Lamond A.I. & Spector D.L. 2003. Nuclear speckles: a model for nuclear organelles. Nat. Rev. Mol. Cell Biol. 4: 605-612.

Lager, I., Fehr M., Frommer W.B. & Lalonde S. 2003. Development of a fluorescent nanosensor for ribose. FEBS Lett. 553: 85-89.

Leary, D. & Huang S. 2001. Regulation of ribosome biogenesis within the nucleus. FEBS Letters 509: 145-150.

Lee, Y.-S. 2010. Radiopharmaceuticals for Molecular Imaging. The Open Nucl. Med. J. 2: 178-185.

Leonhardt, H., Rahn H.-P. Weinzierl P., Sporbert A., CremerT., Zink D. & Cardoso M.C. 2000. Dynamics of DNA replication factories in living cells. J. Cell Biol. 149: 271-279.

Liu, R.S., Chang C.P., Chu L.S., Chu Y.K., Hsieh H.J., Chang C.W., Yang B.H., Yen S.H., Huang M.C., Liao S.Q. & Yeh S.H. 2006. PET imaging of brain astrocytoma with 1-^{11}C-acetate. Eur. J. Nucl. Med. Mol. Imaging. 33: 420–427.

Locasale, J.W., Cantley L.C. & Vander Heiden M.G. 2009. Cancer's insatiable appetite. Nat. Biotech. 27: 916-917.

Lucas, I. & Feng W. 2003. The essences of replication timing: determinants and significance. Cell Cycle.2: 560-563.

Luo, J., Solimini N.L. & Elledge S.J. 2009. Principles of cancer therapy: oncogene and non-oncogene addiction. Cell. 136: 823-837.

Manders, E.M.M., Kimura H. & Cook P.R. 1999. Direct imaging of DNA in living cells reveals the dynamics of chromosome formation. J. Cell Biol. 144: 813-822.

Marik, J., Ogasawara A., Martin-McNulty B., Ross J., Flores J.E., Gill H.S., Tinianow J.N., Vanderbilt A.N., Nishimura M., Peale F., Pastuskovas C., Greve J.M., van Bruggen N. & Williams S.P. 2009 PET of glial metabolism using 2-^{18}F-fluoroacetate. J. Nucl. Med. 50: 982-90.

McNally, J.G., Muller W.G., Walker D., Wolford R. and Hager G.L. 2000. The glucocorticoid receptor: rapid exchange with regulatory sites in living cells. Science. 287: 1262-1265.

McNally, J.G. 2009. Transcription, chromatin condensation and gene migration. J. Cell Biol. 185: 7-9.

Mintz, P.J., Patterson S.D., Neuwald A.F., Spahr C.S. & Spector, D.L. 1999. Purification and biochemical characterization of interchromatin granule clusters. EMBO J. 18: 4308–4320.

Mir, M., Wang Z., Shen Z., Bednarz M., Bashir R., Golding I., Prasanth S.G. & Popescu G. 2011. Optical measurement of cell cycle-dependent cell growth. Proc. Nat. Acad. Sci. 108: 13124-13129.

Miyawaki, A., Llopis J., Heim R., McCaffery J.M., Adams J.A., Ikura M. & Tsien R.Y.. 1997. Fluorescent indicators for Ca2+ based on green fluorescent proteins and calmodulin. Nature. 388: 882-887.

Muller, W.G., Walker D., Hager G.L., and McNally J.G. 2001. Large-scale chromatin decondensation and recondensation regulated by transcription from a natural promoter. J. Cell Biol. 154:33-48.

Murayama, C., Harada N., Kakiuchi T., Fukumoto D., Kamijo A., Kawaguchi A.T. & Tsukada H. 2009. Evaluation of D-18F-FMT, 18F-FDG, L-11C-MET, and 18F-FLT for monitoring the response of tumors to radiotherapy in mice. J. Nucl. Med. 50: 290-5.

Nakamura, H., Morita T. & Sato C. 1986. Sructural organizations of replicon domains during DNA synthetic phase in the mammalian nucleus. Exp. Cell Res. 165: 291-297.

Nakayasu, H. & Berezney R. 1989. Mapping replicational sites in the eukaryotic cell nucleus. J. Cell Biol. 108: 1-11.

Nanni, C., Fantini L., Nicolini S. & Fanti S. 2010. Non FDG PET. Clin. Radiol. 65: 536-48.

Norton, J.T., Pollock C.B., Wang, C., Schink, J.C. Kim, J.J. & Huang S. 2008. Perinucleolar compartment prevalence is a phenotypic pancancer marker of malignancy. Cancer. 113:861-869.

Norton, J.T., Titus S.A., Dexter D., Austin C.P., Zheng W. & Huang S. 2009. Automated high-content screening for compounds that disassemble the perinucleolar compartment. J. Biomol. Screen. 14:1045-1053.

O'Keefe, R.T., Henderson S.C. & Spector D.L. 1992. Dynamic organization of DNA replication in mammalian cell nuclei: spatially and temporally defined replication of chromosome-specific alpha-satellite DNA sequences. J. Cell Biol. 116:1095-1110.

Paige, J.S., Wu K.Y. & Jaffrey S.R. 2011. RNA mimics of green fluorescent protein. Science. 333: 642-646.

Piert, M., Park H., Khan A., Siddiqui J., Hussain H., Chenevert T., Wood D., Johnson T., Shah R.B., Meyer C. 2009. Detection of aggressive primary prostate cancer with ^{11}C-choline PET/CT using multimodality fusion techniques. J. Nucl. Med. 50: 1585-93.

Plathow, C. & Weber W.A. 2008. Tumor cell metabolism imaging. J. Nucl. Med. 49 Suppl 2:43S-63S.

Pombo, A., Jackson D.A., Hollingshead M., Wang Z., Roeder R.G. & Cook P.R. 1999. Regional specialization in human nuclei: visualization of discrete sites of transcription by RNA polymerase III. EMBO J. 18: 2241-2253.

Prasanth, K.V., Sacco-Bubulya P.A., Prasanth S.G. & Spector D.L. 2003. Sequential entry of components of gene expression machinery into daughter nuclei. Mol. Biol. Cell. 14: 1043-1057.

Prasanth, K.V., Prasanth S.G., Xuan Z., Hearn S., Freier S.M., Bennett C.F., Zhang M.Q. & Spector D.L. 2005. Regulating gene expression through RNA nuclear retention. Cell. 123: 249-263.

Prasanth, K.V., Camiolo, M., Chan G., Tripathi V., Denis L., Nakamura T., Hubner M.R., & Spector, D.L. 2010. Nuclear organization and dynamics of 7SK RNA in regulating gene expression. Mol. Biol. Cell. 21: 4184-4196.

Rafalska-Metcalf, I-U. & Janicki S.M. 2007. Show and Tell: visualizing gene expression in living cells. J. Cell Sci. 120:2301-2307.

Rajapaske, I. & Groudine M. 2011. On emerging nuclear order. J. Cell Biol. 192: 711-721.

Ramirez de Molina, A., Gutierrez R., Ramos M.A., Silva J.M., Silva J., Bonilla F., Rosell R. & Lacal J. 2002a. Increased choline kinase activity in human breast carcinomas: clinical evidence for a potential novel antitumor strategy. Oncogene 21: 4317–4322.

Ramirez de Molina, A., Rodriguez-Gonzalez A., Gutierrez R., Martinez-Pineiro L., Sanchez J., Bonilla F., Rosell R. & Lacal J. 2002b. Overexpression of choline kinase is a frequent feature in human tumor-derived cell lines and in lung, prostate, and colorectal human cancers. Biochem. Biophys. Res. Commun. 296: 580–583.

Rasey, J.S., Grierson J.R., Wiens L.W., Kolb P.D. & Schwartz J.L. 2002. Validation of FLT uptake as a measure of thymidine kinase-1 activity in A549 carcinoma cells. J. Nucl. Med. 43: 1210-7.

Sacco-Bubulya, P. & Spector D.L. 2002. Disassembly of interchromatin granule clusters alters the coordination of transcription and pre-mRNA splicing. J. Cell Biol. 156: 425–436.

Saitoh, N., Spahr C.S., Patterson S.D., Bubulya P., Neuwald A.F. & Spector D.L. 2004. Proteomic analysis of interchromatin granule clusters. Mol. Biol. Cell. 15: 3876-3890.

Salic, A. & Mitchison T. 2008. A chemical method for fast and sensitive detection of DNA synthesis in vivo. Proc. Nat. Acad. Sci. 105: 2415-2420.

Sharma, A., Takata H., Shibahara K., Bubulya A. & Bubulya P.A. 2010. Son is essential for nuclear speckle organization and cell cycle progression. Mol Biol. Cell. 21: 650-663.

Sharma, A., Markey M., Torres-Munoz K., Varia S., Bubulya A. & Bubulya P.A. 2011. Son maintains accurate splicing for a subset of human pre-mRNAs J. Cell Sci. 124: *in press.*

Shav-Tal, Y., Darzacq X. Shenoy S.M., Fusco D., Janicki S.M., Spector, D.L. & Singer, R.H. 2004. Dynamics of single mRNPs in nuclei of living cells. Science. 304: 1797-1800.

Sokoloff, L., Reivich M., Kennedy C., Des Rosiers M.H., Patlak C.S., Pettigrew K.D., Sakurada O. & Shinohara M.J. 1977. The [^{14}C]deoxyglucose method for the measurement of local cerebral glucose utilization: theory, procedure, and normal values in the conscious and anesthetized albino rat. Neurochem. 28: 897-916.

Spector, D. 2003. The dynamics of chromosome organization and gene regulation. Ann. Rev. Biochem. 72: 573-608.

Spector, D.L. & Goldman R.D. (Eds). 2006. Basic Methods in Microscopy. Cold Spring Harbor: Cold Spring Harbor Press.

Spector D.L. & Lamond A.I. 2010. Nuclear speckles. Cold Spring Harb. Perspect. Biol. doi: 10.1101/cshperspect.a000646.

Talbot, J.N., Gutman F., Fartoux L., Grange J.D., Ganne N., Kerrou K., Grahek D., Montravers F., Poupon R., & Rosmorduc O. 2006. PET/CT in patients with hepatocellular carcinoma using [(^{18})F]fluorocholine: preliminary comparison with [(^{18})F]FDG PET/CT. Eur. J. Nucl. Med. Mol. Imaging. 33: 1285–1289.

Ting, A.Y., K.H. Kain, R.L. Klemke & R.Y. Tsien. 2001. Genetically encoded fluorescent reporters of protein tyrosine kinase activities in living cells. Proc. Natl Acad. Sci. 98: 15003–15008.

Tripathi, V., Ellis, J. D., Shen, Z., Song, D. Y., Pan, Q., Watt, A. T., Freier, S. M., Bennett, C. F., Sharma, A., Bubulya, P. A., Blencowe, B. J., Prasanth, S. G., & Prasanth, K. V. 2010. The nuclear-retained noncoding RNA MALAT1 regulates alternative splicing by modulating SR splicing factor phosphorylation. Mol. Cell, 39: 925-938.

Tsien, C.I., Brown D., Normolle D., Schipper M., Morand P., Junck L., Heth J.A., Gomez-Hassan D., Ten- Haken R., Chenevert T.L., Cao Y. & Lawrence T.S. 2011. Concurrent Temozolomide and Dose-Escalated Intensity Modulated Radiation Therapy in Newly Diagnosed Glioblastoma. Clin. Cancer Res. *In press.* doi: 10.1158/1078-0432.CCR-11-2073

Tsuchida T., Takeuchi H., Okazawa H., Tsujikawa T. & Fujibayashi Y. 2008. Grading of brain glioma with 1-^{11}C-acetate PET: comparison with ^{18}F-FDG PET. Nucl. Med. Biol. 35: 171–176.Visa, N., Puvion-Dutilleul F., Harper F., Bachellerie J.P. & Puvion E. 1993. Intranuclear distribution of poly(A) RNA determined by electron microscope in situ hybridization. Exp. Cell Res. 208: 19–34.

Tsukamoto , T., Hashiguchi N., Janicki S.M., Tumbar T., Belmont A.S., & Spector D.L. 2000. Visualization of gene activity in living cells. Nat. Cell Biol. 2: 871–878 .

Visa N., Puvion-Dutilleul F., Harper F., Bachellerie J.-P. & Puvion E. 1993. Intranuclear distribution of poly A RNA determined by electron microscope in situ hybridization. Exp. Cell Res. 208: 19–34.

Walter, J., Schermelleh L., Cremer M., Tashiro S. & Cremer T. 2003. Chromosome order in HeLa cells changes during mitosis and early G1, but is stably maintained during subsequent interphase stages. J. Cell Biol. 160: 685-697.

Wang, C., Politz J.C., Pederson T. & Huang S. 2003. RNA polymerase III transcripts and the PTB protein are essential for the integrity of the perinucleolar compartment. Mol Biol. Cell. 14: 2425-2435.

Warburg, O. 1956. On the origin of cancer cells. Science 123: 309–314.

Zaheer, A., Cho S.Y. & Pomper M.G. 2009. New agents and techniques for imaging prostate cancer. J. Nucl. Med. 50: 1387-90.

Zhang, J., Y. Ma, Taylor S.S. & Tsien R.Y. 2001. Genetically encoded reporters of protein kinase A activity reveal impact of substrate tethering. Proc. Natl Acad. Sci. 98: 14997–15002.

Zhang, J., Campbell R.E., Ting A.Y. & Tsien R.Y.. 2002. Creating new fluorescent probes for cell biology. Nat. Rev. Mol. Cell Biol. 3: 906-918.

Zhao, R., Nakamura T., Fu Y., Lazar Z., & Spector D.L. 2011. Gene bookmarking accelerates the kinetic of post-mitotic transcriptional re-activation. Nat. Cell Biol. 13: 1295-1304.

Zitzmann-Kolbe, S., Strube A., Frisk A.L., Kakonen S.M., Tsukada H., Hauff P., Berndorff D. & Graham K. 2010. D-18F-fluoromethyl tyrosine imaging of bone metastases in a mouse model. J. Nucl. Med. 51: 1632-6.